QUANTITATIVE METHODS IN NEUROSCIENCE

QUANTITATIVE METHODS IN NEUROSCIENCE
A NEUROANATOMICAL APPROACH

Edited by

Stephen M. Evans

Director of Clinical Research, Pharmaimaging Limited,
Grangemouth, Scotland, UK

Ann Marie Janson

Department of Neuroscience, Karolinska Institute, Stockholm, Sweden

and

Jens Randel Nyengaard

Stereological Research Laboratory & Electron Microscopy Laboratory,
Aarhus University Hospital, Aarhus, Denmark

OXFORD
UNIVERSITY PRESS

OXFORD

UNIVERSITY PRESS

Great Clarendon Street, Oxford OX2 6DP

Oxford University Press is a department of the University of Oxford.
It furthers the University's objective of excellence in research, scholarship,
and education by publishing worldwide in

Oxford   New York
Auckland   Bangkok   Buenos Aires   Cape Town   Chennai
Dar es Salaam   Delhi   Hong Kong   Istanbul   Karachi   Kolkata
Kuala Lumpur   Madrid   Melbourne   Mexico City   Mumbai   Nairobi
São Paulo   Shanghai   Taipei   Tokyo   Toronto

Oxford is a registered trade mark of Oxford University Press
in the UK and in certain other countries

Published in the United States
by Oxford University Press Inc., New York

© Oxford University Press 2004

British Library Cataloging in Publication Data

Data available

ISBN 0 19 850528 0

10 9 8 7 6 5 4 3 2 1

Typeset by EXPO Holdings, Malaysia
Printed in Great Britain
on acid-free paper by Biddles Ltd., King's Lynn, UK

CONTENTS

FOREWORD

To most biologists, stereology comprises a set of very efficient methods or experimental tools for obtaining precise and accurate quantitative information about any structure, anywhere. However, it probably does not appear to them that *all* of stereology is really about sampling: among animals, in the organ, in the sections, among all directions, etc.

Sampling may be the most useful concept in science since calculus. No experimental science is possible without sampling, and we all do it, every day. Biologists do it more than most scientists, but, curiously, often at an almost unconscious level. In this state, they are therefore prone to overlook the fact that in order to fulfil its sole purpose (to make scientifically valid statements about the reality), sampling must be *uniformly* random.

To many, these are just principles—and this may sound like a boring lecture, but these principles are, nevertheless, the only foundation common to all experimental sciences. As a much more immediately practical consequence of the above, the methods, the wet nitty-gritty of handling animals and tissue and paraffin and stains, must be direct and faithful implementations of these principles. Such careful implementation requires a lot of specific knowledge, hard work, attention to many details, and experience.

This book is mainly about good advice, from experienced people, in a large number of experimental situations of special interest to neuroscientists.

Jacuzzi, United Nations, net-dating, and unbiased estimation based on just uniform samples (from unknown distributions) are all concepts from our generation. The latter is about 50 years old, and was the foundation of the disector, which I described just 20 years ago. Since then, essentially all of the remarkable developments in stereological methods have been based on uniform sampling in all aspects of 3D space. To be precise, several well-known estimators actually have unnoticed steps of non-uniform sampling (with known probability) embedded in them. However, with the general sampling principle of the smooth fractionator, we may have reached the limit of how efficient unbiased estimators from plain uniform samples may ever become.

Still waiting for automatic image analysis to fulfil its promise (I have waited for more than 25 years so far), there are fortunately a number of radically new concepts lurking in the horizon in which the computer will do the hard work and make most decisions (some non-trivial) in spatial sampling...automatically. Most of these new, unbiased stereological estimators will be based on markedly non-uniform, random sampling.

D.C. Sterio
March 2004

ABBREVIATIONS

ABC	avidin–biotin–peroxidase complex
ANOVA	analysis of variance
AP	alkaline phosphatase
asf	area sampling fraction
BCIP	5-bromo-4-chloro-3-indolyl phosphate
BSA	bovine serum albumin
Cb	calbindin
cdf	cumulative distribution function
CE	coefficient of error
CNS	central nervous system
CT	computerized tomography
CV	coefficient of variation
DAB	diaminobenzidine
DEPC	diethylpyrocarbonate
d.f.	degrees of freedom
DG	dentate gyrus
DMEM	Dulbecco's modified Eagle's medium
DR	dorsal raphe
DSS	disuccinimidyl suberate
ECL2	entorhinal cortex layer II
EDTA	ethylene diaminetetraacetic acid
EM	electron microscope
E-PTA	ethanol phosphotungstic acid
FISH	fluorescent *in situ* hybridization
FITC	fluorescein isothiocyanate
GAPDH	glyceraldehyde-3-phosphate dehydrogenase
GC	guanine–cytosine (content)
GFAP	glial fibrillary acidic protein
GFP	green fluorescent protein
H_0, H_1	null and alternative hypotheses, respectively
HPLC	high-performance liquid chromatography
hsf	height sampling fraction
ir	immunoreactive
ISH	*in situ* hybridization
IUR	isotropic uniform random (sections)
LDH	lactate dehydrogenase
MANOVA	multivariate analysis of variance
MAP2′	microtubule-associated protein 2
MBP	myelin basic protein
MCAO	medial cerebral artery occlusion
MRI	magnetic resonance imaging
NA	numerical aperture
nAChR	nicotinic acetylcholine receptor

NADP	NADPH-diaphorase
NBT	nitroblue tetrazoleum
ncp	non-centrality parameter
OCV	observed coefficient of variation
PB	phosphate buffer
PBS	phosphate buffered saline
PET	positron emission tomography
PFA	paraformaldehyde
PI	propidium iodide
PMI	post-mortem interval
psd	probe sampling density
p.s.u.	primary sampling unit
Pv	parvalbumin
PVA–DABCO	polyvinyl acetate–1,4-diazabicyclo[2.2.2]octane
PVP	polyvinylpyrrolidone
RT	room temperature
RT-PCR	reverse transcriptase polymerase chain reaction
SRS	systematic random sampling
SURS	systematic uniform random sampling
SSC	standard sodium citrate
SS	striatal somastatin
ssf	section sampling fraction
TBS	tris-buffered saline
TE	tris–EDTA
TMN	tumor/metastases/nodes
TNF	tumor necrosis factor
tsf	thickness sampling fractions
UFAPP	unbiased for all practical purposes
UV	ultraviolet
Var(SURS)	variance due to systematic uniform random sampling of sections
VUR	vertical uniform random (sections)

CONTRIBUTORS

Dámaso Crespo
Department of Anatomy and Cell Biology, Faculty of Medicine, University of Cantabria, E-39011 Santander, Spain

Luis M. Cruz-Orive
Department of Mathematics, Statistics, and Computation, Faculty of Sciences, University of Cantabria, Avenida Los Castros, E-39005 Santander, Spain

Karl-Anton Dorph-Petersen
Departments of Neuroscience and Psychiatry, University of Pittsburgh, 3811 O'Hara Street, WI651 BST Pittsburgh, Pennsylvania 15213, USA

Stephen M. Evans
Pharmaimaging Limited, Chambers Building, Earls Road, Grangemouth FK3 8XG, Scotland, UK

Bente Finsen
Department of Anatomy and Neurobiology, Institute of Medical Biology, University of Southern Denmark, Winsløwsparken 21, DK-5000 OdenseC, Denmark

Rikke Gregersen
Department of Anatomy and Neurobiology, Institute of Medical Biology, University of Southern Denmark, Winsløwsparken 21, DK-5000 OdenseC, Denmark

Ana M. Insausti
Neural Plasticity and Regeneration Laboratory Research Unit, Navarra Health Department Irunlarrea S/n, E-31007 Pamplona, Spain

Ricardo Insausti
Human Neuroanatomy Laboratory, Department of Health Sciences, School of Medicine, University of Custilla-la Muncha, Avda. de Almansa s/n, E-02006, Albacete, Spain

Ann Marie Janson
Department of Neuroscience, Karolinska Institute, Doktorsringen 17, S-171 77 Stockholm, Sweden

A. Keller
Department II of Anatomy and Neuroanatomy, University of Cologne, Joseph Stelzmann Strasse 9, D-50931 Cologne, Germany

Christoph Köhler
Department II of Anatomy and Neuroanatomy, University of Cologne, Joseph Stelzmann Strasse 9, D-50931 Cologne, Germany

E. Lain
Department II of Anatomy and Neuroanatomy, University of Cologne, Joseph Stelzmann Strasse 9, D-50931 Cologne, Germany

Jytte Overgaard Larsen
Department of Medical Anatomy, The
Panum Institute, University of
Copenhagen, DK-2200 Copenhagen,
Denmark and Stereological Research
Laboratory, University of Aarhus,
Building 185, DK-8000 Aarhus C,
Denmark

Elin Lehrmann
Department of Anatomy and Cell
Biology, Institute of Medical Biology,
University of Odense, Winsløwsparken
19, DK-5000 Odense, Denmark and
Maryland Psychiatric Research Center,
University of Maryland, Baltimore,
Maryland 221228, USA

Annemette Løkkegaard
Department of Neurology, Bispebjerg
Hospital, Bispebjerg Bakke 23,
DK-2400 Copenhagen NV, Denmark

Steen Lovmand
DNA Technology A/S, Science Park
Aarhus, DK-8000 Aarhus C, Denmark

Jens R. Nyengaard
Stereological Research Laboratory and
Electron Microscopy Laboratory,
Aarhus University Hospital, Aarhus
Kommunehospital, DK-8000 Aarhus
C, Denmark

Daniel A. Peterson
Neural Repair and Neurogenesis
Laboratory, Department of
Neuroscience, The Chicago Medical
School, 3333 Green Bay Rd, North
Chicago, Illinois 60064, USA

Raben Rosenberg
Department of Biological Psychiatry,
University Psychiatric Hospital in
Aarhus, DK-8000 Aarhus, Denmark

Hannsjörg Schröder
Department II of Anatomy and
Neuroanatomy, University of Cologne,
Joseph Stelzmann Strasse 9, D-50931
Cologne, Germany

W. Selberis
Department II of Anatomy and
Neuroanatomy, University of Cologne,
Joseph Stelzmann Strasse 9, D-50931
Cologne, Germany

Trine Tandrup
Department of Neurology, University
Hospital of Aarhus, DK-8000 Aarhus,
Denmark

Yong Tang
Neurobiology of Aging Laboratories,
Mount Sinai School of Medicine, One
Gustave L. Levy Place, New York, New
York 10029, USA

Mia von Euler
Division of Geriatric Medicine,
Huddinge University Hospital,
Karolinska Institute, S-171 77
Stockholm, Sweden

Mark J. West Department of
Neurobiology, University of Aarhus,
DK-8000 Aarhus C, Denmark

Brigitte Witter Department II of
Anatomy and Neuroanatomy,
University of Cologne, Joseph
Stelzmann Strasse 9, D-50931 Cologne,
Germany

CHAPTER 1

GENERAL INTRODUCTION

STEPHEN M. EVANS AND JENS R. NYENGAARD

1 Introduction

There is a natural progression from a qualitative description of a structure and its components to the quantification of parameters such as number, size, connectivity, and the spatial arrangement of the different components of the structure under study. Quantitative analysis provides more descriptive information about the structure under study, thus facilitating comparisons and the determination of how changes under different conditions will affect function. However, there is a danger of becoming obsessed with numbers and stereological methodology at the expense of losing sight of the biological problem.

This is not meant to be a reference book for stereologists but rather a 'cookbook' of stereological methods for neuroscientists. We have tried to limit the amount of formulae and in depth discussion of stereological theory. Also, in the simplification of the description of the methods, some of the mathematical rigor of the description has had to be reduced. For those who wish to pursue detailed mathematical discussion on stereology, adequate references are provided.

The first thing we would like to explain about using stereology in neuroscience is when not to use it. Do not quantify things indiscriminately, that is, number for the sake of using numbers. Beware of seeking some form of quantitation even though the biological effect is clear just from simple observation of the material. Sometimes, however, the important biological effects are not so striking. For example, the search for new therapies in the early stages of human neurodegenerative disorders requires observation of subtle changes in the pathological lesions, including analysis of the time course of neurodegeneration and dose–response curves of possible neuroprotective strategies. In such situations quantitative studies are relevant.

Consider a neuroscientist's interest in how nerve cell number and size are related to exposure to a toxic substance. Objects that are processed so that they can be viewed through a microscope are seen as two-dimensional images. One way to obtain the three-dimensional information would be to do a reconstruction of the object but this is rather time-consuming. In addition, the reconstruction can only use the information that is contained in the set of two-dimensional images and assumptions have to be made as to how the object varies in the three-dimensional space between the two-dimensional images. Sometimes these assumptions are not valid: consider the step-like features of many three-dimensional reconstructions of supposedly smoothly shaped objects.

A more efficient and unbiased way to obtain three-dimensional quantitative information would be to assume that information contained in the sections is a two-dimensional sample of the object and then to use a series of methods based on well proven mathematical theory to extract this information. This second way of thinking is known as stereology and the methodology can be used in a variety of scientific disciplines, for example, biology, material science, metallurgy, and geology. A more formal definition of stereology comes from Weibel (1992), namely, a 'set of methods which all deal with the problem of measuring structures with geometrically defined probes'. The geometrical probes used can be points, volumes, areas, and lines arranged in regular arrays to form a test system. They are introduced into the object by superimposing a test system on to a section taken at random through the object.

Early stereological methods were very imprecise (the sample average did not rapidly converge to a stable value, required thousands of measurements to be made) and were model-based, (they required assumptions to be made regarding an object's size, shape, and orientation in a section). These assumptions were nearly always unjustifiable and led to a biased result, i.e., there was a systematic difference between the sample average and the population average no matter how large the sample size. Some of these biases could be as great as 50% (Pakkenberg *et al.* 1991). Worse still, the extent of the bias was unknown and variable.

In 1961 Hans Elias brought together a group of morphologists from many different scientific disciplines to discuss the formation of an international society to increase the communication between interested scientists. The first international congress of the International Society of Stereology was held in Vienna in 1963 and since then international congresses have been held every 4 years. Although stereology has been developed for centuries, both in theory and in practice, it is only in recent years that many of the new unbiased stereological methods have been discovered, and in some cases rediscovered. Many of these new methods use two parallel sections, one reference and one 'look-up', instead of using the more traditional approach of making measurements on just one single section. In 1983

DeHoff published a table that showed that very few three-dimensional geometric properties could be estimated without a geometric shape assumption. By the end of the decade nearly all of the parameters listed could be estimated unbiasedly, that is, without any assumptions about geometric shape and size. Some reviews can be found, including those of Gundersen *et al.* (1988*a,b*), Cruz-Orive and Weibel (1990), Mayhew (1992), and Nyengaard (1999).

This explosion in stereology from the early 1980s, which saw a series of new unbiased methods being developed—the 'new stereology'—is now impacting in neuroscience. These newer methods are not only unbiased, that is, free from systematic error, but they also are very efficient, that is, they produce very precise estimates requiring at most only 200 measurements. For example the total number of neocortical neurons in a human brain, 25×10^9, can be estimated in less than 2 hours by counting only 200 neurons with a coefficient of error (standard error of the mean divided by the mean) of around 5% (Evans *et al.* 1989).

2 Image analysis and computer-assisted devices

Although the 'new stereology' has produced a set of fast and efficient methods, the efficiency could be further increased if automatic image analysis could be used. However, there are two reasons why this approach has not been completely success-ful. The first is that automatic image segmentation is not always possible, especially in biology, and the second is that very often the visual interpretation of the image is so complex that it can only be done by an experienced investigator (see Gundersen *et al.* (1981) and Mathieu *et al.* (1980)). However, it is possible to increase method-ological efficiency by using a semi-automatic computer-assisted system that produces a series of stereological test systems and probes that can be superimposed on an image of interest. The computer allows the user to make a wide selection of user-designed grids and measurements, to calculate the relevant estimate, for example, a volume, surface area, etc., from measurements made with the aid of the computer, and to transfer the data directly to other software packages, for example, word processors, spreadsheets, etc. A computer can also be used to control the movement of the specimen stage thus automating many of the sampling procedures.

3 Stereology

Modern stereology is built on the mathematical foundation of sampling theory (Miles and Davy 1976; Baddeley and Jensen 2002). In fact, stereology is often compared to an opinion poll. Both require some form of sampling followed by a question being put to the suitably sampled individual. To get an unbiased opinion, it is a requirement that both the sampling and the question are unbiased.

For example, suppose an alien was sent to Earth to find out who was the great-est story-teller who ever lived on the planet. The alien does not have time to ask every single person's opinion so he chooses 'typical' populations to sample. Also, the alien has been studying Earth culture for many years, which is one of the reasons he was chosen for his task, and has a particular fondness for the story of the little mermaid. So when the alien stands outside the H.C. Andersen cafe in Copenhagen and asks the question, 'I think that Hans Christian Andersen is the greatest story-teller who ever lived, don't you?', not only is the question biased ('loaded') but so is the sampling. An unbiased question would be 'What do you think of Hans Christian Andersen?' Notice how both the sampling and the question must be unbiased and the interdependence between them. Although the procedures required to make unbiased and efficient samples will only be discussed very briefly, it is important to realize that a well planned sampling design will often be the critical element in deciding if an experiment is feasible.

Stereology asks questions about the three-dimensional parameters of an object, for example, number, volume, surface area, and length. The way in which the questions are asked is to 'throw' geometric probes, such as frames, lines, and points, at the object and see how the probe and the object interact. Although this manual will mainly concentrate on the kind of 'questions' that can be asked using stereological methods, it is essential to understand that for the method to be unbiased both the probing ('questioning') and the sampling must be unbiased.

The sampling should be randomized to give every item of a population the same chance of being sampled. In stereology systematic random sampling is used, that is, after a random start the population of interest is sampled at regular intervals, because this is more efficient than simple random sampling. It is assumed both that the population can be well defined and that individuals can be identified in the sample. This may present a problem in the central nervous system, for example, accurately outlining the borders of sectioned human substantia nigra in all sections, or identifying all neurons in human cerebral neocortex, especially the small interneurons in the visual cortex. If a population cannot be defined, it is almost impossible to sample from it. Also, if an individual cannot be identified in the sample, then the sample may also be invalid.

An effective sampling strategy makes it possible to minimize the workload whilst still obtaining reliable, quantitative information about the whole structure of interest. Sampling is so important to stereology that Chapter 2 is devoted to it, but some of the different sampling strategies used in stereology are briefly discussed below. In all of these sampling strategies all parts of the structure being studied must be sampled with the same probability. This is done by using some form of random start which is obtained from a random number generator.

4 An introduction to sampling as applied to stereology

A sample is a set of individuals that have been selected from a well defined population about which well defined numerical quantities, that is, parameters such as number, size, etc., are being sought. In an unbiased sample the mean of the sample will be the same as the mean of the population from which the sample was taken. This requires that every individual in the population has the same probability of being sampled and this is achieved by randomly sampling the population. An excellent short introduction to sampling theory can be found in Stuart (1984).

It is unacceptable to choose a 'typical' feature as representative of a population without knowing just how 'typical' the feature is. For example, a scientist wants to quantify changes in neuron volume in a particular region of the brain after a new drug treatment. He dissects out his region of interest, cuts it into smaller blocks, chooses some of the blocks as 'typically' representative of the tissue, and prepares them for microscopy. The scientist chooses a 'typical' section and then chooses a 'typical' field of view and measures the volume of all of the neurons in the field of view using a stereological probe such as the rotator (see the chapters in Part 2 of this book). In reality the scientist has estimated the neuronal volume in one field of view, in one section from one block of tissue from one brain. It may be that the scientist with his years of experience has chosen wisely and his 'typical field of view' is representative of the population, but how does he know? More importantly, what happens if the effect is seen outside the 'typical field of view'? One way would be to measure the size of every neuron in the tissue of interest. Apart from being very time-consuming and possibly impractical, how many brains should be used to know if the treatment is having an effect? Thus, some form of sampling has to be used. The tissue has to be sampled uniformly randomly so that all neurons have an equal chance of being sampled. This then begs the question as to how many brains, blocks, sections, fields of view, and measurements are needed.

To give every object in the sample an equal chance of being sampled, every measurement made in every field of view from every section from every tissue block from all areas of the tissue must be chosen at random. Unfortunately, humans are notoriously bad at choosing random events and some form of random number generator is a prerequisite for an unbiased sampling scheme. In stereology systematic random sampling is used. This means that the first object in a series is chosen at random and the subsequent objects chosen at predetermined intervals. Notice how once the first object has been chosen at random all the subsequent objects will also be chosen at random. It has the advantage of being more efficient than simple random sampling provided the predetermined interval does not coincide with a natural occurring periodicity in the object(s) of interest (see Gundersen and Jensen (1987) for more details).

5 Sampling strategies

Envision that a brain is cut into a number, N, of arbitrary slices.

5.1 Independent, uniformly random sampling

The slices are sampled by chosing a random number between 1 and N and the corresponding slice is sampled. This continues until a fixed sample size (number) of slices is chosen. If the same random number is chosen twice, it is ignored and a new random number is chosen.

5.2 Systematic, uniformly random sampling

A fixed sample size (number), m, is decided upon. The first of the slices is taken at random and from then on every N/mth slice is chosen from the set of slices. Note the slices can be arranged in either their natural order or in a unique order, for example, a smooth order.

5.3 Cluster sampling of structures

The brain slices are divided into three parts, for example, dorsal, middle, and ventral. The definition of each cluster is completely arbitrary. Within clusters, sampling is independent, uniform random or systematic, uniform random (if not both are used). All items in the clusters are studied (possibly after further subsampling). The more the clusters are alike or equal and the more the clusters are internally inhomogeneous, the more efficient cluster sampling is.

5.4 Stratified random sampling

Stratified random sampling means that the slices are divided into strata and every stratum is studied using the above methods.

5.5 Variance, bias, precision, and accuracy

Consider a problem faced by some football officials who wish to know how many people go to a stadium to watch a game of football. The stadium can be divided into four stands: North, South East, and West. Each stand can be divided into 20 sections with each section holding 25 rows of seats and each row holding 20 seats. They try several methods for estimating spectator number.

5.5.1 The 'flag method' or 'typical sample'

The first is used by an official who from his years of experience knows that, for every flag seen in the crowd, there are 'typically' 20 spectators. Therefore, he counts all of the flags and multiplies this number by 20 to get an estimate of the total number of spectators. However, he realizes that he does not have time to count all of the flags in the stadium. So he counts all of the flags in a section of the stadium that he has chosen because he believes its population to be representative of the crowd in the stadium. After counting each row he multiplies the number of flags by 20 to get the total number of spectators in this section, Ω. He then estimates the total number of spectators in the stadium to be $\Omega \times 20 \times 4$. He repeats this several times.

5.5.2 Simple random sampling

Another official tries another method that requires that samples are taken at random so that all individuals have an equal chance of appearing in the sample. There are $4 \times 25 \times 20 = 2000$ rows in the stadium. Using a random number generator he chooses a number between 1 and 2000. He then counts all of the people in this row of seats, n. Therefore an unbiased estimate of the total number of spectators at a match is $n \times 25 \times 20 \times 4$. If the crowd is homogeneously dispersed through the stadium, this one estimate of the sample mean is very likely to be close to the population mean. If, however, this is not the case, for example, the row chosen was empty, then further samples must be taken. For simple random sampling the official would repeat his process again choosing the stand, section, and row at random, rejecting it if it had been already chosen. This will be unbiased, because all rows have an equal probability of being sampled and eventually the mean of the sample will equal the mean of the population. However, with simple random sampling there is a risk of sampling rows directly next to each other which is not very efficient, that is, it may take many estimates before the sample mean approaches the population mean.

5.5.3 The results of the 'flag method' and simple random sampling compared

The officials use these methods for several games. They find that the means from the two methods are different. So they ask the Danish football authorities to provide the true value of the average number of spectators going to the stadium. The results are summarized in Fig. 1.1. The assumption that for every Danish flag seen in the crowd there were 20 spectators was wrong; actually there were only 15 spectators for each flag. The official can correct for this bias if it is known but in real life the bias is often unknown. Also notice that in the beginning the bias from the flag method is hidden by the variance. We shall return to this point later in the chapter.

The result from the simple random sampling method is an accurate method in that the mean of the sample equals the mean of the population (compared to the flag method). It is not a very precise value because it has a high sampling variance. In an ideal world we would like to have a precise accurate estimate but would settle for an accurate estimate (since by doing more work the precision can be

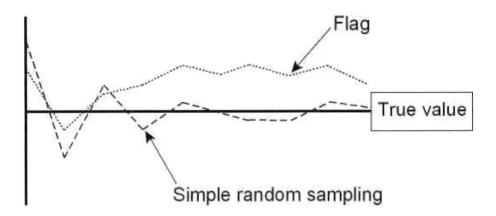

Fig. 1.1 Bias and variance. Initially there is a high degree of variance associated with both methods but as more observations are made the means of both methods settle to a stable value. The mean obtained from the simple random sampling method approaches the true value as the number of observations increases because it is unbiased. However, the mean from the flag method does not approach the true value as the number of observations increases because it was biased.

increased). The worst possible scenario is to have a precise inaccurate estimate where the bias is not known, since no matter how much extra work is done the result will always be an inaccurate estimate.

Notice how the sampling variance decreases as the number of observations increases. For a simple variable, x, with an arithmetic mean, $\bar{x}$, the sampling variance, expressed as the coefficient of error, CE, in stereology, will depend on the natural variation of the population under study, expressed as the coefficient of variation, CV, and the number of observations made, n. The two are related by

$$CE(x) = \frac{CV}{\sqrt{n}} = \frac{\text{standard error of the mean }(x)}{\bar{x}}$$

$$CV(x) = \frac{\text{standard deviation }(x)}{\bar{x}}.$$

The coefficient of error will only reduce by the square root of the number of observations made. This means that, after about 200 observations, this reduction is disproportionately small compared with the extra work required to make the observations.

To make the sampling more efficient, that is, reduce the variance some modifications to the simple random sampling can be used, for example, systematic sampling.

5.5.4 Systematic sampling

Stereological sampling uses systematic random sampling, that is, after a random start the population of interest is sampled at regular intervals, which is more efficient than simple random sampling where the population of interest is repeatedly sampled at random intervals (Gundersen and Jensen 1987). With systematic sampling the coefficient of error will be reduced by a factor greater than the square root of number of observations made, but the exact reduction and how to calculate the coefficient of error CE without bias have only recently been developed. The method was originally developed by Gundersen and Jensen (1987) and then further modified (Gundersen *et al.* 1999). It is commonly used for stereological methods applied to neuroscience and examples of its use can be found throughout the book.

In this example the initial procedure of choosing one row at random is the same as for simple random sampling but, if systematic sampling is used, the subsequent choices will be at a predetermined interval. If there are 2000 rows the official might want to make 10 samples and so would choose a row 200 from the first. To avoid repeatedly sampling the same rows it is best to make the interval a prime number so in this case it would be 197. Thus the estimates will be made from samples dispersed throughout the sample.

In a population that is large and widely dispersed, gathering a simple or a systematic random sample may pose a problem. The population may need to be divided, for example, into clusters or stratified. In cluster sampling, items are grouped into arbitrary 'clusters'. An unbiased sampling method is used to select the clusters that are then sampled without bias. In stratified sampling the population is divided into subpopulations or strata and every stratum is sampled with an unbiased sampling rule.

Finally, it is important to remember that the most important factor in any sampling design is the natural variation between individuals. It is therefore much more efficient to spend extra time sampling more individuals rather than concentrating on making more precise measurements within an individual. The 'Do more less well paper' by Gundersen and Østerby (1981) showed that, in order to reduce the sampling variance in a biological experiment, the measurement of rat glomerular basement membrane thickness, the greatest contribution to sampling variance came from the highest level of sampling, that is, the individual animals. Therefore it was more efficient to study more individuals rather than spending more time taking extra tissue blocks, or making extra sections from the tissue blocks, or making more precise measurements in more fields of view. Intuitively, this is a very simple observation. If the basement membrane thickness is measured very precisely in one field of view, on one section, from one tissue block from one kidney from one individual, a lot of information has been obtained about that one field of view and but not much about the basement membrane thickness of the population of rats under study.

A general rule used in stereology is to take at least five items from each sampling level, that is, animals, blocks, sections, fields of view, and make a total of 200 measurements per animal. This rule can then be used to set up a pilot experiment whose results can be used to calculate the variances at the different levels of sampling. This can then be used to check the feasibility of an experiment and to optimize the experimental conditions (see Chapter 2, this book).

Geometric probes

In a stereological sampling poll the questions are asked using geometric probes, that is, points, lines, areas, and volumes.

Consider a room filled with a regular array of points. If an object is placed in the room, a potato, for example, the number of points that fall inside the potato will be proportional to its volume, that is, the larger the volume of the object the greater the probability of a point hitting the object and the more points that will fall inside it. If the volume associated with each point is known, an estimate of the volume of the object is simply the number of points falling inside the object multiplied by the volume associated with each point, provided that the initial position of the regular array of points was randomized. The purpose of choosing a uniform random start is to give all parts of the object an equal probability of being sampled, that is, to be unbiased; this is a general requirement of all almost all stereological methods. In practice the points are introduced into the object, the potato, by sectioning it and superimposing a test grid of points on to the section. The estimate of the object's volume is the total number of points hitting the object multiplied by

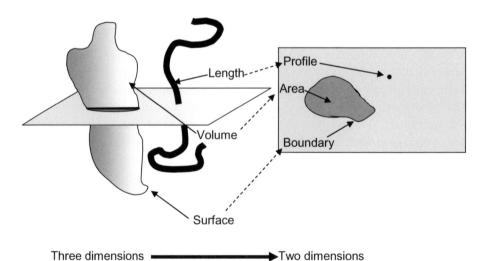

Fig. 1.2 Three dimensions to two dimensions. When three-dimensional objects are reduced to two-dimensional representations their physical properties also change. Volumes become areas, surfaces become boundaries, and lengths become profiles.

Table 1.1 Stereological probe geometric and the characteristics of three-dimensional objects

Stereological probe	Dimension	Geometric characteristic	Dimension
Point	0	Volume	3
Line	1	Surface	2
Plane	2	Curve	1
Disector	3	Number	0

volume associated with each point, that is, the area associated with each point multiplied by the mean section thickness. Notice that, in moving from a three-dimensional object to a two-dimensional section, the potato's volume is reduced to an area, its surface to a boundary, and, if it is a very long potato, its length will become a profile (see Fig. 1.2).

A test point will fall inside an object with a probability related to its volume: hence its use to estimate volumes. A test line will hit an object's surface with a probability related to the object's surface, and an area will contain an object's profile with a probability related to the object's length. These relationships are clarified in Table 1.1. From the table it can be seen that a pattern emerges: the dimensions of the probe plus the dimension of the geometric characteristic of the object equals three so that the geometrical characteristic that is being estimated is meaningful in *three-dimensional space*. These are the principal relations when estimates are obtained by counting events. As will be seen later, intercept measurements can also be used to obtain estimates of volume and spatial distributions using the nucleator principle.

The stereological methods that estimate length and surface area require that isotropy be considered. That is to say, the orientation distribution of either the objects being investigated or the probe itself must be uniform. The stereological methods that estimate number and those that use point counting (i.e. where either the number of dimensions of the probe or of the geometric characteristic equals three) do not have this requirement.

6 Isotropy

Consider a set of test lines used to estimate the surface area of an object, for example, a disk. The number of lines that will intersect the disk's surface will be dependent on the orientation of the disk with respect to the test lines. Therefore, for the estimate to be unbiased, either the orientation of the disk or that of the test lines must be chosen at random, that is, the interaction between test line and disc surface can occur with equal likelihood in all possible directions in three-dimensional space. This is the concept of isotropy.

Therefore either the method employed must take into account isotropy, for example, by making measurement in an isotropic direction in three-dimensional space, for example, using isotropic uniform random (IUR) sections, or the structure itself must be isotropic. The 'new stereology' uses the former approach so that no assumptions need be made about the object under study. Mattfeldt *et al.* (1990) has described a method for producing IUR sections, the orientator. Another practical approach to producing IUR sections is the isector (Nyengaard and Gundersen 1992).

Another elegant solution to the problem of isotropy, which can be applied, for example, to generate isotropic test lines, is the use of so-called vertical sections (Baddeley *et al.* 1986). Vertical uniform random (VUR) sections allow an object to be sectioned in any preferred direction, which may be crucial in order to identify various structures on the section, while still satisfying the requirement for spatially isotropic test lines. Vertical sections can be used to estimate surface area and for local volume estimators, for example, the rotator or nucleator. Estimates of length, when objects have been sectioned in a preferred direction, can be obtained by using vertical slices (Gokhale 1990).

In some situations it may be difficult to produce an IUR/vertical section by rotating the tissue prior to sectioning. A typical example occurs when working with animal models of spinal cord injury. The lesion-induced cavity collapses completely and the lesion interior is lost when the tissue is sectioned in certain directions. To obtain isotropy an alternative strategy may be used—to rotate the geometrical probe through virtual planes within thick physical sections (see also Chapter 13, this book).

Figure 1.3 gives examples of ways of making IUR sections.

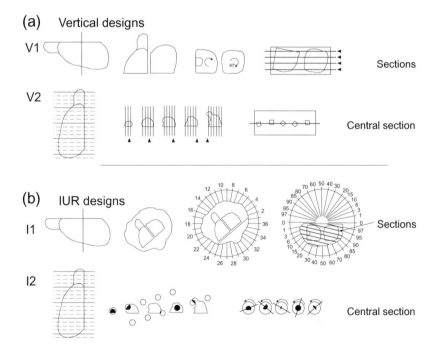

7 A cautionary note on density estimators

Whenever density estimators, for example, N_v, V_v, S_v, L_v etc., are used it is vital to measure the reference volume, (for example using Cavalieri's principle, see Chapter 8 for more details[1]) the volume of the neocortex in the above example, because it is impossible to know if the changes in the density are due to a change in the total quantity, for example, number of neurons, or a change in the reference volume. Investigators who do not measure the reference volume run the risk of falling into the *'reference trap'*. Many examples exist of false conclusions based on density estimates relating to total quantities and some can be found in Brændgaard and Gundersen (1986) and also Casley-Smith (1988).

The majority of these of these ideas are considered in more detail in subsequent chapters.

Fig. 1.3 How to make IUR sections. The figure shows how sections can be produced to solve the problem of making isotropic measurements using vertical IUR sectioning. Two examples of (a) vertical designs and (b) isotropic, uniformly random (IUR) designs, respectively, from a rat brain are illustrated. (a) In the first vertical section design, V1, the brain is divided in two halves and both halves are placed on the cut surface. The first half is rotated randomly and the other half is rotated 90° with respect to the first. Both halves are embedded in a block. Four sections are systematic uniformly random sampled and these are all vertical sections. In V2, the brain is sectioned into, for example, 15 slabs, and five slabs (full drawn lines) are sampled systematic uniformly random. The slabs are cut into bars and these are randomly rotated around their long axes. The rotated bars are embedded in a block and only one central section is used. Some of the bars may only provide a grazing section or none but the central section is still a uniform random vertical section because all bars have uniform random position in the brain. (b) In I1, the brain is cut into two halves. The first half is not moved and the other half is placed on the cut surface and randomly rotated around its axis. The two brain halves are embedded in, for example, agar, with their original axes approximately perpendicular to each other, and a two-step orientator (16) is performed. In the first step, the two embedded brain halves are placed at the center of the circle with 36 (360°) equidistant divisions along the perimeter. A random number between 0 and 36 is looked up (here 4). A cut parallel to the line in the direction 4 to 22 is performed in the embedding media (full drawn line). A straight edge is produced between the cut surface and the bottom surface of the agar. In the last step the embedded brain halves are placed on the cut agar surface with the straight edge parallel to a line between 0 and 0 on the second circle with non-equidistant (sine-weighted) divisions along the perimeter. A random number is looked up between 0 and 97 (here 95). Sections are cut parallel to a direction of 95 on the second circle to generate isotropic, uniform and random sections. In I2, the brain is cut into ~15 slabs and five slabs (solid lines) are sampled systematic uniformly random. Using a template with holes placed randomly over the slabs sampled, small brain blocks are sampled uniformly random and embedded in spherical agar isectors (17). The isectors are embedded in the final embedding media and the central section will be isotropic, uniform, and random. Note that several isectors can be embedded in a single block as with the bars under V2.

References

Baddeley, A.J and **Jensen, E.B.V**. (2002). *Stereology—sampling in three dimensions*, Research report no. 22. Laboratory for Computational Stochastics, Department of Mathematical Sciences, University of Aarhus, Denmark.

Baddeley. A.J., Gundersen, H.J.G., and **Cruz-Orive, L.M.** (1986). Estimation of surface area from vertical sections. *J. Microsc.* **142**, 259–76.

Braendgaard, H. and **Gundersen, H.J.G.** (1986). The impact of recent stereological advances on quantitative studies of the nervous system. *J. Neurosci. Meth.* **18**, 39–78.

Casley-Smith, J.R. (1988). Expressing stereological results 'per cm³' is not enough. *J. Pathol.* **156**, 263–5.

Cruz-Orive, L.M. and **Weibel, E.R.** (1990). Recent stereological methods for cell biology: a brief survey. *Am. J. Physiol.* **258**, L148–56.

DeHoff, R.T. (1983). Quantitative serial sectioning analysis: preview. *J. Microsc.* **131** (3), 259–63.

Evans, S.M., Howard C.V., and **Gunderson, H.J.G.** (1989). An Unbiased estimate of the total number of human neocortical neurons. *J. Anat.* **167**, 249–50.

Gokhale, A.M. (1990). Unbiased estimation of curve length in 3-D using vertical slices. *J. Microsc.* **159**, 133–41.

Gundersen, H.J.G. and **Jensen, E.B.** (1987). The efficiency of systematic sampling in stereology and its prediction. *J. Microsc.* **147**, 229–63.

Gundersen, H.J.G. and **Østerby, R.** (1981). Optimizing sampling efficiency of stereological studies in biology: or "Do more less well!". *J. Microsc.* **121**, 65–73.

Gundersen, H.J.G., Boysen, M., and **Reith, A.** (1981). Comparison of semiautomatic digitizer-table and simple point counting performance in morphometry. *Virchows Arch.* (*Cell. Pathol.*) **37**, 3–45.

Gundersen, H.J.G., Bendtsen, T.F., Korbo, L., Marcussen, N., Møller, A., Nielsen, K., Nyengaard, J.R., Pakkenberg, B., Sørensen, F.B., Vesterby, A., and **West, M.J.** (1988*a*). Some new, simple and efficient stereological methods and their use in pathological research and diagnosis. *Acta Pathol. Microbiol. Immunol. Scand.* **96**, 379–94.

Gundersen H.J.G., Bagger, P., Bendtsen, T.F., Evans, S.M., Korbo, L., Marcussen, N., Møller, A., Nielsen, K., Nyengaard, J.R., Pakkenberg, B., Sørensen, F.B., Vesterby, A., and **West, M.J.** (1988*b*). The new stereological tools: disector, fractionator, nucleator and point sampled intercepts and their use in pathological research and diagnosis. *Acta Pathol. Microbiol. Immunol. Scand.* **96**, 857–81.

Gundersen, H.J., Jensen, E.B., Kieu, K., and **Nielsen, J.** (1999). The efficiency of systematic sampling in stereology—reconsidered. *J. Microsc.* **193**, 199–211.

Mathieu, O., Cruz-Orive, L.M., Hoppeler, H., and **Weibel, E.R.** (1980). Measuring error and sampling variation in stereology: comparison of the efficiency of various methods for planar image analysis. *J. Microsc.* **121**, 75–88.

Mattfeldt, T., Mall, G., Gharehbaghi, H., and **Müller, P.** (1990). Estimation of surface area and length with the orientator. *J.Microsc.* **159**, 301–17.

Mayhew, T.M. (1992). A review of recent advances in stereology for quantifying neural structure. *J. Neurocytol.* **21**, 313–28.

Miles, R.E. and **Davy, P.J.** (1976). Precise and general conditions for the validity of a comprehensive set of stereological fundamental formulae. *J. Microsc.* **107**, 211–26.

Nyengaard, J.R. (1999). Stereologic methods and their application in kidney research. *J. Am. Soc. Nephrol.* **10**, 1100–23, 1999.

Nyengaard, J.R. and **Gundersen, H.J.G.** (1992). The Isector: a simple and direct method for generating isotropic, uniform random sections from small specimens. *J. Microsc.* **165**, 427–31.

Pakkenberg, B., Møller, A., Dam, A.M., Gundersen, H.J.G., and **Pakkenberg, H.** (1991). The absolute number of nerve cells in substantia nigra in normal subjects and in patients with Parkinson's disease estimated unbiasedly with a stereological method. *J. Neurol. Neurosurg. Psychiatry* **54**, 30–33.

Stuart, A. (1984). *Basic ideas of sampling.* Griffin & Co, London.

Weibel, E.R. (1992). Stereology in perspective: a mature science evolves. *Acta Stereol.* **11** (suppl. I), 1–13.

CHAPTER 2

A CASE STUDY FROM NEUROSCIENCE INVOLVING STEREOLOGY AND MULTIVARIATE ANALYSIS

LUIS M. CRUZ-ORIVE, ANA M. INSAUSTI, RICARDO INSAUSTI, AND DÁMASO CRESPO

1 Introduction

Down syndrome is characterized by a triplication of chromosome 21 (HSA-21), and is the most frequent cause of mental retardation in the newborn; other known chromosome abnormalities related to Down syndrome are translocation and mosaicism (Bersu 1980). Various brain abnormalities have been reported in humans (see Insausti *et al.* (1998) for details and references). A special trisomic mouse (Ts65Dn) is a well known animal model for Down syndrome. Motivated by the memory impairment and other behavioral alterations observed in these mice, Insausti *et al.* (1998) carried out a stereological study that revealed significant changes in total neuron numbers in the denate gyrus (DG) and in the pyramidal layer (CA3) of the hippocampus of Ts65Dn mice with respect to controls. We have therefore carried out a new study to try to detect possible differences between the two mice groups in the total neuron number for each of four subpopulations, defined as expressing, respectively:

- calbindin (Cb) D-28k (Freund and Buzsaki 1996);
- parvalbumin (Pv);
- glial fibrillary acidic protein (GFAP) in astrocytes;
- NADPH-diaphorase (NADP; Dun *et al.* 1994);

within each of three well defined compartments of the hippocampus, namely:

- dentate gyrus (DG);
- union of CA1 and CA2;
- CA3 (pyramidal layer).

The problem is first formulated in statistical terms in Section 2.2. Reasons are given there to favor a multivariate approach: for each cell subpopulation the three cell number estimates from each animal are taken together to constitute a three-dimensional random vector.

Section 3 is devoted to the stereology required to estimate the input data (namely, the cell numbers) for the subsequent statistical analyses. It is emphasized that the unbiasedness of the geometric sampling depends only on the sampling design, and not on cell arrangement, object shape, etc. The latter may affect the error variance, but this can be controlled by properly tuning the sampling intensities (this is examined in Section 5). Biases arising in practice from sources other than the geometric design itself are currently a source of debate (e.g. von Bartheld 2001); some comments are devoted to this in Section 8.

The importance of a proper graphical display for a preliminary examination of the data is also emphasized (as illustrated by Figs 2.6, 2.8, and 2.9). The choice of a statistical test (notably between parametric and nonparametric) is discussed in Section 4.2 with particular reference to graphical displays and suitable tests of normality. A fairly detailed description of the univariate parametric case (Section 4.3) and a worked example (Section 4.4) are included.

The results indicated biologically important differences, which nonetheless were not statistically significant due to the small numbers of animals used, to the biological variance among animals, and to the stereological error variance within animals. These facts offered us the opportunity to illustrate how to predict the number of animals required in each case to detect a desired difference between groups for a fixed significance level and a desired power (Section 5). The case in which the stereological workload per animal is kept unchanged is considered in Section 5.1, whereas in Section 5.2 we expand on the possibility of increasing the workload per animal in order to decrease the within-animal (and hence the total) variance. The latter analysis requires the splitting the total variance among the animals of a group into the biological variance and the stereological error variance. A worked example is given in Section 5.4. The multivariate approach is described in Section 6, and illustrated with another example. In the numerical examples, a constant reference is made, with concrete details, to the convenience of using the S-PLUS software package (Becker *et al.* 1988; Venables and Ripley 1999). The results of the complete analyses following validation of the models, and the sample sizes required to detect a given difference between the two groups, are summarized in Section 7. Finally, the main conclusions and a number of remarks pertaining to the stereology and the statistics used are discussed in Section 8. Section 9 comprises a glossary of sampling and statistics and a list of notation used in this chapter.

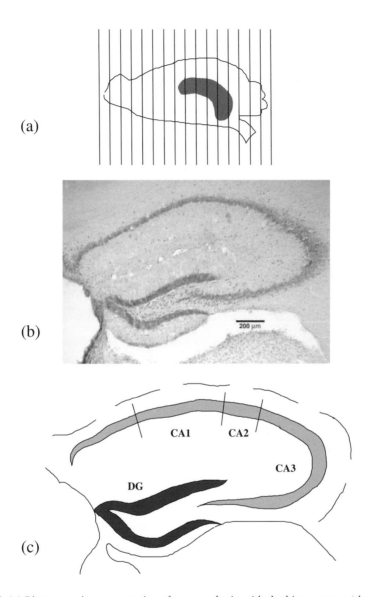

Fig. 2.1 (a) Diagrammatic representation of a mouse brain with the hippocampus (shaded region) inside it. The lines represent the boundaries of adjacent slices (whose thickness has been exaggerated for clarity), which constitute the basis of the stereological design. (b) Micrograph of a light microscopical (LM) slice of 50 μm nominal thickness through the hippocampus of one of the mice used in this study. (c) Diagrammatic version of the image in (b) showing the approximate definitions of the relevant compartments. Within each compartment we considered four cell subpopulations (see Section 1)

2 Problem, material, and statistical framework

2.1 Problem and material

The quantity of interest was the total number of cells—which we denote generically by N(neu)—in each of the three compartments DG, CA3, and the union of CA1 and CA2 (see Fig. 2.1). The problem was to compare these numbers in control and trisomic mice for each of the four cell subpopulations Cb, Pv, GFAP, and NADP mentioned in Section 1.

The trisomic group consisted of $n_2 = 4$ mice (except for the Cb subpopulation, in which $n_2 = 3$) and the control group of $n_1 = 5$ littermates, supplied by Jackson Laboratory (Bar Harbor, ME, USA). The group sizes were unequal owing to the high mortality among trisomic mice. The animals had been used in a previous study (Insausti *et al.* 1998); as described there, they were sacrificed at 20–28 weeks of age with a barbiturate overdose and, after a brief wash with saline, perfused transcardially with 4% paraformaldehyde, 0.1% glutaraldehyde, and 15% saturated picric acid in 0.1 M phosphate buffer (pH 7.3). The brains were removed and placed in 30% sucrose in phosphate buffer prior to slicing for stereology (see Section 3.1).

2.2 Statistical framework

For each subpopulation, the cell numbers in the relevant compartments of a given hippocampus will tend to be correlated, because they are somewhat predetermined by the total number of cells in the same hippocampus. Therefore, repeated statistical tests comparing control and trisomic groups separately for each compartment will not be independent, and the adopted significant levels will tend to be underrated. To avoid this shortcoming, for each subpopulation the estimated cell numbers in the three relevant compartments within an animal may be integrated together into a three-dimensional random vector, whereby the subsequent statistical tests become multivariate instead of univariate.

The control group consists of n_1 animals, which may be regarded as n_1 independent replications of the three-dimensional random vector

$$\mathbf{X}_1 = (\hat{N}(\text{neu}) \text{ in DG}; \hat{N}(\text{neu}) \text{ in CA1 and CA2}; \hat{N}(\text{neu}) \text{ in CA3}), \quad (2.1)$$

where $\hat{N}$(neu) denotes the estimator (obtained by stereology; see Section 3), of the corresponding true N (neu) in each case. For the trisomic group we define the corresponding vector $\mathbf{X}_2$, of which we observe n_2 independent replications. The data are displayed in Table 2.2 which is as located and discussed in Section 6.3.

For each subpopulation, let μ_1, μ_2 denote the means of the vectors $\mathbf{X}_1$, $\mathbf{X}_2$, respectively. The problem is reduced to test the null (H_0) and alternative (H_1) hypotheses

$$H_0: \mu_1 = \mu_2 \quad \text{versus} \quad H_1: \mu_1 \neq \mu_2 \quad (2.2)$$

at a specified level of significance. If the null hypothesis H_0 is rejected, then we can formally identify which compartments are responsible for the detected difference between the two groups.

3 Estimation of neuron number by the classical Cavalieri-disector stereological design

3.1 Design and method

Uniform sampling guarantees estimation unbiasedness (notably for population totals) and it means sampling of the population units with identical probabilities. There are many kinds of uniform sampling, which may differ widely in their efficiency: the best known are simple random sampling, and many variants of systematic sampling. Stereology uses systematic sampling (in a geometric context) quite extensively, first because it is easy to implement, and second because it is often much more efficient than simple random sampling.

The designs described below are variants of geometric systematic sampling, and therefore they guarantee estimation unbiasedness irrespective of the size and shape of the reference space, and of the size, shape, and spatial arrangement of the cells in the reference space. The populations of interest are assumed to be non-random, and the necessary mechanism of randomness is carried by the probes. To estimate the error variance predictors of total cell number estimators (Section 3.3) mild assumptions may occasionally be made about the spatial arrangement of the cells (notably the Poisson assumption).

Diverse artifacts may nonetheless interfere with the correct implementation of the aforementioned sampling mechanisms in practice; this aspect is briefly discussed in Section 8.2.

The total number $N(\text{neu})$ of cells of each subpopulation within each compartment of a given hippocampus may be estimated indirectly via the corresponding number density $N_v(\text{neu, ref}) = N(\text{neu})/V(\text{ref})$ using the identity

$$N(\text{neu}) = V(\text{ref}) \cdot N_v(\text{neu, ref}) \tag{3.1}$$

first proposed by Evans *et al.* (1989) (see also Braendgaard *et al.* 1990; West and Gundersen 1990), where $V(\text{ref})$ represents the volume of the reference space (namely, the compartment of interest in this case). To estimate $V(\text{ref})$ and $N_v(\text{neu, ref})$ a Cavalieri design was adopted. The identity (3.1) is a particular case of the classical multilevel or 'cascade' design (see Weibel 1979; Cruz-Orive and Weibel 1981).

In the present study each brain was cut coronally into slices of $t = 50$ μm thickness (Fig. 2.1(a)). Every fifth slice (chosen with a random start) was stained with 0.25% thionin with a double purpose: to estimate $V(\text{ref})$ and to help in the identification of the reference space when quantifying the cell subpopulations.

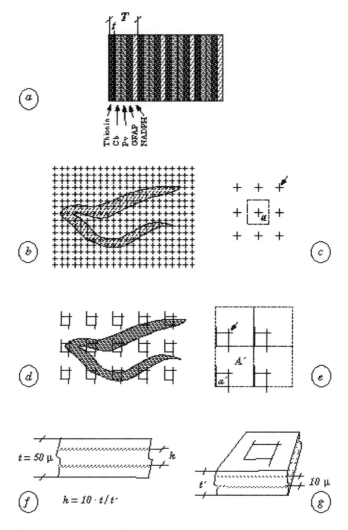

Fig. 2.2 (a) Slice treatment protocol. The thionin slices were used to estimate the relevant compartment volumes of the hippocampus, and to help identifying their boundaries when analysing the remaining slices. The latter were used to reveal the relevant cell populations. Each of the five series was a Cavalieri series of period $T = 5t$. (b), (c) Estimation of the area of a section (of DG here) by point counting for eqn. (3.2). The arrowed corner defines a test point. (d), (e) Subsampling of the section by systematic quadrats. The quadrat boundaries are kept in focus while the slice is scanned from top to bottom through a height h (see (f)). A neuron is counted if its nucleus (or the whole cell in some cases) appears in focus for the first time between both faces of the disector provided it hits neither the upper face of the disector, nor the forbidden lines of the disector frames (thicker lines in (e), (g)). (g) Perspective view of the slice in (f) after shrinkage. The nominal disector thickness was always 10 μm, and the proper h had to be estimated as indicated in (f) before plugging it into eqn. (3.3).

Adjacent slices (Fig. 2.2(a)) were handled by immunohistochemical techniques to reveal the Cb, Pv, GFAP, and NADP subpopulations, as described elsewhere (Hsu et al. 1981). In this way, five different Cavalieri series were obtained to suit our purposes. The period of each series therefore was $T = 5t = 250$ μm $= 0.250$ mm.

For the reasons explained in Insausti et al. (1998), for each animal only one hemisphere—either left or right, with probability 1/2—was analysed. Thus, the estimate of V(ref) or of N(neu) reported for each brain was twice that computed for the available hemisphere. This operation does not introduce any bias, but it doubles the estimation variance within each brain, and it adds a term to the observed variance of the estimators among animals which depends on the hemisphere asymmetry (see Section 5.3 for more details).

The relevant compartment volume was estimated in each case via

$$\hat{V}(\text{ref}) = T \cdot a \cdot \sum P(\text{ref}) \quad \text{mm}^3 \qquad (3.2)$$

(e.g. Gundersen and Jensen 1987; Howard and Reed 1998; and references therein). For the sake of simplicity we omit summation limits and variable subscripts, and we understand that the summations run over the set of sampled sections or slices. For instance, in the preceding formula $\sum P(\text{ref})$ denotes the total number of test points counted in the region of interest with a test system superimposed on each section uniformly at random (Fig. 2.2(b)). The constant a represents the area per point (corrected for magnification) for the test system used (Fig. 2.2(c)), and it was different for each subdivision in the hippocampus to keep $\sum P(\text{ref})$ between 100 and 400. The latter prescription is intended only to keep the error variance under reasonable limits, and it does not affect the estimation unbiasedness. We adopted $a = 11525$ μm^2 for DG, $a = 8004$ μm^2 for CA1 and CA2, and $a = 25931$ μm^2 for CA3, respectively. The procedure was carried out on a monitor screen with the aid of the CAST Grid system (Olympus Denmark). The $\hat{V}(\text{ref})$ results were reported in Insausti et al. (1998).

The cell number densities were estimated on the specially treated adjacent series (Fig. 2.2(a)) by systematically subsampled optical disectors (Gundersen et al. 1988; West and Gundersen 1990), as illustrated in Fig. 2.2(d). The estimator used was the ratio of the total number of neurons counted in all disectors (as illustrated in Fig. 2.3), $\sum Q^-(\text{neu})$, to the corresponding total disector volume in the reference space, namely,

$$\hat{N}_V(\text{neu, ref}) = \frac{\sum Q^-(\text{neu})}{a' \cdot \sum \left(h(\text{ref}) \cdot P'(\text{ref}) \right)} \quad \text{mm}^{-3} \qquad (3.3)$$

where $h(\text{ref})$ represents the corrected optical disector height for a slice (Fig. 2.2(f)), $P'(\text{ref})$ the number of upper right corners of the sampling frames (Fig. 2.2(e)) hitting the relevant reference space in the upper optical face within the same slice, and a' the area of the counting frame (corrected for magnification; Fig. 2.2(e)). It

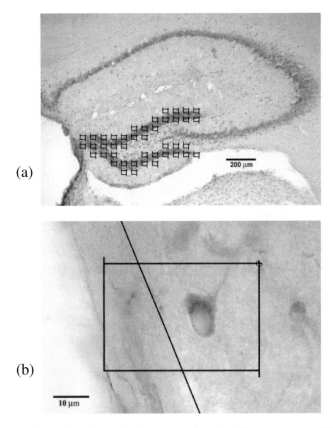

(a)

(b)

Figure 2.3 (a) Subsampling of a section by systematic optical disectors, as a real illustration of Fig. 2.2(d). (b) Sampling frame (see area a' in Fig. 2.2 (e)) of one of the optical disectors illustrated in (a). A cell was counted if it was captured by the frame (as explained in Fig. 2.2 (e)) and its nucleus was observed in focus within a pre-established disector thickness controlled by an electronic microcator (see Fig. 2.2 (g)) and within the reference space. The crossing line is a fragment of the boundary of the reference space, drawn before subsampling with the aid of the CAST Grid system at a low magnification. The disector would contribute to P'(ref) in the denominator of eqn. (3.3) because its associated point at the upper right-hand corner hits the reference space.

is important to realize that the relevant reference area in a section is estimated by P'(ref) $\cdot$ a' (= $4a'$ in Fig. 2.2(d)), and not by the number of quadrats hitting the reference area times a' (= $8a'$ in Fig. 2.2(d)). The actual disector height, measured with the microcator of the CAST Grid system, was 10 μm, (Fig. 2.2(g)). The nominal slice thickness was $t = 50$ μm, but the final slice thickness t' after shrinkage varied from one slice to another; each slice thickness was measured at several points in the slice with the aid of the microcator, and the corresponding average was used to represent t' for that slice. The corresponding corrected disector thickness was h(ref) $= 10 \cdot t/t'$ μm, (Fig. 2.2(f)). The total optical disector volume sampled from the slice therefore was estimated by $a' \cdot h(\text{ref}) \cdot P'(\text{ref})$. Summation over slices constitutes the denominator of eqn. (3.3). The area a' was different for each region, to keep $\Sigma Q^-(\text{neu})$ and $\Sigma P'(\text{ref})$ within the limits 100–400 in each region. For the Cb subpopulation we adopted $a' = 1107$, 886, and 1550 μm^2 for DG, CA1–CA2, and CA3, respectively; for the Pv subpopulation, $a' = 1240$, 1107, and 974 μm^2; for the GFAP subpopulation, $a' = 1328$, 1151, and 1151 μm^2, and for the NADP subpopulation, $a' = 1328$ μm^2, for all three compartments, respectively.

3.2 Preliminary results

The preceding stereological procedures were applied to each animal of each of the two groups described in Section 2.1. The results are presented graphically by means of dot plots in Fig. 2.4. Dot plots enjoy a number of advantages, and they constitute a good choice for neurostereology. In particular:

- The vertical axes describe a global quantity of biological meaning. The most common descriptors in the stereological context are number, length, surface area, and volume for features of dimension 0, 1, 2, 3, respectively. The numerical scales start at zero.

- Each dot represents a primary sampling unit (henceforth abbreviated 'p.s.u.'), namely, an animal in the present context. The sample mean is represented by a horizontal trace. The dot patterns allow a cursory comparison between two groups at a glance, as well as a fair assessment of the variability within a group, and also a possible departure from normality, the presence of outliers, etc. If the number of animals per group exceeds, say, 20, then box plots may be more appropriate (e.g. Venables and Ripley 1999).

The type of graphs displayed in Fig. 2.4 incorporate the relevant information supplied by the experiment in a rather comprehensive manner. Numerical tables may nonetheless be considered to complement the information. For instance, the first three columns of Table 2.4 (see Section 7.1.) incorporate estimates of the relative group mean differences, defined for each cell subpopulation in each hippocampus compartment as follows,

$$\Delta\% = 100 \cdot \left(\frac{\text{Mean } \hat{N}(\text{neu}) \text{ in trisomic group}}{\text{Mean } \hat{N}(\text{neu}) \text{ in control group}} - 1 \right). \tag{3.4}$$

A glance at Fig. 2.4 reveals that the only significant change is likely to correspond to the number of Cb neurons in the CA3 compartment. Often, a formal statistical test will merely confirm a proper scientific assessment based on a proper graph: this study was no exception.

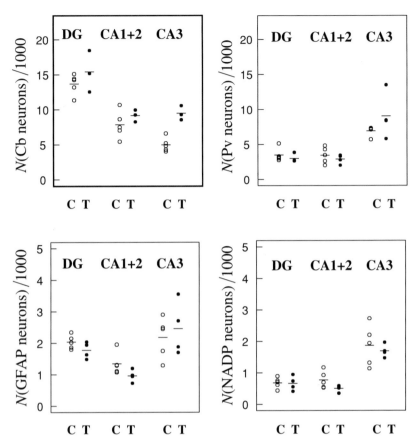

Figure 2.4 Cell numbers estimated for the relevant subpopulations in the hippocampal compartments considered in this chapter. Each open circle corresponds to a control, and each filled dot to a Ts65Dn mouse. The horizontal segments represent group means. C, control group; T, trisomic group.

As pointed out in Section 1, one of our goals was to illustrate the statistical treatment of the data to detect group mean differences. While the approach of choice here is multivariate (Section 2.2), for didactic purposes this is deferred to Section 6, and the univariate approach is described first (Section 4).

3.3 Error variance of the neuron number estimator

Our purpose is to estimate, within each animal, the error variance (due to stereology only) of the estimator based on eqn. (3.1), namely,

$$\hat{N} = \hat{V} \cdot \hat{N}_v \qquad (3.5)$$

where $\hat{N} \equiv \hat{N}(\text{neu})$, $\hat{V} \equiv \hat{V}(\text{ref})$, and $\hat{N}_V \equiv \hat{N}_V(\text{neu, ref})$, for short. Within each animal the estimators $\hat{V}$ and $\hat{N}_V$ are obtained independently, in which case it can be shown that

$$CE^2(\hat{N}) \approx CE^2(\hat{V}) + CE^2(\hat{N}_V) \qquad (3.6)$$

where $CE^2(\cdot) = \text{Var}(\cdot)/(\text{Mean}(\cdot))^2$ denotes the true square coefficient of error of an estimator. The Cavalieri contribution $CE^2(\hat{V})$ may be estimated as shown in García-Fiñana and Cruz-Orive (2000a) and García-Fiñana et al. (2003). However, the modern refinements incorporated in the latter formulae have not yet been implemented to estimate the contribution $CE^2(\hat{N}_V)$ due to disector sampling from systematic sections; in fact, the only formula hitherto available to estimate the CE^2 of a ratio estimator from systematic sections is formula (16) from Gundersen and Jensen (1987) (see also eqn. (6.9) from Pache et al. 1993), which in the light of the recent developments cannot be regarded as fully satisfactory to account for the local variation within slices (Cruz-Orive 1999).

The preceding difficulties can be circumvented if we note that the right-hand side of eqn. (3.5) is basically equivalent to the alternative expression arising from the fractionator idea, namely,

$$\hat{N} = f_1 \cdot f_2 \cdot f_3 \cdot \sum Q^- \qquad (3.7)$$

where f_1, f_2 and f_3 are known constants called the sampling periods, whereas $\sum Q^-$ is the total number of neurons counted in all the optical disectors (the summation runs over the available slices; see Gundersen 1986; Geiser et al. 1990; West et al. 1991). With reference to Fig. 2.2, here we would have $f_1 = T/t$, $f_2 = A'/a'$, and $f_3 = t/h = t'/10$. The actual values of these sampling periods are not involved in $CE^2(\hat{N})$, however, because in principle they are constant or nearly constant, and therefore we have

$$CE^2\left(\hat{N}\right) = CE^2\left(\sum Q^-\right), \qquad (3.8)$$

at least approximately. Thus, although the fractionator design was not used in the present study, we may use the preceding identity to predict the error of the Cavalieri-disector estimator (3.5). The right-hand side of the identity (3.8) no longer involves the CE^2 of a ratio, but only the sum of cell counts over systematic slices, for which an estimator is available, namely,

$$ce^2\left(\hat{N}\right) = \alpha(\tau) \cdot \frac{3\left(C_0 - \sum Q^-\right) - 4C_1 + C_2}{\left(\sum Q^-\right)^2} + \frac{1}{\sum Q^-} \qquad (3.9)$$

(Cruz-Orive 1999; eqn. (3.12)), where

$$C_k = \sum_{i=1}^{n_s-k} Q_i^- \, Q_{i+k}^-, \qquad k = 0, 1, \ldots, n_s - 1, \qquad n_s \geq 3 \qquad (3.10)$$

(n_s is the number of slices, Q_i^- the total number of cells counted in the ith slice). The first term in the right-hand side of (3.9) estimates the slice contribution and it depends solely on the shape of the reference space, whereas the second term estimates the local variation or 'nugget' within the slices. The latter term is based on the assumption that cells are spread purely at random (Poisson model); we emphasize that the latter assumption is made only to simplify the nugget contribution. Further, the coefficient $\alpha (\tau)$ depends on the ratio $\tau = h/T$ (namely the proportion of space scanned by the optical disectors in a direction normal to the slices), and on the smoothness constant m of the Cavalieri area function of the reference space. The relevant expressions are

$$\alpha(\tau) = \begin{cases} \frac{1}{6}(1-\tau)^2 / (2-\tau), & m=0, \\ \frac{1}{6}(1+2\tau - 2\tau^2)(1-\tau)^2 / (40 - 10\tau^2 + 3\tau^3), & m=1 \end{cases} \qquad (3.11)$$

(Gual-Arnau and Cruz-Orive 1998; Cruz-Orive 1999). If the number of slices is, say, less than 10, then we may choose $m = 1$ if the reference space is fairly regular ('quasi-ellipsoidal'), and $m = 0$ otherwise (García-Fiñana and Cruz-Orive 2000b). Note that eqn. (3.9) constitutes an improvement over earlier ones (e.g. Gundersen et al. 1999) because the inclusion of the coefficient $\alpha(\tau)$ enforces the natural requirement that the slice contribution tends to zero when the ratio $\tau = h/T$ tends to 1 (equivalent to exhaustive sampling by completely observed adjacent slices.)

4 Statistical detection of group mean differences: univariate analysis

4.1 Statement of the problem

In this section we assume that we are interested only in a given cell subpopulation from a given compartment of the hippocampus. The problem is to compare the mean cell number per animal in the control and the trisomic groups.

The variable of interest is the relevant cell number estimator $\hat{N}(\text{neu})$ per animal. We define $X_1 = \hat{N}(\text{neu})$ for a control animal and $X_2 = \hat{N}(\text{neu})$ for a trisomic animal. The population means of X_1 and X_2 are denoted by μ_1 and μ_2, respectively. Define the population mean difference $\delta = \mu_2 - \mu_1$. The problem is reduced to testing the null hypotheses $\delta = 0$ against the two-sided alternative, namely,

$$H_0: \delta = 0 \quad \text{versus} \quad H_1: \delta \neq 0. \quad (4.1)$$

We want to construct a test with a fixed significance level (or type I error) equal to α, which represents the probability of rejecting H_0 when it is true.

4.2 Choice of a statistical test

First we have to distinguish whether we have a two-sample design or a paired design. In the former design an independent sample of different p.s.u.s is chosen from each of the two populations under study. In the paired design, on the contrary, only one sample is used, and two variables (typically a control and a treatment variable) are observed in each p.s.u. of the sample. A variant of the latter design is the 'matched samples' design. A sample is drawn from the treatment population (consisting, for instance, of brains of people who suffered from a degenerative neurological disorder); then, for each member of this sample, a control match is chosen so that it ideally differs only in that the treatment (e.g. the degenerative disorder) is absent, while the remaining characteristics (sex, age, body or organ weight, etc.) are as similar as possible in both subjects. In a paired, or matched, design the relevant variable is the difference between the values observed in each pair. Clearly, in the present study we have a two-sample design.

Once we have identified the type of design, we have to choose among a non-parametric (or distribution-free) test, and a parametric test. The former are popular in biosciences, but the following remarks are opportune.

1 Non-parametric tests are not universally valid: they usually require correct sampling, which guarantees statistical independence among the observations.

2 Non-parametric tests are often chosen to circumvent the need to assume that the observations are normally distributed. This is however not necessary. First, a goodness of fit test for normality such as Kolmogorov–Smirnov's (see Sections 4.4 and 6.4 for numerical illustrations) is fairly adequate and easy to implement. Second, the parametric t-test, for instance, is fairly robust against moderate departures from normality.

3 The price paid for avoiding any distribution assumptions is usually a loss of power (i.e. a smaller probability of detecting a difference when it is there) with respect to a parametric test. Thus, other conditions being equal, a non-parametric test requires larger sample sizes than a parametric test provided that the model assumptions for the latter are correct.

4.3 Univariate two-sample *t*-test

A test based on normality may be adopted if the normality assumption can be verified. A popular, traditional test of normality is the chi-square (χ^2) goodness of fit test (e.g. Snedecor and Cochran 1980), which nevertheless requires moderately large samples. For any sample size no less than 4, there are excellent graphical diagnostic tools and alternative tests, notably the Kolmogorov–Smirnov test. The latter procedures are tedious to implement by hand, but—as suggested in Section 1—with a modern statistical software such as S-PLUS the task is very simple. Among the graphical tools S-PLUS offers cdf.compare and qq.norm (see Figs 2.6, 2.8 and 2.9 later in the chapter) whereas the Kolmogorov–Smirnov test may be implemented with the function ks.gof (Section 4.4).

With few exceptions ($\hat{N}$(Pv neu) in DG, both groups, and in CA3, control group) the *P*-values estimated with ks.gof were generally quite large (and consequently the null hypothesis of normality was not rejectable). In the exceptional cases the Wilcoxon non-parametric test yielded the same conclusions as the *t*-test described below.

We therefore concentrate on the case in which the random variables X_1 and X_2 defined in Section 4.1 are independent and normally distributed. To specify the test we must decide whether X_1 and X_2 have a common population variance (denoted by σ^2), or not (see, for instance, Snedecor and Cochran 1980 or Huntsberger and Billingsley 1989). If the null hypothesis of equality of variances is not rejected, then one may use the exact *t*-test described below; otherwise, one may use Welch's test. To test the null hypothesis of equality of variances one may use the *F*-test (which relies also on the normality assumption). The latter test was not significant for all the samples in Fig. 2.4 with the exception of the CA3 samples of $\hat{N}$(Pv neu).

We draw independent simple random samples from the control group, (sample size n_1, sample mean $\overline{X}_1$, sample variance s_1^2) and from the treatment group ($n_2, \overline{X}_2, s_2^2$, respectively). For instance, for X_1,

$$\overline{X}_1 = \frac{1}{n_1} \sum_{i=1}^{n_1} X_{1i}, \quad s_1^2 = \frac{1}{n_1 - 1} \sum_{i=1}^{n_1} \left(X_{1i} - \overline{X}_1\right)^2, \tag{4.2}$$

and similarly for X_2.

Under the preceding assumptions the sample mean difference is normally distributed with mean δ and variance given as

$$\overline{D} = \overline{X}_2 - \overline{X}_1, \quad \overline{D} \sim N\left(\delta, \sigma^2\left(\frac{1}{n_1} + \frac{1}{n_2}\right)\right) \tag{4.3}$$

where '~' means 'is distributed as', and $N(m, v)$ denotes a normal random variable with mean m and variance v. The estimator of the common variance σ^2 is

$$s^2 = \frac{(n_1 - 1)s_1^2 + (n_2 - 1)s_2^2}{n_1 + n_2 - 2}. \tag{4.4}$$

The optimal strategy consists in rejecting H_0 if $|\bar{D}| > D_c$ (see Fig. 2.5(a)), where D_c is called the critical value, which can be calculated from the condition $\mathbb{P}\left(|\bar{D}| > D_c | H_0 \text{ true}\right) = \alpha$, where '$\mathbb{P}$' denotes 'probability' and '|' denotes 'given that' or 'conditional upon'. It is more convenient to consider the statistic

$$t = \frac{\bar{D}}{s\sqrt{\dfrac{1}{n_1} + \dfrac{1}{n_2}}} \sim t_{n_1 + n_2 - 2} \quad \text{if } H_0 \text{ is true} \tag{4.5}$$

where t_k represents the Student's t distribution with k degrees of freedom (d.f.). The strategy is thus reduced to rejecting H_0 if $|t| > t_c$ (see Fig. 2.5(b)), where $t_c = t_{n_1 + n_2 - 2}(1 - \alpha/2)$ is the $100(1 - \alpha/2)\%$ percentile of $t_{n_1 + n_2 - 2}$. A popular alternative is to calculate the P-value (Fig. 2.5(c)), which intuitively represents the *a priori* probability of observing a mean difference between groups at least as large as ours when H_0 is true. Then H_0 is rejected if $P \leq \alpha$.

The power of the test is defined as

$$\text{Power} = 1 - \beta = \mathbb{P}(\text{rejecting } H_0 \text{ when } H_1 \text{ is true}), \tag{4.6}$$

(see Fig. 2.5(a)) where β is the 'type II error'. The following will help to better understand the multivariate case. For simplicity suppose that σ^2 is known. Then,

$$\begin{aligned}\beta &= \mathbb{P}\left(|\bar{D}| \leq D_c | H_1 \text{ true}\right) \\ &= \mathbb{P}\left(|\bar{D}| \leq z(1 - \alpha/2) \cdot \sigma \sqrt{\frac{1}{n_1} + \frac{1}{n_2}} \, \middle| \, H_1 \text{ true}\right) \\ &= \mathbb{P}\left(|Z^*| \leq z(1 - \alpha/2)\right), \end{aligned} \tag{4.7}$$

where $z(1 - \alpha/2)$ is the $100(1 - \alpha/2)\%$ percentile of the standard normal $Z \sim N(0,1)$, whereas

$$Z^* = \frac{\bar{D}}{\sigma \sqrt{\dfrac{1}{n_1} + \dfrac{1}{n_2}}} \sim N\left(\frac{\delta}{\sigma \sqrt{\dfrac{1}{n_1} + \dfrac{1}{n_2}}}, 1\right) \tag{4.8}$$

can be regarded as a 'non-central standard normal' with non-centrality parameter $\delta/\sigma\sqrt{1/n_1 + 1/n_2}$.

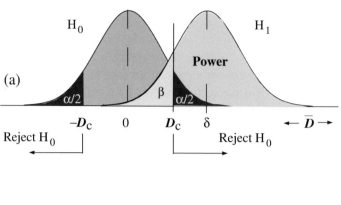

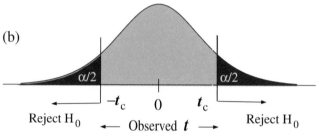

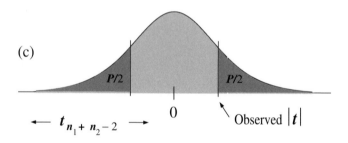

Fig. 2.5 (a) Illustration for the univeriate testing of the hypotheses (4.1). The real difference between the two population means is δ, and the observed difference is the random variable $\bar{D}$, which may lie anywhere along the abscissa. The curve on the left represents the probability density of $\bar{D}$ when H_0 is true ($\delta = 0$), whereas the curve on the right represents a case in which H_1 is true. The significance level α is fixed and it determines the critical value D_c. H_0 is rejected if $|\bar{D}| > D_c$. (b) Equivalently, H_0 is rejected if $|t| > t_c$, where t is given by (4.5) and t_c is the critical value (see Section 4.3). (c) Illustration of the concept of P-value, namely, $P = 2(1 - \mathbb{P}(t_{n_1 + n_2 - 2} \leq |t|))$.

In practice σ is not known and it is estimated by s. If we want to compute the power for $\delta = \bar{D}$, namely, for the observed difference, then the non-centrality parameter becomes precisely the t statistic computed via eqn. (4.5), and

$$\text{Power} = 1 - \beta = 1 - \mathbb{P}\left(\left|t^*\right| \le t_{n_1 + n_2 - 2}\left(1 - \alpha/2\right)\right)$$

(4.9)

where t^* follows the non-central t distribution with $n_1 + n_2 - 2$ d.f. and the afore-mentioned non-centrality parameter. A convenient alternative is to use eqn. (6.6) (see Section 6.1) with $p = 1$ and non-centrality parameter given by the square of the t statistic (4.5) (this is because the square of a t_k random variable is a $F_{1,k}$ random variable). For an illustration see the S-PLUS command (4.15) below.

4.4 Worked numerical example

To illustrate the foregoing test procedures we have chosen the GFAP neuron subpopulation in the union of the CA1 and CA2 layers of the hippocampus, first, because the normality goodness of fit test was non-significant for the

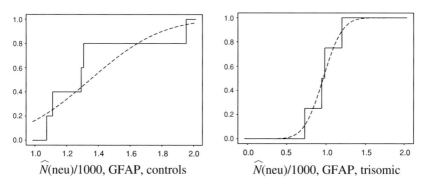

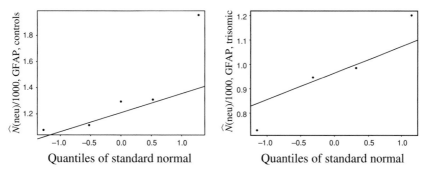

Figure 2.6 Graphical diagnoses of normality for the samples in Table 2.1 (Upper panels) Empirical (solid lines) and hypothesized normal (dotted lines) cumulative distribution functions. (Lower panels) Corresponding Q–Q plots.

corresponding samples and second, because the data are also useful to illustrate the sample size prediction procedures (Section 5.3). The sample data (control and trisomic) are displayed in Tables 2.1 (in Section 5.2) and 2.2 (in Section 6.3).

As a first step we checked the normality of the samples. In Fig. 2.6 (upper panels) the empirical cumulative distribution function (cdf) of each sample is plotted against the normal cumulative distribution function with the estimated mean and variance. The graphs were obtained directly with the command cdf. compare of S-PLUS. Quantile plots (or 'Q-Q plots') obtained with the functions qq.norm and qq.line are also displayed in Fig. 2.6 (lower panels). Normality was checked by Kolmogorov–Smirnov's test using the ks.gof function of S-PLUS, which yielded a *P*-value of 0.051 for the control group, and greater than 0.50 for the trisomic group.

The *t* statistic (4.5) requires that the null hypothesis of equality of variances is not rejected against the two-sided alternative. To test this we compute the statistic

$$F = \frac{\text{larger } s^2}{\text{smaller } s^2} = \frac{0.3543^2}{0.1934^2} \approx 3.36. \tag{4.10}$$

The critical value at the level $\alpha = 0.05$ is $F_{4,3}(0.975) \approx 15.10$ and, since this exceeds 3.36, the null hypothesis is not rejected. Alternatively, the *P*-value is $2(1 - \mathbb{P}\ (F_{4,3} \leq 3.36)) \approx 0.35$ and, since this exceeds 0.05, the null hypothesis is not rejected. This test may be carried out directly with the S-PLUS function var.test. By eqn. (4.4) the estimate of the common variance is $s^2 = 0.08774828$, and the *t* statistic (4.5) becomes,

$$t = \frac{-0.38415}{\sqrt{0.08774828\left(\dfrac{1}{5} + \dfrac{1}{4}\right)}} \approx -1.933. \tag{4.11}$$

The critical value at the level $\alpha = 0.05$ is $t_7(0.975) \approx 2.36$ and, since this exceeds 1.93, the null hypothesis of equality of the population means is not rejected. Alternatively, the *P*-value is $2(1 - \mathbb{P}(t_7 \leq 1.933)) \approx 0.095$ and, since this exceeds 0.05, the conclusion is the same. The S-PLUS function for the whole test is t.test.

Since the normality of the control sample is at the borderline of significance and the sample sizes are very small, we may try also the non-parametric Wilcoxon test with, say, the S-PLUS command wilcox.test. This yields a *P*-value of 0.064, which is also greater than 0.05.

The conclusion is therefore that, in spite of the fact that the estimate of the group mean relative difference defined by Eq. (3.4) is

$$\hat{\Delta} = (-0.38415/1.34840) \cdot 100 \approx -28.5\%, \tag{4.12}$$

namely, relatively large, this difference could not be declared significant by a formal statistical test. The reason for this is that the standard error of the group mean difference, namely, the denominator of eqn. (4.5), is too large. Equivalently, the power of the t-test to detect as significant the group mean difference $\delta = -0.38415$ is, according to eqn. (6.6),

$$\text{Power} = 1 - \beta = 1 - \mathbb{P} \left(F^* \leq F_{1,7}(0.95) \right) \approx 0.386, \tag{4.13}$$

which is too small. Here F^* follows a non-central $F_{1,7}$ distribution whose non-centrality parameter (ncp) is the square of the t statistic computed above (as discussed at the end of Section 4.3), namely,

$$\text{ncp} = t^2 \approx (-1.933)^2 \approx 3.736. \tag{4.14}$$

With S-PLUS, the above power may be computed with the command

$$1 - \text{pf}(\text{qf}(0.95, 1, 7), 1, 7, \text{ncp} = 3.736). \tag{4.15}$$

The purpose of the next section is to discuss how to increase the power.

5 Sample size predictions for further action

5.1 Sample size prediction when the observed variance among animals is not altered

We follow the notation and assumptions adopted in Section 4.3. If the true mean absolute difference $|\delta|$ between the two population means is large, then the observed mean absolute difference $|\bar{D}|$ is likely to be also large. However, if the error variance of $\bar{D}$, namely, $\text{Var}(\bar{D}) = \sigma^2(1/n_1 + 1/n_2)$, is large, then the power $1 - \beta$ of the test will be low (see Fig. 2.7(a)). In such cases it is unlikely that δ can be detected, that is, it is unlikely that H_0 can be rejected at the specified level α. If $\bar{D}$ is nonetheless biologically important, then we may try to redress matters (namely, to decrease $\text{Var}(\bar{D})$), either by increasing n_1 and n_2 (see below and Fig. 2.7(b)) or by decreasing σ^2 (Section 5.2). The equivalent goal is to reduce the denominator of the t statistic (4.5) so that the statistic itself increases and statistical significance is reached with probability $1 - \beta$ given α and δ.

Here we show how to predict the minimum sample size n per group so that H_0 can be rejected for a given true difference δ in favour of the two-sided alternative H_1 (see (4.1)) with a specified level of significance α and a desired power $1 - \beta$, when the variance σ^2 among animal estimates is not altered. From the definitions of significance level and of power (Fig. 2.5(a)) we obtain the equations

$$\frac{D_c}{\sigma\sqrt{2/n}} = z(1 - \alpha/2) \quad \text{and} \quad \frac{|\delta| - D_c}{\sigma\sqrt{2/n}} \approx z(1 - \beta), \tag{5.1}$$

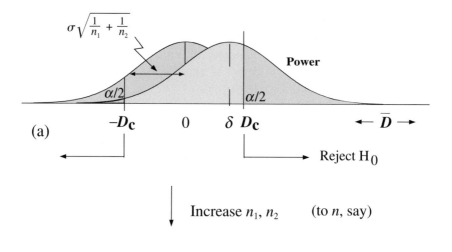

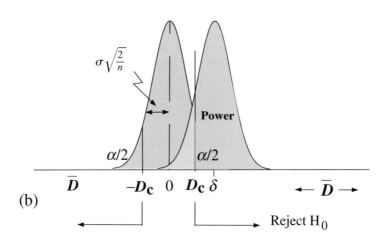

Figure 2.7 (a) The power of the test to detect the true difference δ between two population means is low because $\text{SE}(\overline{D})$ is large (either because the SD among animals, σ, is large, or because the sample sizes n_1 and n_2 are small, or both). (b) Everything else remaining equal, the power of the test can be arbitrarily increased by increasing the sample size per group (n) accordingly.

respectively (we recall that $z(q)$ is the q-percentile of the standard normal). Eliminating D_c we obtain the required number of animals per group, namely,

$$n = \left(z(1-\alpha/2) + z(1-\beta)\right)^2 \cdot \frac{2\sigma^2}{\delta^2}. \tag{5.2}$$

Equation (5.2) is logical: the numerator includes the observable variation among individuals σ^2, which is mainly responsible for the 'noise of the system', whereas the denominator includes the mean difference δ we want to detect, which is in a way a 'measure of our ambition'. Snedecor and Cochran (1980, §6.14) tabulate the multiplication factor for a few values of α and β. For one-sided alternative hypotheses, $z(1 - \alpha/2)$ should be replaced with $z(1 - \alpha)$. For the paired two-sample test $2\sigma^2$ should be replaced with σ_D^2 (namely, the variance of the difference $X_1 - X_2$).

Equation (5.2) could be refined by replacing σ^2 with its estimator s^2 (see eqn. (4.4)) and the normal percentiles with the corresponding t percentiles, but this is not worthwhile because we only seek an orientative idea of the required sample size. By the same token, the following sample size predictor is suggested by eqn. (5.2) and adopted because it is convenient and useful for practical purposes, but it is not rigorously obtained from eqn. (5.2):

$$\hat{n} = 10.5 \cdot \frac{\mathrm{cv}_a^2\left(\hat{N}\text{ controls}\right) + \mathrm{cv}_a^2\left(\hat{N}\text{ trisomic}\right)}{\Delta^2} \tag{5.3}$$

where

- $10.5 \approx (z(0.975) + z(0.90))^2$, which corresponds to a significance level $\alpha = 0.05$ and a desired power $1 - \beta = 0.90$;
- $\mathrm{cv}_a^2(\hat{N}) = \mathrm{var}_a(\hat{N})/(\mathrm{mean}_a(\hat{N}))^2$, estimate of the square coefficient of variation of the neuron number estimate among the animals (hence the subscript 'a') of a group;
- $\Delta = \mathrm{Mean}_a(\hat{N}\text{ trisomic})/\mathrm{Mean}_a(\hat{N}\text{ controls}) - 1$, the true relative difference we want to detect between the two group means.

The estimator (5.3) makes sense only if the cell number estimator $\hat{N}$ is unbiased for all animals, as warranted (at least as far as geometric sampling is concerned) by the designs described in Section 3.1).

5.2 Reduction of the observed variance among animals: variance splitting

The preceding subsection describes the strategy of tuning the number of animals per group to attain a given power without acting upon the variance among the animal estimates in each group. Here we address the latter possibility.

We recall that true means, variances, coefficients of variation, and coefficients of error are denoted by Mean($\cdot$), Var($\cdot$), CV($\cdot$), and CE($\cdot$), respectively, whereas the corresponding estimators are denoted by mean($\cdot$), var($\cdot$), cv($\cdot$), and ce($\cdot$), respectively.

An important identity in the present context is the following,

$$\mathrm{Var}_a(\hat{N}) = \mathrm{Var}_a(N) + \mathrm{Mean}_a(\mathrm{Var}_w(\hat{N})) \tag{5.4}$$

where:

- $Var_a(\hat{N})$ = true variance of the cell number estimates among animals (represented by σ^2 in the right-hand side of eqn. (5.2));
- $Var_a(N)$ = true variance of the true cell number among animals, (or 'biological variance'),
- $Mean_a (Var_w(\hat{N}))$ = true mean among animals (hence the subscript 'a') of the true error variance, due to stereology only, within an animal (hence the subscript 'w').

The possibility is now open to reduce the numerator of eqn. (5.2) by reducing the error variance $Var_w(\hat{N})$ within each animal, that is, by acting upon the stereological sampling protocol (note that, in principle, $Var_a(N)$ cannot be altered). It is convenient to work with the following estimation version, based on (5.4) but not a rigorous consequence of it,

$$cv_a^2(\hat{N}) = cv_a^2(N) + mean_a(ce_w^2(\hat{N})). \tag{5.5}$$

For practical purposes it is sufficient to concentrate on the control group. The computation of $cv_a^2(\hat{N})$ is straightforward (see Table 2.1 and Section 5.4). Next we only need a special formula for the estimator $ce_w^2(\hat{N})$ of the square coefficient of error within an animal, whereby the biological square coefficient of variation can be estimated from (5.5) by difference, that is,

$$cv_a^2(N) = \max\left(0,\ cv_a^2(\hat{N}) - mean_a(ce_w^2(\hat{N}))\right) \tag{5.6}$$

(note that the difference in the right-hand side of the preceding equation may be negative, in which case we take $cv_a^2(N) = 0$ (see e.g. Table 2.6 in Section 7.2, Pv in CA3 case).

Table 2.1   Data (thousands of cells) and preliminary calculations for the example described in Section 4.4 and 5.4. The biological $cv_a^2 (N)$ is computed via eqn. (5.6) for each group

	$\hat{N}$(neu)/1 000			$ce_w^2(\hat{N}) \approx 1/\Sigma \bar{Q}^-$	
	Controls	**Trisomic**		**Controls**	**Trisomic**
Animal 1	1.954	1.200		0.0115	0.0200
Animal 2	1.112	0.945		0.0213	0.0357
Animal 3	1.308	0.984		0.0286	0.0232
Animal 4	1.075	0.728		0.0233	0.0333
Animal 5	1.293			0.0233	
$mean_a(\hat{N})$	1.3484	0.9643	$mean_a(ce_w^2(\hat{N}))$	0.0216	0.0281
s_a	0.3543	0.1934			
$cv_a^2(\hat{N})$	0.0690	0.0402			
$cv_a^2(N)$	0.0474	0.0121			

The idea is as follows. Suppose that we can predict a new value of the coefficient of error, $ce_{w,new}^2(\hat{N})$ when we change say the number of slices, or the number of disectors sampled per animal. Since the biological variation is unchanged and we have calculated an estimator $cv_a^2(N)$ of it, we can predict the new total variation in each group by

$$cv_{a,new}^2(\hat{N}) = cv_a^2(N) + mean_a(ce_{w,new}^2(\hat{N})). \tag{5.7}$$

Finally, we can predict the required number of animals per group with the new protocol by replacing $cv_a^2(\hat{N})$ with $cv_{a,new}^2(\hat{N})$ for each group in the right-hand side of (5.3).

The most efficient way to decrease the true $CE_w^2(\hat{N})$ within each animal would be to increase the number of slices. For practical reasons this would be precluded in the used brains, but not in new ones. In the used brains one may, however, consider, say, doubling the number of optical disectors in the available slices. In the present study, and probably in many others in the present context, the main component of $ce_w^2(\hat{N})$ is the nugget component, namely, the second term in the right-hand side of eqn. (3.9). Thus, for practical purposes we may take

$$ce_w^2\left(\hat{N}\right) \approx \frac{1}{\sum Q^-}, \tag{5.8}$$

and therefore, if ΣQ^- is doubled, then one can expect $ce_w^2(\hat{N})$ to be approximately halved. The new total coefficient of variation (see (5.7)) will then be expected to be

$$\begin{aligned} cv_{a,new}^2(\hat{N}) &= cv_a^2(N) + mean_a(ce_{w,new}^2(\hat{N})) \\ &= cv_a^2(N) + \tfrac{1}{2} \cdot mean_a(ce_w^2(\hat{N})), \end{aligned} \tag{5.9}$$

and the new number of animals per group can be predicted using eqn. (5.3) with $cv_{a,new}^2(\hat{N})$ in the place of $cv_a^2(\hat{N})$ for each group.

Finally, it may be interesting to know the number of animals required per group in the hypothetical case in which the amount of labor per animal was 'infinite', that is, when the relevant number of neurons was known exactly. In this extreme case $CE_w^2(\hat{N}) = 0$ for all animals, and therefore

$$cv_{a,new}^2(\hat{N}) = cv_a^2(N). \tag{5.10}$$

Equation (5.3) would now give some idea of the 'very minimum' number of animals required per group.

5.3 The 'half-organ design'

In the present study—and in many others in a similar context—one of the two hemispheres of each brain was chosen for analysis with probability $\tfrac{1}{2}$ (Section 3.1). In such cases the variance splitting (5.4) is not exact; although the practical relevance of this may be expected to be usually small, a clarification may be in order.

Let N_L, N_R denote the true total cell numbers in the left and right hemisphere of a brain, respectively, for a cell subpopulation in a compartment of interest: thus $N_L + N_R = N$, the total number of cells. In the first stage sampling, either the left, or the right hemisphere is chosen with probability $\frac{1}{2}$. The first stage unbiased estimator of N would therefore be $\hat{N} = 2N_L$ with probability $\frac{1}{2}$, and $\hat{N} = 2N_R$ with probability $\frac{1}{2}$. This estimator is not practical, however, because neither N_L nor N_R are known. In a second sampling stage, if the left hemisphere is chosen, then N_L is estimated with stereology by the unbiased estimator $\hat{N}_L$ with a within-hemisphere error variance σ_L^2, and similarly for the right hemisphere. The final unbiased estimator of N is therefore $\tilde{N} = 2\hat{N}_L$ with probability $\frac{1}{2}$, and $\tilde{N} = 2\hat{N}_R$ with probability $\frac{1}{2}$. Two-stage sampling theory (e.g. Cochran 1977) shows that

$$\text{Var}_a(\tilde{N}) = \text{Var}_a(N) + \text{Mean}_a(N_L - N_R)^2 + 2\,\text{Mean}_a(\sigma_L^2 + \sigma_R^2) \qquad (5.11)$$

instead of (5.4). The last term in the right-hand side of the preceding equation can be estimated with one-hemisphere data, but the second term can unfortunately not be so estimated. As a consequence, the right-hand side of eqn. (5.6) will tend to overestimate $CV_a^2(N)$ when there is an asymmetry between the right and left brain hemispheres.

5.4 Worked numerical example (contd.)

We retake the example considered in Section 4.4. The group mean relative difference $\Delta \approx -28.5\%$ (see (4.12)) may be biologically important, but it could not be detected statistically due to the low power of the test (see (4.13)). In order to increase the power to 0.90 with a level of significance of 0.05 we may apply eqn. (5.3) using the $\text{cv}_a^2(\hat{N})$ displayed in Table 2.1. We obtain

$$\hat{n} = 10.5 \cdot \frac{0.0690 + 0.0402}{0.285^2} \approx 15 \text{ animals}. \qquad (5.12)$$

If we wish to save animals we may try to see what would happen if we counted twice as many neurons per animal (this could be achieved by doubling the number of disectors—unfortunately the number of slices could not be changed in the processed brains). Application of eqn. (5.8) yields the $\text{ce}_w^2(\hat{N})$ displayed in Table 2.1. By eqn. (5.9) we obtain, for the control group,

$$\text{cv}_{a,\text{new}}^2(\hat{N}) = 0.0474 + 0.0216/2 = 0.0582 \qquad (5.13)$$

and likewise $\text{cv}_{a,\text{new}}^2(\hat{N}) = 0.02615$ for the trisomic group. Now eqn. (5.3) yields

$$\hat{n}_{\text{new}} = 10.5 \cdot \frac{0.0582 + 0.02615}{0.285^2} \approx 11 \text{ animals}, \qquad (5.14)$$

as given in Table 2.5 (Section 7.2)

6 Statistical detection of group mean differences: multivariate analysis

6.1 Problem and analysis: Hotelling's T^2 test

For each cell subpopulation the estimators of the cell numbers in the three relevant compartments of the hippocampus are regarded as the components of a three-dimensional random vector (see eqn. (2.1)). The problem is to compare the population means of these vectors for the control and the trisomic groups (see eqn. (2.2)). For convenience we define the mean vector difference $\delta = \mu_2 - \mu_1$, whereby the statistical problem is reduced to test

$$H_0: \delta = 0 \text{ versus } H_1: \delta \neq 0. \tag{6.1}$$

(Here we follow Chatfield and Collins (1980), specially subsection 7.5.) To make the notation compatible with the software S-PLUS, however, here a vector is by default a row vector.

Under normality of the samples (Section 4.2) and homogeneity of their variances (Section 4.3) the control and treatment vectors, $\mathbf{X}_1$ and $\mathbf{X}_2$, respectively, are assumed to be p—variate normal ($p = 3$ here), with a common $p \times p$ covariance matrix Σ, that is, $\mathbf{X}_1 \sim N_p(\mu_1, \Sigma)$ and $\mathbf{X}_2 \sim N_p(\mu_2, \Sigma)$. To test the hypotheses (6.1) we draw independent simple random samples from the control vector (sample size n_1, sample mean sample covariance matrix $\mathbf{S}_1$) and from the treatment group (n_2, $\overline{\mathbf{X}}_2$, $\mathbf{S}_2$, respectively).

Under the preceding assumptions,

$$\mathbf{D} = \overline{\mathbf{X}}_2 - \overline{\mathbf{X}}_1, \quad \overline{\mathbf{D}} \sim N_p\left(\delta, \left(\frac{1}{n_1} + \frac{1}{n_2}\right)\Sigma\right) \tag{6.2}$$

The estimator of the common covariance matrix Σ is

$$\mathbf{S} = \frac{(n_1 - 1)\mathbf{S}_1 + (n_2 - 1)\mathbf{S}_2}{n_1 + n_2 - 2}. \tag{6.3}$$

By analogy with (4.5),

$$T^2 = \frac{\overline{\mathbf{D}}\mathbf{S}^{-1}\overline{\mathbf{D}}^T}{\frac{1}{n_1} + \frac{1}{n_2}} \sim \text{Hotelling's } T^2_{p, n_1 + n_2 - 2} \text{ if } H_0 \text{ is true} \tag{6.4}$$

where $\mathbf{x}^T$ represents the transposed form of a vector $\mathbf{x}$. It can also be shown that

$$F = \frac{n_1 + n_2 - p - 1}{p(n_1 + n_2 - 2)} \cdot T^2 \sim F_{p, n_1 + n_2 - p - 1} \text{ if } H_0 \text{ is true.} \tag{6.5}$$

The optimal strategy is: reject H_0 if $F > F_c$, where $F_c = F_{p,n_1 + n_2 - p - 1} (1 - \alpha)$ is the $100(1 - \alpha)\%$ percentile of Snedecor's F distribution with p degrees of freedom in the numerator and $n_1 + n_2 - p - 1$ in the denominator.

By analogy with eqn. (4.9) the power is computed for given α, β, and for a desired δ as

$$\text{Power} = 1 - \beta = 1 - \mathbb{P}(F^* \leq F_{p,n_1 + n_2 - p - 1} (1 - \alpha)) \tag{6.6}$$

where F^* is the non-central F distribution with p and $n_1 + n_2 - p - 1$ degrees of freedom and non-centrality parameter $\delta^T \Sigma^{-1} \delta / (1/n_1 + 1/n_2)$ for a desired δ. In practice we replace Σ with its estimator S. If we want to compute the power for the observed mean vector difference, namely, for $\bar{\delta} = \bar{D}$, then the estimated non-centrality parameter is precisely the statistic T^2 given by eqn. (6.4).

6.2 Multivariate analysis of variance (MANOVA)

In the same way as the two-sample t-test is a particular case of the one-way analysis of variance ('ANOVA') with two groups to compare, Hotelling's T^2 test is a particular case of one-way multivariate analysis of variance ('MANOVA') with two groups to compare. Either ANOVA or MANOVA analyses can be easily carried out in S-PLUS with the functions aov, manova, respectively. The latter is illustrated in Section 6.4.

If the null hypothesis H_0 of equality of control and trisomic vector means is rejected for a given subpopulation of cells, then we may identify which compartment(s) is(are) responsible for the difference. This is easily carried out with the option univariate = T, which returns the p univariate ANOVAs (Section 6.4).

The necessary verification of the normality assumption is also an easy task for S-PLUS. It may be done by plotting and testing the residuals (briefly, these are the random deviations of the observed cell numbers from the corresponding means); see Section 6.4 and Figs 2.8 and 2.9.

6.3 Worked numerical example: Hotelling's T^2 test

For the sake of illustration we compare the control and trisomic population means of the number of GFAP cells in the three hippocampal compartments DG, CA12, and CA3 simultaneously, namely, we compare the hypotheses (6.1) at the level $\alpha = 0.05$. For each animal we observe a three-dimensional row vector (see (2.1)) with $n_1 = 5$ independent animals (i.e. replications) for the control vector X_1, and $n_2 = 4$ independent replications for the trisomic vector X_2. The data (thousands of cells) are displayed in Table 2.2.

As indicated in the preceding subsection the diagnostics of normality of the residuals are easily made with the aid of S-PLUS in the context of the more general

Table 2.2 Data (thousands of GFAP cells) represented by trivariate row vectors with 5 and 4 replications (animals) in control and trisomic groups, respectively. The mean vectors and covariance matrices are also given, as well as the pooled covariance matrix. The latter are used in Section 6.3 to test the null hypothesis of equality of vector means against the two-sided alternative (see (6.1)) by means of Hotelling's T^2 test

Control group (Gfap.c)

	DG	CA12	CA3
[1,]	2.344	1.954	2.493
[2,]	2.149	1.112	2.448
[3,]	1.798	1.308	2.902
[4,]	2.022	1.075	1.752
[5,]	1.865	1.293	1.294

Mean

	DG	CA12	CA3
	2.0356	1.3484	2.1778

Covariance matrix

	DG	CA12	CA3
DG	0.04842830	0.04568195	0.0280864
CA12	0.04568195	0.12552130	0.0657821
CA3	0.02808640	0.06578210	0.4148082

Pooled covariance matrix **S**

	DG	CA12	CA3
DG	0.05627574	0.02327990	0.03285294
CA12	0.02327990	0.08774828	0.10226906
CA3	0.03285294	0.10226906	0.54660211

Trisomic group (Gfap.t)

	DG	CA12	CA3
[1,]	1.943	1.200	3.555
[2,]	1.487	0.945	1.880
[3,]	1.634	0.984	2.705
[4,]	2.038	0.728	1.704

Mean

	DG	CA12	CA3
	1.7755	0.96425	2.461

Covariance matrix

	DG	CA12	CA3
DG	0.06673900	-0.00658950	0.03920833
CA12	-0.00658950	0.03738425	0.15091833
CA3	0.03920833	0.15091833	0.72232733

MANOVA (see Section 6.4 and Figs 2.8 and 2.9.) Further, variance comparison tests based on the F-statistic (4.10) are non-significant. The pooled sample co-variance matrix (6.3) is also given in Table 2.2. The T^2 test is known to be little affected by moderate differences in the covariance matrices, however, provided that the sample sizes are nearly the same (Chatfield and Collins 1980, p. 125). The T^2 statistic (6.4) becomes

$$T^2 = \frac{3.495713}{1/5 + 1/4} \approx 7.768 \tag{6.7}$$

and now eqn. (6.5) yields the following value for the F-statistic,

$$F = \frac{5 + 4 - 3 - 1}{3(5 + 4 - 2)} \cdot 7.768 \approx 1.85, \tag{6.8}$$

which is less than the critical value $F_{3,5}(0.95) = 5.41$ and, therefore, H_0 is not rejectable. The P-value is $P = 1 - \mathbb{P}(F_{3,5} \le 1.85) \approx 0.26$, as displayed in Table 2.4 (Section 7.2).

Finally, the power of the test when the true mean vector difference δ is the observed one, namely, $\bar{\mathbf{D}} = (0.2601, 0.3842, -0.2832)$ (see eqn. (6.6)) may be computed with S-PLUS via

$$1 - pf(qf(0.95, 3, 5), 3, 5, ncp = 7.768) \tag{6.9}$$

which yields the result 0.34 (Table 2.4, & Section 7.2).

6.4 Worked numerical example: MANOVA

As pointed out in Section 6.2, the foregoing analysis can be carried out easily in S-PLUS by regarding the analysis as a MANOVA. For convenience the control and trisomic matrices displayed in Table 2.2 are named Gfap.c and Gfap.t, respectively. The necessary steps are shown in Table 2.3. First we have to construct a data frame (named Gfap.frame) which incorporates a factor vector Gfap.fac with $n_1 = 5$ one's and $n_2 = 4$ two's, and the sample data (Gfap.dat) made by assembling the rows of Gfap.c and of Gfap.t together as indicated. Now we apply the manova function and we name the resulting object Gfap.man. As expected we obtain the same P-value as in Section 6.3, (≈ 0.26).

Although no significant difference is detected ($P = 0.25$) we may see what happens for each of the compartments by disclosing the univariate ANOVAs with the command summary and the option univariate = T (see Table 2.3). We see that the closest compartment to exhibit a difference is CA12 ($P = 0.09$), which, together with the fact that the observed difference (−28.5%) may be considered to be important, prompted us to carry out the further analysis of Sections 5.4 and 7.2.

Table 2.3 Procedure to carry out a MANOVA for the GFAP data given in Table 2.2, with the aid of S-PLUS. For explanation see Section 6.4

```
> Gfap.fac_factor (c(1,1,1,1,1,2,2,2,2))
> Gfap.dat_rbind (Gfap.c, Gfap.t)
> Gfap.frame_data.frame (Gfap.fac,Gfap.dat)

> Gfap.frame
```

Gfap.fac		DG	CA12	CA3
1	1	2.344	1.954	2.493
2	1	2.149	1.112	2.448
3	1	1.798	1.308	2.902
4	1	2.022	1.075	1.752
5	1	1.865	1.293	1.294
6	2	1.943	1.200	3.555
7	2	1.487	0.945	1.880
8	2	1.634	0.984	2.705
9	2	2.038	0.728	1.704

```
> Gfap.man_manova (Gfap.dat~Gfap.fac,Gfap.frame)
> summary (Gfap.man)
```

	Df	Pillai Trace	approx.F	num df	den df	P-value
Gfap.fac	1	0.52601	1.849583	3	5	0.255552
Residuals	7					

```
> summary (Gfap.man, univariate = T)
```

Response: DG

	Df	Sum of Sq	Mean Sq	F Value	Pr(F)
Gfap.fac	1	150 337.8	150 337.8	2.671449	0.1461791
Residuals	7	393 930.2	56 275.7		

Response: CA12

	Df	Sum of Sq	Mean Sq	F Value	Pr(F)
Gfap.fac	1	327 936.1	327 936.1	3.737236	0.09447714
Residuals	7	614 237.9	87 748.3		

Response: CA3

	Df	Sum of Sq	Mean Sq	F Value	Pr(F)
Gfap.fac	1	178 227	178 227.2	0.3260639	0.5858429
Residuals	7	3 826 215	546 602.1		

Finally, the graphical diagnoses of normality

```
> x_as.vector (resid (Gfap.man))
> cdf.compare (x, dist = 'normal', mean = mean (x), sd = sqrt (var (x)))
```

and

```
> qqnorm (x)
> qqline (x)
```

namely, the Q–Q plots to compare the residuals against the normal distribution, were carried out for each cell subpopulation (see Figs 2.8 and 2.9, respectively). The graphs visually suggest a fair degree of normality of the data. The formal Kolmororov–Smirnov test carried out for instance for the GFAP data gives us (with simplified output),

> ks.gof (x)
ks = 0.1213, p-value = 0.382
alternative hypothesis:
    True cdf is not the normal distn. with estimated parameters

which confirms that the normality hypothesis was not rejectable, and similarly for the remaining three cell subpopulations.

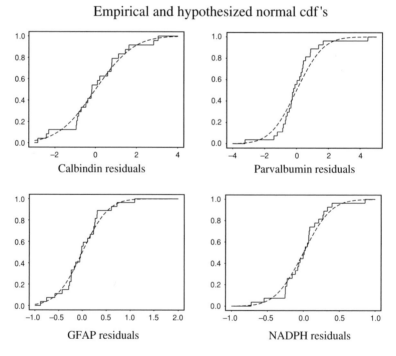

Empirical and hypothesized normal cdf's

Figure 2.8 Empirical (solid lines) and hypothesized normal (dotted curves) cumulative distribution functions for the residuals of the estimated cell numbers in each of the four subpopulations of interest. There are 24 calbindin residuals and 27 residuals in each of the remaining three cases. See Section 6.4.

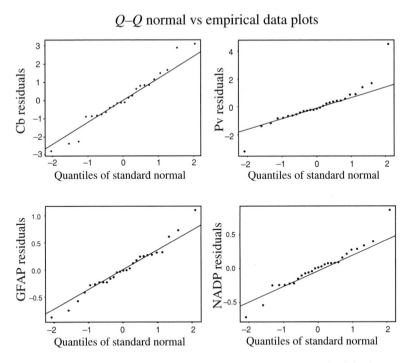

Figure 2.9 *Q–Q* plots for the residuals of the estimated cell numbers in each of the four subpopulations of interest. There are 24 calbindin residuals and 27 residuals in each of the remaining three cases. See Section 6.4.

7 Results

7.1 Multivariate comparisons of the group means

The procedures described in the preceding section were applied to the sample data obtained for each of the four cell subpopulations of interest (Section 3.2 and Fig. 2.4). The results are summarized in Table 2.4.

As anticipated in Section 3.2, the only difference (*P*-value less than 0.05) detected by the univariate ANOVA method (see Section 6.4), corresponded to the Cb subpopulation and, within this, the responsible compartment was CA3. For the remaining cell subpopulations the Hotelling T^2-test revealed no statistically significant differences, which was not surprising because, due to the small numbers of animals used (and also to the variation among animals), the powers of the tests were too low (Table 2.4).

Table 2.4 Estimated relative mean group differences (i.e. estimates $\hat{\Delta}\%$ of $\Delta\%$; see eqn. (3.4)) of cell number estimates in the relevant subpopulations and hippocampal compartments defined in the Introduction and in Section 2.1. The remaining items are explained in Sections 4.3 and 6

Cell subpop.	$\hat{\Delta}\%$			P-value	Power $(1 - \beta)$ $(\alpha = 0.05)$	p	df of F-statistic $n_1 + n_2 - p - 1$
	DG	CA1 + 2	CA3				
Cb	12.6	16.4	90.1	0.03	0.89	3	4
Pv	−13.9	−16.2	30.6	0.52	0.18	3	5
GFAP	−12.8	−28.5	13.0	0.26	0.34	3	5
NADP	−2.5	−35.0	−9.2	0.42	0.22	3	5

Table 2.5 Numbers of animals/group required to declare statistically significant the observed relative mean group differences in cell numbers ($\Delta\%$) considered to be of potential biological significance. The powers correspond to individual two-sample t-tests. For further details see Section 5

Subpopulation	$\Delta\%$	Power with current sample sizes, $n_1 = 5, n_2 = 4$	Required number of animals/group with $\alpha = 0.05$ and $1 - \beta = 0.90$		
			Same work/animal	Double work/animal	∞ work/animal
Pv in CA3	30.6	0.36	16	15	13
GFAP in CA1 + 2	−28.5	0.34	15	11	8
NADP in CA1 + 2	−35.0	0.32	15	13	11

Table 2.6 Splitting of the total coefficient of variation of total cell number estimates among animals into the biological component and the (stereological) error component, according to eqn. (5.5), for each of the cell subpopulations considered in Table 2.5. All cv's and (mean) ce's are given as percentage of the group mean

Subpopulation	$\Delta\%$	Controls			Trisomic		
		$cv_a(\hat{N})$	$cv_a(N)$	$ce_w(\hat{N})$	$cv_a(N)$	$cv_a(N)$	$ce_w(\hat{N})$
Pv in CA3	30.6	9.8	0.0	10.0	35.7	33.6	12.1
GFAP in CA1 + 2	−28.5	26.3	21.4	15.3	20.1	9.7	17.6
NADP in CA1 + 2	−35.0	35.4	32.4	14.4	21.3	15.2	14.8

7.2 Sample size predictions to increase the power

Although most group mean differences were not statistically significant, as indicated in the preceding subsection, some of the observed differences could be judged to be biologically important in at least the three cases collected in Table 2.5. First, the corresponding powers were computed via (6.6) for the observed differences, with the initial sample sizes and with $p = 1$ (univariate case). The powers were found to be too low (see Table 2.5, third column). Thus, in these three cases the situation was similar to that illustrated in Fig. 2.7(a). To attain a power of 0.90 with different workloads per animal the required sample sizes, computed as explained in Section 5, are displayed in Table 2.5, last three columns. The auxiliary decompositions of the total variation into the biological and the stereological ones, computed as indicated in Section 5.2, are shown in Table 2.6.

8 Conclusions and remarks

8.1 Pilot experiment

- The only mean group difference found statistically significant concerned the increase in mean cell number for the Cb subpopulation in the CA3 compartment (Table 2.4); this can already be appreciated in Fig. 2.4.

- Three other mean group differences were greater than 25% of the control means, and they were therefore deemed to be of potential biological importance. Consequently, a deeper analysis was carried out to establish how many animals, depending upon the amount of labor invested per animal, would be required to report such differences as significant (at the 5% level) with a probability of 90%, (the power). The results are displayed in Table 2.5. It is seen that the sample size requirements are similar in all three cases. In particular, even if the number of neurons counted per animal were doubled, the required number of animals per group would still lie between 11 and 15. This is mainly a consequence of the generally large variation observed among animals.

8.2 Remarks on the stereology

- The choice of five adjacent Cavalieri series (Fig. 2.2(a)) to carry out the stereology was only imposed by the need to stain the reference space and the different cell subpopulations in different ways. In the absence of such restrictions, all measurements may be made in a single series.

- Cavalieri type sampling from a reference space guarantees unbiasedness of cell number estimators irrespective of the spatial arrangement of the cells in that reference space. The error variance may then be arbitrarily reduced by increasing the sampling intensity.

- Unbiasedness refers, however, only to the geometric sampling. The present study revealed the difficulty in preserving unbiasedness in practice at all stereological steps, mainly because the boundaries of the different reference compartments (specially CA1 and CA2) were difficult to identify for some neuron subpopulations.

- The preceding remarks should not cause us to yield to the temptation of renouncing geometrically unbiased designs. Likewise, trying to 'validate' a method against a 'golden standard' such as exhaustive cell counting through 'complete reconstruction' (e.g. von Bartheld 2001, and some references therein) is futile. First, the latter will never yield general conclusions because it can be carried out only in very few special cases. Second, the typical artifacts (cell misrecognition, slice compression, loss of grazing cell fragments at the slice faces, etc.) remain intact when serial slices (as opposed to a few systematic slices) are analysed. A better direction may be to preserve geometric unbiasedness, and then try to remove artifacts by improving the instrumentation and the laboratory manipulations.

- The fractionator estimator (3.7) was invoked only to borrow the error predictor formula (3.9) for the adopted estimator (eqn. (3.5)). Only the local error contribution in eqn. (3.9) uses a special assumption for the arrangement of the cells (approximately Poisson).

8.3 Remarks on the statistics

- The above conclusions are based on the multivariate Hotelling's T^2-test (conveniently carried out by means of a MANOVA analysis, see Sections 6.2 and 6.4), which seems a more reasonable alternative than repeated t-tests in the present context.

- It might be questioned why parametric tests are favored whenever possible (Section 4.2) whereas the usual stereological designs are generally non-parametric (Section 3). A technical reason for the latter choice can be found in Baddeley and Cruz-Orive (1995; section 5), but the issue is not entirely closed.

- The predicted sample sizes are tailored to detect the corresponding observed differences between group means with a given probability $1 - \beta$ (the power). It should be borne in mind, however, that in eqn. (5.3) we are replacing Δ with an estimator $\hat{\Delta}$ of it, and the latter is a random variable that will vary from an experiment to another. Thus, Δ is really unknown, and a new experiment with the calculated sample sizes and with $\alpha = 0.05$, $1 - \beta = 0.90$ will have a power less than 0.90 if $\Delta < \hat{\Delta}$.

- In some cases the observed variation among animals was much larger in one group than in the other; for practical purposes one may then increase the sample size more in the more variable group than in the other using the results in Table 2.5, albeit then the theory behind the methods becomes difficult and inexact.

- It should be borne in mind that neither P-values nor the 'magic' threshold of the '5% level' encapsulate either all the science or the only reason to make a decision. The main reason for the 5% choice was that in the celebrated statistical tables of R.A. Fisher and F. Yates—quite prior to the computer era!—the relevant statistical distributions were computed for convenience at levels such as 1, 5, and 10%.

8.4 How many animals?

Technical steps to address the above question in practice are detailed in Section 2.5, but nonetheless a few additional remarks may be opportune.

The question should be better formulated as follows: 'How many animals per group should we use to detect a biologically meaningful difference δ between population means with a probability $1 - \beta$?' To obtain a proper answer the scientist should be committed to fix at least the meaningful constant δ (for eqn. (5.2)), or Δ (for eqn. (5.3)), at the outset. This choice depends on the context of the experiment, and it cannot be produced by the statistician alone. Failure to fix δ in advance may lead to apparent paradoxes, for instance: 'For each experiment it is always possible to choose a design so that the null hypothesis $H_0 : \delta = 0$ is rejectable!' To see this look for instance at eqn. (4.5): no matter how small $\bar{D}$ is, we can always choose n_1, n_2 so that t is as large as we wish.

Modern software for stereology offers the user the possibility to design stereological test grids, etc. The user, however, may not always see clearly which is the ideal grid for a given experiment or, more importantly, how many sections should be cut, how many disectors should be subsampled per section (Fig. 2.3(a)), etc. Here it might help if the system asked the user for the following information at the outset.

1 Fix the constants Δ, α, and $1 - \beta$ (with defaults 0.15, 0.05, and 0.90, say, respectively).

2 Give a guess of the maximum number of animals n per group you could afford to use to detect the relative difference Δ with the given probability $1 - \beta$.

3 Give a guess $cv_a^2 (N)$ of the biological coefficient of variation among animals for the target quantity. (With time, databases should incorporate this kind of information for an increasing number of relevant quantities in different animal species).

With the data 1–3 an estimate $cv_a^2(\hat{N})$ of the total observable square coefficient of variation among animal estimates can be obtained from eqn. (5.3). From this and $cv_a^2(N)$ an estimate $\text{mean}_a(ce_w^2(\hat{N}))$ of the tolerable mean square coefficient of error due to stereology can be calculated by difference using eqn. (5.5). A negative difference would mean that the given data were inconsistent, and a new set of data 1–3 would have to be re-entered. If the difference was positive, then eqn. (5.8) would give an idea of the total number of neurons to count per animal, and with this the neuroscientist would have a solid ground to proceed. One can also lower the number of animals and see whether the resulting amount of work required per animal remains tolerable, and so forth.

Given all the preceding considerations, however, we should never forget that they will lose their meaning if sampling is not correct (see the fourth point in Section 8.2).

8.5 Final conclusions

Relatively complex designs involving several compartments within a given organ, as described in this chapter, are commonplace in neuroscience and elsewhere. Details are provided:

(1) to carry out the stereology required to acquire the data;

(2) to represent the data graphically in a simple but comprehensive manner (Fig. 2.4) prior to any statistical analysis;

(3) to choose a statistical test. The neuroscientist should no longer hesitate to choose a multivariate design wherever appropriate, because nowadays statistical software such as S-PLUS makes the task very simple (Table 2.3). The mentioned software also makes it easy to verify, for instance, the underlying normality assumptions (Figs 2.8 and 2.9);

(4) to predict, at least approximately, the number of animals per group, with a desired sampling intensity within animals, required to detect a given difference between group means with a desired probability (the power) for a given level of significance (Section 5).

9 Appendix to Chapter 2

9.1 Glossary of sampling and statistics, with worked numerical examples

The examples refer to the population data P1 and P2 given in the entry 'Population'. Not all terms included in this glossary appear in the main text. The symbols may be checked in Section 9.2. Since the terms are listed alphabetically, breaking the conceptual order is unfortunately unavoidable—for instance, $\text{Var}(\overline{X})$ is given by eqn. (9.7) whereas an estimator of it is given by eqn. (9.1). The reader can hopefully overcome this.

Bias of an estimator $\hat{\theta}$ of a parameter θ: Mean error of $\hat{\theta}$, namely, the difference between the mean value of all possible realizations of $\hat{\theta}$ and θ. Thus, $\text{Bias}(\hat{\theta}) = \mathbb{E}(\hat{\theta}) - \theta$, where $\mathbb{E}(\hat{\theta})$ is calculated from the sampling distribution of $\hat{\theta}$.

Examples

Under systematic sampling of period $k = 3$ from P2 we have $\mathbb{E}(\tau) = 29 = \tau_2$ (see 'Estimator of the population total') and therefore Bias$(\hat{\tau}) = 0$ in this case. However, under the same design, from the corresponding sampling distribution of $\bar{X}$ (see 'Sampling distribution, Examples') we obtain $\mathbb{E}(\bar{X}) = (11/3) \cdot (1/3) + 4 \cdot (1/3) + 5 \cdot (1/3) = 38/9 = 4.\dot{2}$, whereas $\mu_2 = 29/7 \approx 4.14$, and therefore Bias$(\bar{X}) = 38/9 - 29/7 = 5/63 \approx 0.079$. See also 'Estimator of the population mean'.

Coefficient of error. Synonym of coefficient of variation when it refers to an estimator, that is, $\mathrm{CE}(\cdot) = \mathrm{SE}(\cdot)/\mathbb{E}(\cdot)$.

Coefficient of variation, $\mathrm{CV}(\cdot) = \mathrm{SD}(\cdot)/\mathbb{E}(\cdot)$. For a variable X (such as neuron number) defined on a natural population, we can write $\mathrm{CV}(X) = \sigma/\mu$.

Consistency. An estimator $\hat{\theta}$ is consistent for a parameter θ if it converges to θ (in a well defined manner) as the sample size increases. If an estimator is unbiased and its variance is finite and it decreases when the sample size increases, then it is also consistent. The reciprocal does not need to be true.

Deviation of an estimator $\hat{\theta}$ is the difference $\hat{\theta} - \mathbb{E}(\hat{\theta})$.

Efficiency of an estimator $\hat{\theta}_1$ with respect to another estimator $\hat{\theta}_2$ (obtained with the same sample size) of a given parameter θ is the ratio Var$(\hat{\theta}_2)$/Var$(\hat{\theta}_1)$.

Example (see 'Variance of the estimator of the population total'). Under simple random sampling of size $n = 3$ from P2, Var$(\hat{\tau}_1) = 54.\dot{2}$. Under systematic sampling of period $k = 3$ from P2, however, Var$(\hat{\tau}_2) = 14$. Thus the efficiency of $\hat{\tau}_2$ with respect to $\hat{\tau}_1$ is $54.\dot{2}/14 \approx 3.87$ (even though the sample size for $\hat{\tau}_1$ is 3, whereas the mean sample size for $\hat{\tau}_2$ is $7/3 < 3$).

Error of an estimator $\hat{\theta}$ of a parameter θ is the difference $\hat{\theta} - \theta$.

Estimate. A particular numerical value of an estimator.

Estimator of a parameter θ. A function $\hat{\theta}(X_1, \ldots, X_n)$ of the elements of a sample, namely, a rule that is used to compute numerical estimates of θ from concrete samples.

> **Estimator of the population mean**. If sampling is simple random, then the sample mean $\bar{X}$ is an unbiased estimator of μ. If sampling is systematic with period k, then $\bar{X}$ does *not* need to be an unbiased estimator of μ.
>
> *Example*
>
> Under simple random sampling of size $n = 3$ from P1, the mean value of $\bar{X}$, computed from its sampling distribution, is: $\mathbb{E}(\bar{X}) = (10/3) \cdot (1/4) + \ldots + (16/3) \cdot (1/4) = 4.5 = \mu_1$, which verifies that $\bar{X}$ is unbiased for μ_1.
>
> **Estimator of the population total**. If sampling is simple random, then $\hat{\tau} = N\bar{X}$ is an unbiased estimator of τ. If sampling is systematic with period k, then $\hat{\tau} = kQ$, where Q is the sample total, is an unbiased estimator of τ.

Example

Under systematic sampling of period $k = 3$ from P2, the mean value of $\hat{\tau}$, computed from its sampling distribution, is: $\mathbb{E}(\hat{\tau}) = 33 \cdot (1/3) + 24 \cdot (1/3) + 30 \cdot (1/3) = 29 = \tau_2$, which verifies that $\hat{\tau}$ is unbiased for τ_2.

Estimator of the population variance. If sampling is simple random, then $\hat{\sigma}^2 = (1 - 1/N) \cdot s^2$ ($\approx s^2$ when N is large) is an unbiased estimator of σ^2. If sampling is systematic with period k, then there is no simple relation between the sample variance and the population variance.

Example

Under simple random sampling of size $n = 3$ from P1, the mean value of $\hat{\sigma}^2$, computed from its sampling distribution (see 'Sampling distribution, Examples') is $\mathbb{E}(\hat{\sigma}^2) = 1.75 \cdot (1/4) + \ldots + 4.75 \cdot (1/4) = 5.25 = \sigma_1^2$, which verifies that $\hat{\sigma}^2$ is unbiased for σ_1^2.

Estimator of the variance of the sample mean (error variance). If sampling is simple random, then

$$\mathrm{var}(\overline{X}) = \left(1 - \frac{n}{N}\right) \cdot \frac{s^2}{n} \approx \frac{s^2}{n} \text{ when } N \text{ is large} \tag{9.1}$$

is an unbiased estimator of $\mathrm{Var}(\overline{X})$ (it suffices to replace σ^2 with its unbiased estimator $\hat{\sigma}^2 = (1 - 1/N) s^2$ in eqn. (9.7)).

Example

Under simple random sampling of size $n = 3$ from P1, the mean value of $\mathrm{var}(\overline{X})$, computed from its sampling distribution (see 'Sampling distribution, Examples') is: $\mathbb{E}[\mathrm{var}(\overline{X})] = 0.19\dot{4} \cdot (1/4) + \ldots + 0.52\dot{7} \cdot (1/4) = 0.58\dot{3} = \mathrm{Var}(\overline{X})$, which verifies that $\mathrm{var}(\overline{X})$ is unbiased for $\mathrm{Var}(\overline{X})$.

- **Expectation (mathematical).** Weighted mean of a random variable, using the probabilities as weights. For instance, if the sampling distribution of the random variable X consists of r values $\{x_1, x_2, \ldots, x_r\}$ with corresponding probabilities $\{p_1, p_2, \ldots, p_r\}$, then

$$\mathbb{E}(X) = x_1 p_1 + x_2 p_2 + \ldots + x_r p_r \tag{9.2}$$

Remark. In a population $\{x_1, x_2, \ldots, x_N\}$ the individual weights are usually assumed to be identical and equal to $1/N$. If an element is drawn with probability $1/N$, the outcome is a random variable X whose expectation $\mathbb{E}(X)$ is precisely the population mean μ, and whose variance is σ^2 (see eqn. (9.3)).

Fractionator, Systematic sampling, typically in various stages. The estimator of the total τ_2 of the population P2 under systematic sampling is a fractionator estimator with one stage. Particularly interesting is the smooth fractionator (Gundersen 2002).

Example

As given, population P2 is arranged smoothly, and $\mathrm{Var}(\hat{\tau}_2) = 14$ (see 'Variance of the estimator of the population total'). For a less favorable arrangement of P2 such as {2, 8, 6, 3, 5, 1, 4}, however, $\hat{\tau}_2$ remains of course unbiased, but its variance now rises to 56 (worse than under simple random sampling!)

Mean square error (MSE) of an estimator $\hat{\theta}$ of a parameter θ. $\mathrm{MSE}(\hat{\theta}) = \mathbb{E}(\hat{\theta} - \theta)^2 = \mathrm{Var}(\hat{\theta}) + [\mathrm{Bias}(\hat{\theta})]^2$.

Noise. See 'nugget'. The term 'noise' is a synonym of 'pure random error' accompanying an observed 'signal', or a measurement in general.

Nugget. Typically the second stage sampling variance of an estimator. It may be regarded as noise error in the context of geometric sampling.

Examples

In the Cavalieri estimator of a volume the first stage variance is due to the fact that the section areas are different (and their number is generally random), whereas the second stage variance, or nugget component, arises from the use of test points to estimate each area. In the Cavalieri slices estimator of particle number, the first stage variance is due to the fact that the slice volumes are different and their number random, whereas the nugget component arises from the use of optical disectors to subsample each slice. The handling of this case is simplified if the particles ('nuggets') are assumed to be 'Poisson scattered' within the slices.

Parameter. Fixed quantity θ defined on a population (e.g. μ, τ, σ^2, etc.).

  Target parameter. Parameter of interest that we want to estimate.

Percentile of a random variable X, corresponding to a probability p, is a value $x(p)$ of X such that $\mathbb{P}(X \le x(p)) = p$.

Poisson model. Suppose that points are distributed uniformly and independently at random in space, with a numerical density of λ points per unit volume. Then, the number X of points in a domain of volume V with arbitrary shape and connectivity, is a Poisson random variable of mean λV. The variance of X is also λV, and therefore $\mathrm{CV}(X) = 1/(\lambda V)$. If we count Q^- points in a domain (such as a collection of optical disectors) of total volume V, then Q^- is an observation of X; hence Q^- is an unbiased estimator of $\mathbb{E}(X) = \lambda V$. Moreover $\mathrm{CE}(Q^-) = 1/(\lambda V)$ and an estimator of this is $\mathrm{ce}(Q^-) = 1/Q^-$. This is the model used in eqn. (3.9) for the nugget contribution.

Population (finite of size N). A well-defined set of elements $\{e_1, e_2, \ldots, e_N\}$. Each element bears a numerical value x of a variable X of interest, and the population is then regarded as the corresponding set of numerical values $\{x_1, x_2, \ldots, x_N\}$.

Examples:

- P1 = {2, 3, 5, 8} (thousands of neurons) corresponding to a well defined compartment of the hippocampus of four rats.

- P2 = {2, 3, 6, 8, 5, 4, 1} (hundreds of neurons) corresponding to 7 consecutive, adjacent slices of a well defined compartment of a rat brain.
  Population mean. $\mu = (x_1 + \ldots + x_N)/N$. *Examples.* $\mu_1 = 4.5$, $\mu_2 = 29/7 \approx 4.14$.
  Population total. $\tau = x_1 + \ldots + x_N$. *Examples.* $\tau_1 = 18$, $\tau_2 = 29$.
  Population variance.

$$\sigma^2 = \frac{1}{N} \sum_{i=1}^{N} (x_i - \mu)^2 = \frac{1}{N} \sum_{i=1}^{N} x_i^2 - \mu^2. \tag{9.3}$$

Examples. $\sigma^2_1 = 5.25$, $\sigma^2_2 \approx 4.98$.

Quantile. See 'percentile'.

Sample. A collection of elements taken from a population to infer properties of the latter. A sample of size n is denoted by $\{X_1, X_2, \ldots, X_n\}$, where each X represents some element x of the population.

- **Sample mean.** $\bar{X} = (X_1 + \ldots + X_n)/n$.
- **Sample total.** $Q = X_1 + \ldots + X_n$.
- **Sample variance.**

$$s^2 = \frac{1}{n-1} \sum_{i=1}^{n} (X_i - \bar{X})^2 = \frac{1}{n-1} \left[\sum_{i=1}^{n} X_i^2 - \frac{1}{n} \left(\sum_{i=1}^{n} X_i \right)^2 \right]. \tag{9.4}$$

One reason for dividing by $n-1$ is that as such s^2 is an unbiased estimator of σ^2 under simple random sampling when N is large: see 'Estimator of the population variance'.

Sampling. The act of drawing a sample.

  Random sampling. Any kind of sampling involving randomness. Some times it is used as a synonym of simple random sampling.

  Sampling design. Collection of rules that define the random mechanism to draw a sample.

  Sampling distribution of an estimator. Collection of all possible realizations of the estimator, together with the corresponding probabilities, which are determined by the sampling design.

  Examples

  Under simple random sampling of size $n = 3$ from P1 the possible samples are the following four:

$$\{\{2, 3, 5\}, \quad \{2, 3, 8\}, \quad \{2, 5, 8\}, \quad \{3, 5, 8\}\} \tag{9.5}$$

each with probability 1/4. The sampling distribution of $\bar{X}$ consists of the corresponding values $\{10/3, 13/3, 15/3, 16/3\}$, each with probability 1/4. The sampling distribution of $\hat{\sigma}^2 = (1 - 1/N) s^2$ consists of the values $\{1.75, 7.75, 6.75, 4.75\}$, each with probability 1/4. The sampling distribution of $\text{var}(\bar{X})$ (see eqn. (9.1)) is $\{0.19\dot{4}, 0.86\dot{1}, 0.750, 0.52\dot{7}\}$ each with probability 1/4.

Under systematic sampling of period $k = 3$ from P2, the possible samples are the following three:

$$\{\{2, 8, 1\}, \{3, 5\}, \{6, 4\}\} \tag{9.6}$$

each with probability 1/3. The sampling distribution of $\hat{\tau}$ consists of the corresponding values $\{33, 24, 30\}$, each with probability 1/3. The sampling distribution of $\bar{X}$ consists of the corresponding values $\{11/3, 4, 5\}$, each with probability 1/3.

Simple random sampling of fixed size n, without replacement ($n \leq N$). Sampling by the lottery method. Formally: Number all sampling units in the population from 1 to N. Draw successive independent random numbers among $\{1, 2, \ldots, N\}$ and keep drawing the corresponding units if they have not been drawn before. Stop when n units have been drawn. Properties: (1) the probability that any particular unit is sampled is n/N, constant—thus it is uniform sampling (it is, however, not 'independent sampling' because the covariance between pairs of sampled units is small and negative); (2) there are $\binom{N}{n}$ possible samples, all equally likely.

Systematic sampling (of fixed period k). Procedure: Number all sampling units in the population from 1 to N, in the order we like. Particular orderings such as the 'smooth' one (see 'Fractionator') may yield a very low estimation variance. Draw a single random number Z among $\{1, 2, \ldots, k\}$. The sample consists of the units bearing order numbers $\{Z, Z + k, Z + 2k, \ldots\}$ until an order number larger than N is reached. Properties: (1) the probability that any particular unit is sampled is $1/k$, constant—hence it is uniform sampling; (2) there are precisely k possible samples, all equally likely; (3) the sample size is random (unless N/k is a whole number).

Uniform sampling. Any sampling procedure with the property that all sampling units have identical probabilities to be included in the sample.

Standard deviation. Square root of the variance. See SD($\cdot$) in Section 9.2.

Standard error. Standard deviation of an estimator, namely the square root of its (error) variance. See SE($\cdot$) in Section 9.2.

Unbiasedness. An estimator $\hat{\theta}$ of a parameter θ is unbiased if its mean error, or bias, is zero for any fixed sample size.

Remarks. In general there is no formula available to estimate the bias of an estimator from a sample of data. Unbiasedness is warranted only if the sampling procedure used is correct. Note also that an unbiased estimator is so for any sample size, whether big or small (unbiasedness should not be confused with 'Consistency'!).

Variance, Var($\cdot$). Mean square deviation of the random variable within parentheses. Formally, $\mathrm{Var}(X) = \mathbb{E}[X - \mathbb{E}(X)]^2 = \mathbb{E}(X^2) - [\mathbb{E}(X)]^2$.

Variance of the population. See 'Population variance'.

Variance of the sample mean (true error variance). If sampling is simple random, then a theorem of sampling theory states that

$$\text{Var}(\overline{X}) = \frac{N-n}{N-1} \cdot \frac{\sigma^2}{n} \approx \frac{\sigma^2}{n} \text{ when } N \text{ is large.} \tag{9.7}$$

Example.

Under simple random sampling of size $n = 3$ from P1,

$$\text{Var}(\overline{X}) = \frac{4-3}{4-1} \cdot \frac{5.25}{3} = 0.58\dot{3}. \tag{9.8}$$

Alternatively, from the sampling distribution of $\overline{X}$,

$$\begin{aligned}
\text{Var}(\overline{X}) &= \mathbb{E}(\overline{X}^2) - [\mathbb{E}(\overline{X})]^2 \\
&= \mathbb{E}(\overline{X}^2) - \mu^2 \\
&= \left(\frac{10}{3}\right)^2 \cdot \frac{1}{4} + \ldots + \left(\frac{16}{3}\right)^2 \cdot \frac{1}{4} - 4.5^2 \\
&= 0.58\dot{3}.
\end{aligned} \tag{9.9}$$

The equivalence of the preceding two approaches is a key concept of sampling theory. The former approach of using a shortcut prediction formula is restricted to a few mathematically tractable designs such as simple random sampling. The second approach is universally applicable, but it requires the knowledge of the sampling distribution. Usually the latter is accessed by computer resampling.

Variance of the estimator of the population total. If sampling is simple random, then $\text{Var}(\hat{\tau}) = N^2 \, \text{Var}(\overline{X})$. If sampling is systematic there is no simple formula (see García-Fiñana and Cruz-Orive 2000a, b for recent developments). Nonetheless, for any sampling design the variance of any estimator can be computed exactly using the second approach in the preceding paragraph.

Examples.

Under simple random sampling of size $n = 3$ from P1, $\text{Var}(\hat{\tau}) = 4^2 \cdot 0.58\dot{3} = 9.\dot{3}$, which can also be obtained from the sampling distribution of $\hat{\tau}$.

Under simple random sampling of size $n = 3$ from P2, $\text{Var}(\hat{\tau}) = 7^2 \cdot \text{Var}(\overline{X}) = 54.\dot{2}$.

Under systematic sampling of period $k = 3$ from P2, we resort to the sampling distribution of $\hat{\tau}$ and we obtain:

$$\begin{aligned}
\text{Var}(\overline{\tau}) &= \mathbb{E}(\hat{\tau}^2) - [\mathbb{E}(\hat{\tau})]^2 \\
&= \mathbb{E}(\hat{\tau}^2) - \tau^2 \\
&= 33^2 \cdot \frac{1}{3} + 24^2 \cdot \frac{1}{3} + 30^2 \cdot \frac{1}{3} - 29^2 \\
&= 14.
\end{aligned} \tag{9.10}$$

9.2 List of notation

α	Level of significance, or type-I error, of a statistical test, namely the probability of rejecting the null hypothesis H_0 when it is true (§4.1)
$\alpha(\tau)$	Coefficient entering eqn. (3.9), defined by eqn. (3.11) (§3.3)
a, a'	Test area per test point (at the specimen scale) (§3.1)
β	Type-II error of a statistical test, namely the probability of not rejecting the null hypothesis H_0 when it is false. The quantity $1 - \beta$ is the power of the test, namely, the probability of rejecting H_0 when it is false (§4.3)
C_k	Quantity entering eqn. (3.9), defined by eqn. (3.10) (§3.3).
CAx	Hippocampus compartment ($x = 1, 2, 3$) (§1).
Cb	Calbindin, (neuron subpopulation) (§1).
$ce^2(\cdot)$	Estimator of $CE^2(\cdot)$, for instance $ce^2(\overline{X}) = \text{var}(\overline{X})/\overline{X}^2 = cv^2(X)/n$ under simple random sampling from a large population (§5.2)
$CE^2(\cdot)$	Square coefficient of error of the estimator within parentheses. It is a synonym of $CV^2(\cdot)$ when the quantity within parentheses is an estimator (§3.3).
$cv^2(\cdot)$	Estimator of $CV^2(\cdot)$, for instance $cv^2(X) = \text{var}(X)/\overline{X}^2 = s^2/\overline{X}^2$ under simple random sampling from a large population, (§5.1, 5.2).
$CV^2(\cdot)$	Square coefficient of variation of the random variable within parentheses. Formally, $CV^2(\cdot) = \text{Var}(\cdot)/[\mathbb{E}(\cdot)]^2$, (§5.2).
δ	True difference between two population means, (§4.1).
$\Delta = \delta/(\text{true control mean})$	True relative difference between two population means, (§3.2).
$\hat{\Delta} = \overline{D}/(\text{sample control mean})$	Observed relative difference between two sample means (it estimates Δ) (§4.4).
$\overline{D}$	Observed difference between two sample means (it estimates δ) (§4.3).
DG	Dentate gyrus (hippocampus compartment) (§1)
$\mathbb{E}(\cdot)$	Mean value (or mathematical expectation) of the random variable within parentheses

F	Statistic (4.10), also Snedecor's F distribution (§4.4).
$F_{u,v}(\cdot)$	Percentile (corresponding to the probability within parentheses) of Snedecor's F distribution with u degrees of freedom in the numerator and v in the denominator (§4.4).
GFAP	Glial fibrillary acidic protein, (astrocyte subpopulation) (§1)
h	Shortcut for $h(\text{ref})$ = disector height, corrected for shrinkage (§3.1)
H_0, H_1	Null and alternative hypotheses, respectively (§2.2, 4.1)
Hotelling's T^2 test	See T^2
μ	True population mean. It is the preferred alternative notation to $\mathbb{E}(X)$ when X is a variable (e.g. neuron number) defined on a natural population (§2.2, 4.1)
MANOVA	Multivariate Analysis of Variance. It is a statistical procedure to compare the population means of two or more groups when each observation has several components (e.g. neuron numbers in different hippocampus compartments) (§6.2)
mean($\cdot$)	Sample mean of the random variable within parentheses. Thus, mean(X) is the same as $\overline{X}$ (§5).
Mean($\cdot$)	True mean of the random variable within parentheses. Thus, Mean(X) is the same as $\mathbb{E}(X)$ (§3.2).
n	Number of animals per group (§2.2, 4.3). In eqn. (3.10), n_s = number of slices per animal. In general n may also denote sample size (see §9.1)
$\hat{n}$	A predictor of n from pilot data, (§5.1).
N	Number of cells, or disjoint particles. Shortcut for $N(\text{neu})$ in this study, namely, the true total number of neurons in a compartment (§3.1). In general N may also denote the size of a population (see §9.1)
$\hat{N}$	Unbiased estimator of N, (§2.2, 3.1).
$N(m, v)$	Normal distribution of mean m and variance v (§4.3)
NADP	NADPH-diaphorase (neuron subpopulation) (§1)
N_V	Shortcut for the ratio $N_V(\text{neu,ref}) = N(\text{neu})/V(\text{ref})$ in this study (§3.1)
$\hat{N}_V$	Estimator of N_V (§3.1)
neu	Neuron(s) (§2.2)

p	Dimension (i.e. number of components) of a random vector (§2.2, 6.1)
$P(\cdot)$, $P'(\cdot)$	Total number of test points counted (per section in this study) in the compartment within parentheses (§3.1)
P-value	Probability of observing an absolute mean difference between groups at least as large as ours when H_0 is true. Thus, the smaller the P-value, the less we trust that H_0 is true (§4.3, Fig. 2.5)
$\mathbb{P}(\cdot)$	Probability of the event within parentheses (§4.3).
p.s.u.	Primary sampling unit. Prior to sampling the population is partitioned into elements, or non-overlapping sets of elements, called primary sampling units, whose union is the whole population. In this study a p.s.u. is a mouse hippocampus. Each p.s.u. may be partitioned into secondary sampling units (e.g. slices), and so forth.
Pv	Parvalbumin (neuron subpopulation) (§1).
Q^-	Shortcut for Q^-(neu) = total number of neurons counted (per slice in this study) (§3.1).
Q	Sample total (see §9.1)
Q–Q plot	Quantile–quantile plot (§4.4, Fig. 2.6)
ref	Reference space (§3.1)
S-PLUS®	Statistical software package used in this study (http://www.splus.com) (§1, 4.3)
σ^2	True population variance. Preferred notation for $\text{Var}(X)$ when X is a variable (e.g. neuron number, or an unbiased estimator of it) defined on a natural population (§4.3).
s^2	Sample variance, usually an unbiased estimator of σ^2 (§4.3 and §9.1, eqn. (9.4)).
SD$(\cdot)$	Standard deviation of the random variable within parentheses, SD$(\cdot) = \sqrt{\text{Var}(\cdot)}$. When the variable (e.g. neuron number) is defined on a natural population, SD$(\cdot) = \sigma$. In the biomedical literature 'SD' is often used to represent SD(X).
sd$(\cdot)$	Estimator of SD$(\cdot)$ (we write $\hat{\sigma}$ for an estimator of σ). For instance, $\hat{\sigma} = s$ under simple random sampling from a large population (see §9.1, 'Estimator of the population variance').
SE$(\cdot)$	Standard error of the estimator within parentheses; synonym of SD$(\cdot)$ when the random variable

$se(\cdot)$	within parentheses is an estimator. For instance, $SE(\bar{X})$ is the same as $SD(\bar{X})$, namely $\sqrt{Var(\bar{X})}$. Under simple random sampling, $Var(\bar{X})$ is given by eqn. (9.7)
	Estimator of $SE(\cdot)$, namely, $se(\cdot) = \sqrt{var(\cdot)}$ when the random variable within parentheses is an estimator. In the biomedical literature 'SEM' is often used to represent $se(\bar{X})$. Under simple random sampling from a large population $se(\bar{X}) = s/\sqrt{n}$ (see §9.1, eqn. (9.1))
$\tau = h/T$ (§3.3)	The total of a population is also commonly denoted by τ (see §9.1)
$\hat{\tau}$	Estimator of the population total (see §9.1)
t, t'	Slice thickness before and after shrinkage, respectively (§3.1). In §4.3 and §4.4 the symbol t is used to denote the statistic (4.5), and the Student's t distribution.
$t_k(\cdot)$	Percentile (corresponding to the probability within parentheses) of the Student's t distribution with k degrees of freedom (§4.3)
T	True distance between parallel systematic sections, or between slice midplanes, in a Cavalieri design (§3.1)
T^2	Hotelling's test. It is a MANOVA with only two groups to compare (§6.3)
V	Shortcut for $V(ref)$ in this study, namely the true total volume of a compartment (§3.1)
$\hat{V}$	Unbiased estimator of V (§3.1)
$var(\cdot)$	Estimator of $Var(\cdot)$ (§5)
$Var(\cdot)$	Mean square deviation of the random variable within parentheses. Formally, $Var(X) = \mathbb{E}[X - \mathbb{E}(X)]^2 = \mathbb{E}(X^2) - [\mathbb{E}(X)]^2$ (§3.3, 4.3)
$\bar{X}$	Sample mean, usually an unbiased estimator of μ, (§4.3).
$z(\cdot)$	Percentile (corresponding to the probability within parentheses) of the standard normal distribution $N(0, 1)$ (§4.3)

Acknowledgments

We thank Drs Steve Evans and Marta García-Fiñana for their constructive comments on earlier drafts of this chapter.

The authors acknowledge support from the Spanish Ministry of Science and Technology research project BSA2001–0803–C02–01.

References

Baddeley, A.J. and **Cruz-Orive, L.M.** (1995). The Rao–Blackwell theorem in stereology and some counterexamples. *Adv. Appl. Probab.* **27**, 2–19.

Becker, R.A., Chambers, J.M., and **Wilks, A.R.** (1988). *The new S language. A programming environment for data analysis and graphics.* Wadsworth & Brooks/Cole Advanced Books & Software, Pacific Grove, CA.

Bersu, E.T. (1980). Anatomical analysis of the developmental effects of aneuploidy in man: the Down syndrome. *Am. J. Med. Genet.* **5**, 399–420.

Braendgaard, H., Evans, S.M., Howard, C.V., and **Gundersen, H.J.G.** (1990). The total number of neurons in the human neocortex unbiasedly estimated using optical disectors. *J. Microsc.* **157**, 285–304.

Chatfield, C. and **Collins, A.J.** (1980). *Introduction to multivariate analysis.* Chapman & Hall, London.

Cochran, W.G. (1977). *Sampling techniques,* 3rd edn. J. Wiley & Sons, New York.

Cruz-Orive, L.M. (1999). Precision of Cavalieri sections and slices with local errors. *J. Microsc.* **193**, 182–98.

Cruz-Orive, L.M. and **Weibel, E.R.** (1981). Sampling designs for stereology. *J. Microsc.* **122**, 235–72.

Dun, N.J., Huang, R., Dun, S.L., and **Forstermann, U.** (1994). Infrequent co-localization of nitric oxidase synthase and calcium binding proteins immuno-reactivity in rat neocortical neurons. *Brain Res.* **666**, 289–94.

Evans S.M., Howard, C.V., and **Gundersen, H.J.G.** (1989). An unbiased estimate of the total number of human neocortical neurons [abstract]. *J. Anat.* **167**, 249–50.

Freund, T.F. and **Buzsaki, G.** (1996). Interneurons of the hippocampus. *Hippocampus* **6**, 345–474.

García-Fiñana, M. and **Cruz-Orive, L.M.** (2000*a*). Fractional trend of the variance in Cavalieri sampling. *Image Anal. Stereol.* **19**, 71–9.

García-Fiñana, M. and **Cruz-Orive, L.M.** (2000*b*). New approximations for the variance in Cavalieri sampling. *J. Microsc.* **199**, 224–38.

Garcia-Fiñana, M., Cruz-Orive, L.M., Mackay, C.E., Pakkenberg, E.B., and **Roberts, N.** (2003). Comparison of MR imaging against physical sectioning to estimate the volume of human cerebral compartments. *NeuroImage* **18**, 505–16.

Geiser, M., Cruz-Orive, L.M., Im Hof, V., and **Gehr, P.** (1990). Assessment of particle retention and clearance in the intrapulmonary conducting airways of hamster lungs with the fractionator. *J. Microsc.* **160**, 75–88.

Gual-Arnau, X. and **Cruz-Orive, L.M.** (1998). Variance prediction under systematic sampling with geometric probes. *Adv. Appl. Probab.* **30**, 889–903.

Gundersen, H.J.G. (1986). Stereology of arbitrary particles. A review of unbiased number and size estimators and the presentation of some new ones, in memory of William R. Thompson. *J. Microsc.* **143**, 3–45.

Gundersen, H.J.G. (2002). The smooth fractionator. *J. Microsc.* **207**, 191–210.

Gundersen, H.J.G. and **Jensen, E.B.** (1987). The efficiency of systematic sampling in stereology and its prediction. *J. Microsc.* **147**, 229–63.

Gundersen, H.J.G., Bagger, P., Bendtsen, Evans, S.M., T.F., Korbo, L., Marcussen, N., Møller, A., Nielsen, K., Nyengaard, J.R., Pakkenberg, B., Sørensen, F.B., Vesterby, A., and **West, M.J.** (1988). The new stereological tools: disector, fractionator, nucleator and point sampled intercepts and their use in pathological research and diagnosis. *Acta Pathol. Micro biol. Immunol. Scand.* **96**, 857–81.

Gundersen, H.J.G., Jensen, E.B.V., Kiêu, K., and **Nielsen, J.** (1999). The efficiency of systematic sampling in stereology—reconsidered. *J. Microsc.* **193**, 199–211.

Howard, C.V. and **Reed, M.G.** (1998). *Unbiased stereology. Three-dimensional measurement in microscopy.* Bios Scientific Publishers Ltd, Oxford.

Hsu, S.M., Raine, L., and **Fanger, H.** (1981). Use of avidin–biotin–peroxidase complex (ABC) in immunoperoxidase techniques: a comparison between ABC and unlabelled antibody (PAP) procedures. *J. Histochem. Cytochem.* **29**, 577–80.

Huntsberger, D.V. and **Billingsley, P.P.** (1989). *Elements of statistical inference,* 6th edn. Wm. C. Brown Publ., Dubuque, Iowa.

Insausti, A.M., Megías, M., Crespo, D., Cruz-Orive, L.M., Insausti, R., and **Flórez, J.** (1998). Hippocampal volume and neuronal number in Ts65Dn mice: a murine model of Down syndrome. *Neurosci. Lett.* **253**, 1–4.

Pache, J.C., Roberts, N., Zimmerman, A., Vock, P., and **Cruz-Orive, L.M.** (1993). Vertical LM sectioning and parallel CT scanning designs for stereology: application to human lung. *J. Microsc.* **170**, 9–24.

Pakkenberg, B. and **Gundersen, H.J.G.** (1988). Total number of neurons and glial cells in human brain nuclei estimated by the disector and the fractionator. *J. Microsc.* **150**, 1–20.

Snedecor, G.W. and **Cochran, W.G.** (1980). *Statistical methods,* 7th edn. The Iowa State University Press, Ames, Iowa.

Venables, W.N. and **Ripley, B.D.** (1999). *Modern applied statistics with S-Plus,* 3rd edn. Springer Verlag, Berlin.

von Bartheld, Ch.S. (2001). Comparison of 2–D and 3–D counting: the need for calibration and common sense. *Trends Neurosci.* **24**, 504–6.

Weibel, E.R. (1979). *Stereological methods.* Vol. 1: *Practical methods for biological morphometry.* Academic Press, London.

West, M.J. and **Gundersen, H.J.G.** (1990). Unbiased stereological estimation of the number of neurons in the human hippocampus. *J. Comp. Neurol.* **296**, 1–22.

West, M.J., Slomianka, L., and **Gundersen, H.J.G.** (1991). Unbiased stereological estimation of the total number of neurons in the subdivisions of the rat hippocampus using the optical fractionator. *Anat. Record* **231**, 482–97.

NUMBER

CHAPTER 3

SECTION INTRODUCTION

STEPHEN M. EVANS AND JENS R. NYENGAARD

1 Introduction

> The number of neurons and their relative abundance in different parts
> of the brain is a determinant of neural function and, consequently, of
> behaviour.
>
> Williams and Herrup (1988)

Most neuroscientists would accept that neurons and synapses form the fundamental building blocks for information processing in the central nervous system.
Although this is a somewhat simplified view, since it neglects the subtle interaction
between other elements of the nervous system such as glial cells and the local
microenvironment, it is not unreasonable to assume that perturbations in their
number will be involved in the pathophysiology of both disease and normal development. This chapter describes the basic principles of number estimation with a
series of examples and exercises to help explain the methodology.

The historical development of stereological estimators of number was described by Bendtsen and Nyengaard (1989). For many years stereologists advised that any one trying to estimate number should try to redefine their problem and avoid 'counting'. This was because number is a zero-dimensional property and will require a three-dimensional counting rule (Fig. 3.1). This was somewhat of a dilemma for neuromorphologists because the number of neurons, the functional units of the central nervous system, was one of the most valuable biological parameters but there was no way to unbiasedly estimate it except to use serial sectioning and reconstruction.

The easiest counting rule in any situation is to count everything. Then all objects have an equal probability of being counted. For example, to count the number of patients who attend a neurologist's out-patient clinic on a particular day is a relatively simple task requiring no sampling, other than choosing the appropriate day and being able to count. However, to count all the neurons in the brain of one of these patients would require that the brain be viewed at a sufficiently high magnification to be able to see the neurons and, more importantly, since it is impractical to count billions of neurons, a sampling scheme. For now it is assumed that an efficient and unbiased sampling scheme has generated a series of smaller sampling units that are to be used to estimate neuron number. The next question is deciding when a neuron is part of the sampling unit.

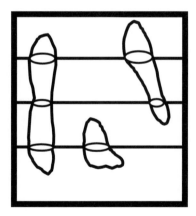

Fig. 3.1 Height bias. To count three-dimensional objects it is necessary to employ a three-dimensional counting rule. For example, 'tall' objects will appear in more sections. Therefore, if one only counts neuronal profiles in one section there will be a bias towards the 'tall' objects.

2 Counting rules

Luis Cruz-Orive (1980) was the first to overcome the barrier for the estimation of number using a volume probe consisting of serial sections cut with a random starting position, but at an arbitrary orientation. However, the big breakthrough came with the description of the disector principle (Sterio 1984), even though the basic principles behind the disector had been known for decades by isolated or unrecognized researchers.

There are three main counting rules in the 'new stereological toolkit':

- associated point (Miles 1972)
- unbiased brick (Gundersen 1981; Howard *et al.* 1985)
- disector (Sterio 1984).

In theory the associated point rule is the easiest to implement because it can be performed using one section and the unbiased brick is the easiest to implement if a three-dimensional reconstruction has been done. In practice the disector is usually the easiest to implement because with the associated point it is necessary to identify a unique point associated with a particle and with the brick it is necessary to obtain complete three-dimensional information about the particle before a decision can be made as to whether or not a particle should be counted. The disector only requires a decision as to when the first identifiable two-dimensional profile of the particle falls within an unbiased counting frame (Gundersen 1977). Notice that this does not have to be the 'top' of the particle but it has to be some unique feature of the particle that occurs in one section and not another.

There was a modified version of the unbiased brick by Williams and Rakic (1988*a*). Unfortunately, the modification was biased, and they made an alteration to their method (Williams and Rakic 1988*b*) converting it to be essentially the same as the unbiased brick. From a historical perspective this was the first time the unbiased brick rule was used in a conventional microscope to study nervous tissue and the paper also contains a number of pertinent references to the history of number estimation.

2.1 The unbiased counting frame and the unbiased brick

2.1.1 Two-dimensional counting

It is easier to start to understand some of the new stereological counting tools using a two-dimensional example. Please note that one of the characteristics of many of the new stereological tools is that they are unbiased in '*n*'-dimensional space.

At high magnification and in one field of view it would be reasonably simple to count all of the neuron profiles but what happens when a neuronal profile falls on the edge of the field of view? Does it belong to this field of view or its neighbor? The worst possible scenario occurs when a profile lies in a corner because it could

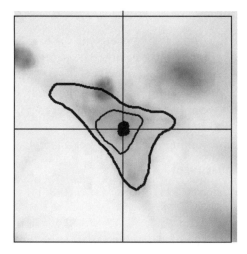

Fig. 3.2 Where is the nucleolus? The neuron's nucleolus could be considered to be part of all four quadrates.

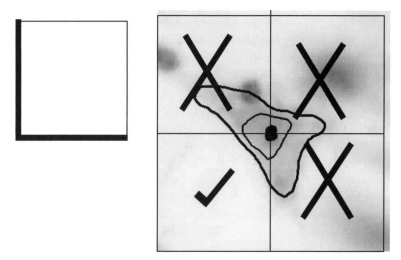

Fig. 3.3 A well established counting rule from hematology. The counting frame is placed over each of the quadrates. The nucleolus is counted in the quadrate when it does not touch any of the forbidden, thickened lines. Hence it is counted once, in the quadrate with the tick.

potentially belong to four fields of view (Fig. 3.2). There are many strategies that have been developed to solve this problem and one of the most common ones comes from the field of hematology (Fig. 3.3). This rule works if the profiles are very spherical but for other shapes it can break down. Consider the neuronal perikaryon in Fig. 3.4. In this example the profile will be counted in two quadrates

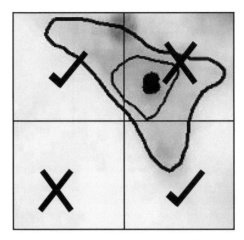

Fig. 3.4 Problems with the well-established counting rule. Using the rule described in Fig. 3.3, the neuron's perikrayon is counted in the quadrate where it does not touch any of the forbidden, thickened, lines. Hence this neuron is counted twice in the quadrates that are ticked.

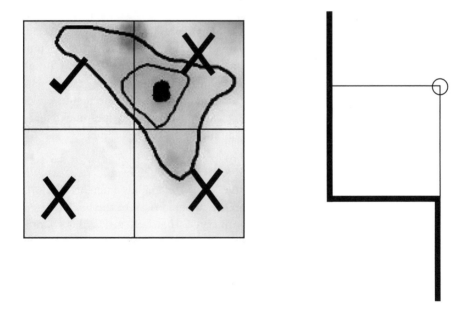

Fig. 3.5 The unbiased counting frame. The neuron is counted in the quadrate so long as it does not touch any of the forbidden, thickened, lines. Hence this neuron is counted once in the quadrate with the tick. Also note that the thickened lines extend to infinity in both directions, effectively dividing two-dimensional space in half.

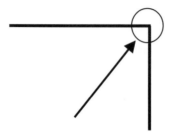

Fig. 3.6 The definition of a point. A point has zero dimensions and therefore cannot be the intersection of the two drawn lines. The lines have a width, otherwise we could not see them, and therefore so will the point created at the intersection of the two drawn lines. So a stricter definition is required. The point is defined as one of the corners of the intersection of the two lines, indicated by the arrow above. Notice also that, if points were being counted, then a simple quadrate could be used since a true point cannot fall on a corner.

and the situation would be worse for non-convex shapes. Therefore, if this rule is used, profiles can be potentially oversampled depending on their shape.

The unbiased counting frame Gundersen (1977) overcomes this bias by extending the forbidden lines to infinity in both directions (Fig. 3.5). Please also observe that one of the corners of the frame has been encircled. This is to help in deciding when the counting frame falls in the tissue of interest. There are times when only part of the frame will fall inside the tissue of interest. The simplest way to make this decision is to have a point associated with the frame—in this case the encircled corner (Fig. 3.6). If the point falls in the tissue then its associated frame does as well. The objects would still be sampled using the rule described above, even if it were decided that the frame itself did fall inside the tissue of interest.

2.1.2 Moving to three dimensions: the unbiased brick

Imagine that the unbiased frame now leaves the page and has a height. The forbidden lines become forbidden sides, as does the upper surface of the brick (Fig. 3.7). Hence a sampling volume is formed. Any object that falls inside the volume is counted so long as it does not touch any of the forbidden sides. The rule is ideal in a three-dimensional reconstruction where the whole of the object can be visualized but its implementation on sectioned material is not always practical. This is because it is necessary to follow the whole object continuously throughout all of the sections to make sure that the object does not touch any of the exclusion planes. In contrast, as is described below, the disector only requires one section to decide whether or not the decisive profile appears in the unbiased counting frame.

As with the unbiased counting frame, a point is associated with the brick to decide when it falls inside the tissue of interest.

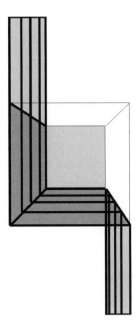

Fig. 3.7 The unbiased brick. Please see the text for more details on this counting rule.

2.1.3 A biased modification of the unbiased frame

Williams and Rakic (1988*a*) also proposed a counting rule. Again the thickened lines are extended to infinity but, as can be seen from Fig. 3.8, the neuron's

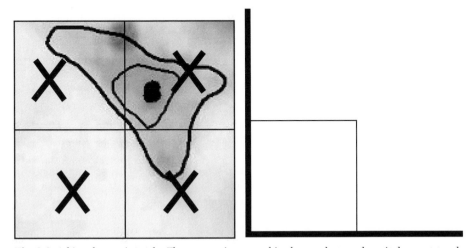

Fig. 3.8 A biased counting rule. The neuron is counted in the quadrate so long it does not touch any of the forbidden, thickened lines. Hence this neuron is never counted in any of the quadrates.

perikaryon will never be counted. This rule is biased in two dimensions and this bias is attenuated when evoked in three-dimensional space.

2.2 The associated point

As described above, a point has zero dimensions. Therefore it cannot fall on the edge of a sampling volume and there is no height bias. Since the point can only occur in one section, the sampling volume is formed by the section height multiplied by the area of the quadrate. In practice nothing in neuroscience can truly be reduced to a point. A nucleolus visualized at high magnification could be considered as a point but it is not zero-dimensional and the bias will eventually become apparent.

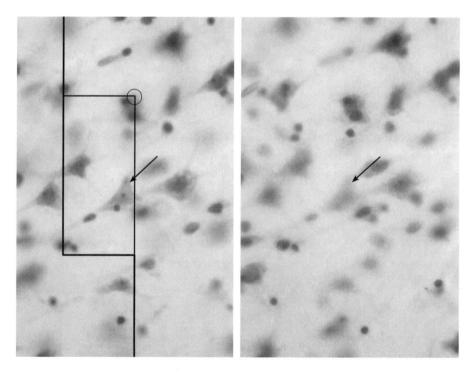

Fig. 3.9 The disector. The figure shows two optical sections. A neuron (arrowed) was counted using the disector principle, that is to say, it is counted if the widest, clearest nuclear profile is seen in the optical section on the left but not in the one in the right and the profile does not touch any of the forbidden lines, (the solid lines of the test frame in the micrograph). This definition is preferred because optical sectioning of whole neurons revealed that some neurons did not have a clearly visible nucleolus. The neuron in the right micrograph, the one indicated by the arrow, is counted because it is in focus and in the sampling frame but it is not in focus in the 'look-up frame' in the left micrograph.

2.3 The disector

In a disector, two sections, either optical or physical, are used to create a sampling volume with a lower, reference section, containing a test frame. The test frame, the unbiased counting frame, is a set of geometrical criteria that ensures that all objects are counted with equal probability (see above). The sampling volume, $v(\text{dis})$ is the area of the test frame multiplied by the distance between the upper surface of the sections. The disector sampling volume is counted as falling in the reference space if the upper right-hand corner of the test frame, the circled point, falls in the reference space. A particle is sampled if it appears in the lower reference section inside the unbiased counting frame and it does not appear anywhere in the upper section (Fig. 3.9). The disector is a three-dimensional probe that samples structures proportional to their number without regard to the size or shape of the structures.

Optical sectioning can be performed using confocal or conventional light microscopes fitted with high numerical aperture oil immersion objective lenses, and it allows many of the stereological methods to be applied more efficiently and facilitates the use of other powerful techniques. A guard volume is normally required above and below the disector probe to facilitate the identification of objects and the decision process as to which objects should be counted. The use of optical sectioning in stereology and some of the equipment required was described in Chapter 1.

3 Sampling design

Estimation of number in the central nervous system is simple, quick, and efficient provided a suitable sampling scheme is employed and the basic requirement of being able to identify the population of objects to be counted can be fulfilled. There are no assumptions about spatial homogeneity or randomness (Poisson distribution). However, it is easy to confuse the disector probe, the disector/Cavalieri, and disector/fractionator. As they are commonly used terms their theory is discussed below. A detailed discussion of sampling techniques used in stereology will be left to a later chapter. A good sampling scheme is essential for estimating number and thus a brief outline of two common techniques that are used for number estimation is given below. However, the aim for both schemes is to produce an efficient and unbiased sampling scheme that generates series of smaller sampling units that are used to estimate number.

A popular misconception, introduced by Benes and Lange (2001), is that the disector technique requires that the objects being counted, for example, neurons in the cerebral cortex, have a Poisson distribution. If this were the case then using a disector/Cavalieri design to estimate neuron number in the cerebral cortex could be done using only one large disector so long as the disector was sufficiently large enough to provide a representative sample and could be placed anywhere in the cerebral cortex. However, neocortical neurons are arranged in layers whose regional composition varies. One large disector could be used to estimate numerical density provided its position was chosen at random and all areas of the neocortex had an equal chance of appearing in the sample. It would be an unbiased estimate but also a very imprecise estimate. Or, put another way, if another large disector is used in the next field of view it will, not surprisingly, provide another estimate of numerical density. To conclude that the disector method is biased would be a misunderstanding of the application of sampling theory to stereological methodology.

The estimator of neuron number is unbiased but the estimators of the efficiency of these methods are complicated, if misused, potentially biased. The estimators of the efficiency of a sampling scheme are used to optimize the sampling scheme, that is, how many individuals, samples, sections, fields of view, disectors, etc., are needed in order to provide an acceptable estimate of neuron number. This is considered in more detail in Chapter 2.

Modern stereology employs a specialized form of sampling theory, systematic random sampling (Fig. 3.10). This is both efficient and unbiased (Gundersen and Jensen 1987; Gundersen *et al.* 1999). Samples must be chosen at random to give

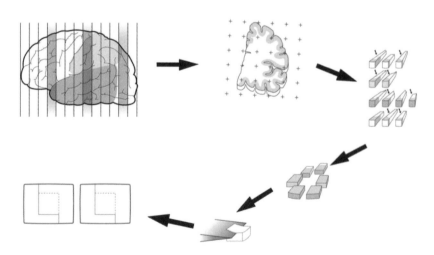

Fig. 3.10 Sampling. In most sampling schemes whole structures, in this case the cerebrum, are reduced to slices from which tissue blocks are sampled. These are themselves sectioned and further sampling occurs at the level of the choice of section and the field of view on which measurements are made. Hence for these measurements to be meaningful an appropriate sampling scheme must be employed. The point grid on the coronal section is estimating the volume of the cerebrum.

all possible tissue an equal chance of being sampled but, after the initial random start, the rest of the sample can be chosen at specific intervals.

Imagine that a brain is cut into a series of slices, for example 20. Every fourth slice is chosen with the first slice being chosen at random. If the first slice was always chosen as the start point and something interesting happened in the third slice it would never ever be detected. So all parts of the tissue under study must have an equal chance of appearing in the sample. Also note that, if the slice thickness is known, then the volume of the brain can be estimated using Cavalieri's estimator.

The sampled slices are now cut into a series of blocks. The blocks can be kept in the sequential order in which they were obtained from the slices (since the position of the first slice was chosen at random) or they can be mixed together. The former approach will have a better sampling efficiency. Every tenth block is chosen, again with a random start, and prepared for histology. These blocks are then cut into sections.

3.1 The Physical Disector/Fractionator design

For the physical disector/fractionator sampling scheme (Gundersen 1986), the block is completely sectioned and every fiftieth section and its adjacent section are chosen, again with the first section being chosen at random. The section thickness should be sufficiently small so that no object can go undetected inside the section. On each of the sections every tenth field of view is sampled using the disector probe. Again the starting position of the first field of view must be chosen at random. The size of the disector's unbiased counting frame is half that of the size of the field of view. The neurons are counted if they fall inside the frame but do not touch any of the exclusion lines and do not occur in the adjacent 'look up' section.

An estimate of the total number of neurons, N(neu), will be the number of neurons counted using the disector probe, n(neu) multiplied by 1/sampling fraction where

$$\text{Sampling fraction} = \tfrac{1}{2} \text{(for the area of the frame)} \times \tfrac{1}{50} \text{(for the sections)} \times \tfrac{1}{10}$$
$$\text{(for the tissue blocks)} \times \tfrac{1}{4} \text{ (for the slices).}$$

The physical disector/fractionator combination is the most robust of the disector designs. It estimates number directly and the estimate is unaffected by tissue deformation. The only requirements are that the objects being counted must not completely disappear during the sampling process and that the sampling fraction must be known. There is no requirement to know either slice or section thickness. However, the sampling must be applied rigorously and all of the tissue block must be cut into sections to obtain the sampling fraction for this sampling level.

3.2 The Optical Disector/Fractionator design

The optical disector/fractionator has one extra sampling step, the measurement of section thickness. The block is again completely sectioned but, instead of taking two sections, only one thick section is chosen. The disector probe will be composed of two optical sections inside a thick physical section and an extra level of sampling occurs, namely, the height of the disector divided by the section thickness. Neither the absolute height of the disector nor the absolute section thickness needs to be known but their ratio, the 'height sampling fraction', must be known. So, for example, if the height of the disector probe is 5 units (as measured using a micro-cator or the units on the microscope's fine focus 'knob') and the section thickness is 10 units then the height sampling fraction $= \frac{5}{10} = \frac{1}{2}$.

An estimate of the total number of neurons, N(neu), will be the number of neurons counted using the disector probe, n(neu), multiplied by 1/sampling fraction,

Sampling fraction $= \frac{1}{2}$ (for the area of the frame) $\times \frac{1}{50}$ (for the sections) $\times \frac{1}{10}$

(for the tissue blocks) $\times \frac{1}{4}$ (for the slices).

This design still estimates number directly but the estimate can be affected by tissue deformation in the z-axis (see Dorph-Petersen *et al.* 2001 and Chapter 10, this book).

3.3 The Physical Disector/Cavalieri design

For the physical disector/Cavalieri sampling scheme the tissue blocks are not completely sectioned and there is no requirement to know the sampling fraction. However, the reference space volume V(ref), in this case the brain volume, V(brain), and the total volume of the disectors falling inside the reference space must be known. The brain volume can be estimated using Cavalieri's principle. The disector sampling volume is counted as falling in the reference space if the upper right-hand corner of the test frame, the circled point, falls in the reference space. The volume of a disector is the area of the unbiased counting frame multiplied by its height, the distance between the sections, that is, the section thickness. The total volume of disectors is then the number of disectors falling in the reference space multiplied by the volume of the disector.

Only one section and its adjacent section are required from anywhere in the tissue block. The section thickness should be sufficiently small so that no object can go undetected inside the section. The section is then sampled systematically and at each field of view neurons are counted if they fall inside the frame but do not touch any of the exclusion lines and do not occur in the adjacent 'look-up' section. The numerical density of neurons in the brain, N_V(neu/brain), can be estimated from the number of neurons counted divided by the total volume of disectors (the number of disectors falling in the reference space multiplied by the volume of the disector).

An estimate of the total number of neurons, N(neu) = N_v(neu/brain) × V(brain).

The physical disector/Cavalieri combination has the advantage that sampling is not as exhaustive as in the fractionator designs but it estimates number indirectly and the estimate can be affected by tissue deformation. In order to verify the assumption that there is no change in the reference space for both estimates, any tissue deformation that occurs between the performance of the Cavalieri estimate of volume and the subsequent numerical density estimate must be quantified.

However, if the estimation of volume by the Cavalieri principle and the estimation of numerical density by the disector can be performed on the same histological sections, then the section thickness t cancels out. The estimate of the number of structures will then be independent of tissue deformation just as in the disector/fractionator combination (Pakkenberg and Gundersen 1988).

3.4 The Optical Disector/Cavalieri design

The optical disector/Cavalieri sampling scheme is exactly the same as the physical version except that only one thick physical section is taken instead of two. Again this can be from anywhere inside the tissue block. The disector is formed by two optical sections a known distance apart, and with a guard volume above and below the disector probe. The volume of a disector is the area of the unbiased counting frame multiplied by its height, the distance between the optical sections. As with the physical design the reference space volume, V(ref), in this case the brain volume, V(brain), and the total volume of the disectors falling inside the reference space must be known. The disector sampling volume is counted as falling in the reference space if the upper right-hand corner of the test frame, the circled point, of the first optical section falls in the reference space.

Only one section and its adjacent section are required from anywhere in the tissue block. The section is then sampled systematically and at each field of view neurons are counted if they fall inside the frame but do not touch any of the exclusion lines and do not occur in the adjacent 'look up' section. The numerical density, N_v(neu/brain), can be estimated from the number of neurons counted divided by the total volume of disectors (the number of disectors falling in the reference space multiplied by the volume of the disector).

An estimate of the total number of neurons, N(neu) = N_v(neu/brain) × V(brain).

The optical disector/Cavalieri combination has the advantage that it can be performed on one section. As with the physical version it estimates number indirectly and the estimate can be affected by tissue deformation. Once again, the assumption that there is no change in the reference space for both estimates must be quantified.

4 Other disectors

4.1 The double disector

This is a modification of the disector to measure the numerical density of different objects, for example, synapses and neurons, when section thickness cannot be determined and it is impractical to exhaustively section a tissue block. A systematic sample has been prepared for electron microscopy and two thin adjacent sections have been obtained. The numerical density of neurons, N_V(neu/brain), is estimated using a frame of area a(neu). The height of the disector is t, the number of disectors falling in the reference tissue is P(neu), and the total number of neurons counted is Q^-(neu). The numerical density of synapses (syn), N_V(syn/brain) is estimated using a frame of area a(syn). The height of the disector is t, the number of disectors falling in the reference tissue is P(syn), and the total number of neurons counted is Q^-(syn). The ratio of synapses to neurons, N(syn)/N(neu), is N_V(syn/brain) divided by N_V(neu/brain) and section thickness cancels out of the equation. If the total number of neurons in the brain has been estimated using one of the methods described above at the light microscopical level then the total number of synapses can be obtained from

$$N(\text{syn}) = N(\text{syn}) / N(\text{neu}) \times N(\text{neu}) = [N_V(\text{syn} / \text{brain}) / N_V(\text{neu} / \text{brain})] \times N(\text{neu})$$
$$= [\{Q^-(\text{syn}) \times P(\text{neu}) \times a(\text{neu})\} / \{Q^-(\text{neu}) \times P(\text{syn}) \times a(\text{syn})\}] \times N(\text{neu}).$$

4.2 Number of connected sets of objects (ConnEulor)

The disector can also be used to estimate the number of connected objects. Using the zero-dimensional Euler–Poincaré characteristic (Euler number), estimation of number with the disector is not restricted to isolated objects but may also be useful in estimating number of elements in a histologically complex network, for example, the number of capillaries in a capillary network (Gundersen *et al.* 1993) For the estimation of Euler number, χ, in a network, the number of isolated parts (islands), the number of extra connections (bridges), the so-called connectivity, and the number of enclosed cavities (holes) in the network have to be taken into account.

χ = number(isolated parts) – number(extra connections) + number(enclosed cavities).

There is one isolated part and no enclosed cavities in a capillary network; thus the equation for a capillary network boils down to $\chi = 1$ – connectivity. The Euler number is a zero-dimensional entity and may then be sampled by the disector for estimating the numerical density of capillaries, W_v, as described in Figure 7.1 in Chapter 7 and using the following equation

$$W_V: \; = \frac{-\sum(\text{number of islands} + \text{number of holes} - \text{number of bridges})}{2 \cdot t \cdot \sum a(\text{frame})}$$

5 Potential problems

There are two main problems in applying quantitative methods in neuroscience. The first is defining the regional boundaries of an area under study. The second is defining what is being sampled. Or, put more simply, the investigator has to be able to define what elements are being sampled and where they can be found.

An example of the first problem is trying to produce anatomical maps of discrete cortical areas in human neocortex. One, the most famous, is based on the cyto-architecture of human neocortex, Brodmann's areas. Apart from the problem that the tissue will be sampled at a macroscopic level but the definition of a boundary will be based on microscopical information, there is a degree of individual subject variability and also intraindividual variability due to the plastic adaptivity of a neocortical area that might make the precise definition of a border ambiguous. Some of these problems and their solutions are further elaborated by Schleicher *et al.* (2000).

The second problem arises when it is difficult to distinguish between the different cell types, for example, trying to distinguish between healthy and dying neurons (Clarke and Oppenheim 1995). Thus, it is important that all of the cells being sampled are identified with an equal probability or at least a known probability. For example, the Golgi stain provides beautiful detailed morphology of neurons and their microstructure (Cajal 1911). However, it is only a select group of neurons that are stained and possibly part of the axonal tree is unstained. It is unknown why these neurons are stained preferentially and, more importantly, what proportion of the total number of neurons this group forms.

The contributors to the various chapters of this section give detailed descriptions of how they have coped with these problems.

5.1 Tissue deformation, lost caps, and overprojection

5.1.1 Tissue deformation during sample preparation

There has been one report of a cutting artifact that might cause a bias if neglected (Hatton and Von Bartheld 1999). The authors report that neuronal nuclei had a bimodal distribution in paraffin and glycolmethacryalate physical sections. They propose that the margins of the tissue in a section are compressed during sectioning causing an increase in neuronal nuclei numerical density at the top and bottom of a section and that, by using only the central volume of a physical section for optical sectioning, it is possible to underestimate the numerical density. While it is always prudent to be aware of, check for, and, if necessary, correct for this potential source of bias, neither the authors of this chapter nor any of our collaborators have so far been able to substantiate these findings. One of the authors was also unable to reproduce these findings in some of the sections used by the researcher in the original article but, nevertheless, we would recommend that it be checked for any stereological study using optical sectioning in thick physical sections.

Tissue deformation and how it affects the disector both in physical and optical sections are extensively discussed in Dorph-Petersen *et al.* (2001). In addition this paper and some of the later chapters of this section give details of strategies to overcome these potential sources of bias.

5.1.2 Lost caps

Loss of fragments of a structure or 'lost caps' means that fragments of the structure actually or apparently disappear. Fragments may be lost due to mechanical action during the cutting of the tissue, or due to a chemical reaction during the histological process. Fragments can also be invisible due to weak tissue contrast or it can be impossible to identify small fragments using strict morphological criteria. The number and size of lost caps are generally unknown. Lost caps will frequently be small and peripheral and will therefore not affect the estimates of different dimensions of structures equally. Lost caps in two dimensions are in general smaller than the average size of profiles, in which case the relative importance of lost caps is greatest for total number of profiles, less for total circumference, and the least for total area. The effect and significance of bias caused by lost caps are unpredictable. However, so long as the object being counted does not completely disappear from the sample, the disector is generally free from this effect.

5.1.3 Overprojection

Overprojection or 'Holmes effect' is caused by the positive thickness, t, of transparent slices because all size measurements of non-transparent structures are greater than or equal to the real size. The disector is again unaffected by this phenomenon.

6 Concluding remarks

The practical implementation of these methods is discussed in more detail in the subsequent chapters of this section. These methods are remarkably efficient and are neither time-consuming nor fatiguing. For example, using an optical disector/Cavalieri design, it takes just over 2 hours to obtain an estimate of the total number of neocortical neurons (Evans *et al.* 1989).

The optical disector can be implemented in a projection microscope. One of the cheapest optical disectors was constructed from a monocular microscope and a UK£0.20 (US$0.30) mirror, which means that these methods are well within the reach of researchers with relatively small amounts of funding. The mental and physical fatigue involved in using such a system is considerably less than that involved in using a dissecting microscope in an operating theatre during an afternoon of surgery. Some of the latest implementations of these methods are computer-aided and this makes them no more fatiguing than playing a video game.

The new 'unbiased' stereological methods for estimating neuron number represent the most modern and efficient methodology available and are a considerable advance over the older model-based methods. The experimenter is still expected to fulfill some basic requirements: it is necessary to know what objects are being measured and where they can be found. To implement these methods it is required that experimental design and sampling theory is understood and sometimes that one realizes that quantitative microscopy might not be the best research tool for a particular biological problem. Of course, some researchers may wish to stay in a two-dimensional world, a flat earth society, but they will not appreciate the extra information that awaits them in a three-dimensional world. Welcome to the third dimension, welcome to the twenty-first century!

References

Bendtsen, T.F. and **Nyengaard, J.R.** (1989). Unbiased estimation of particle number using sections—an historical perspective with special reference to the stereology of glomeruli. *J. Microsc.* **153**, 93–102.

Benes, F.B. and **Lange, N.** (2001). Reconciling theory and practice in cell counting. *Trends Neurosci.* **24**, 378–80.

Cajal, R.Y. (1911). *Histologic du systeme nerveux de I'hemme et des vertebres.* L Azoulay, Paris. [Reprint 1972, Malonie, Paris]

Clarke, P.G.H. and **Oppenheim, R.W.** (1995). Neuron death in vertebrate development: *in vitro* methods [review]. *Methods Cell Biol.* **46**, 277–321.

Cruz-Orive, L.M. (1980). On the estimation of particle number. *J. Microsc.* **120**, 15–27.

Dorph-Petersen, K-A., Nyengaard, J.R., and **Gundersen, H.J.G.** (2001). Tissue shrinkage and unbiased stereological estimation of particle number and size. *J. Microsc.* **204**, 232–46.

Evans, S.M., Howard, C.V., and **Gunderson, H.J.G.** (1989). An unbiased estimate of the total number of human neocortical neurons. *J. Anat.* **167**, 249–50.

Gundersen, H.J.G. (1977). Notes on the estimation of the numerical density of arbitrary profiles: the edge effect. *J. Microsc.* **111**, 219–23.

Gundersen, H.J.G. (1981). *Stereologi: eller hvordan tal for runlig form og indhold opnas ved iagttagelse af struktureer pa snitplaner.* Laegeforeningens Forlag, Copenhagen.

Gundersen, H.J.G. (1986). Stereology of arbitrary particles. A review of unbiased number and size estimators and the presentation of some new ones, in the memory of William R. Thompson. *J. Microsc.* **143**, 3–45.

Gundersen, H.J.G. and **Jensen, E.B.** (1987). The efficiency of systematic sampling in stereology and its prediction. *J. Microsc.* **147**, 229–63.

Gundersen, H.J.G., Boyce, R.W., Nyengaard, J.R., and **Odgaard, A.** (1993). The conneulor: unbiased estimation of connectivity using physical disectors under projection. *Bone* **14**, 217–22.

Gundersen, H.J., Jensen, E.B., Kieu, K., and **Nielsen, J.** (1999). The efficiency of systematic sampling in stereology—reconsidered. *J. Microsc.* **193**, 199–211.

Hatton, W.J. and **von Bartheld, C. S.** (1999). Analysis of cell death in the trochelar nucleus of chick embryos: calibration of the optical disector couting technique reveals systematic bias. *J. Comp. Neurol.* **409**, 169–86.

Howard, C.V., Reid, S., Baddeley, A., and **Boyde, A.** (1985). Unbiased estimation of particle density in the tandem scanning reflected light microscope. *J. Microsc.* **138**, 203–12.

Miles, R.E. (1972). The random division of space. *Adv. Appl.Prob.* **4** (Suppl.), 243–66.

Pakkenberg, B. and **Gundersen, H.J.G**. (1988). Total number of neurons and glial cells in human brain nuclei estimated by the disector and fractionator. *J. Microsc.* **150** (1), 1–20.

Schleicher, A., Amunts, K., *et al.* (2000). A stereological approach to human cortical architecture: identification and delineation of cortical areas. *J. Chem. Neuroanat.* **20** (1), 31–47.

Sterio, D.C. (1984). The unbiased estimation of number and sizes of arbitrary particles using the disector. *J. Microsc.* **134** (2), 127–36.

Williams, R.W. and **Herrup, K.** (1988). The control of neuron number. *Annu. Res. Neurosci.* **11**, 42–53.

Williams, R.W. and **Rakic, P.** (1988*a*). Three-dimensional counting: an accurate and direct method to estimate numbers of cells in sectioned material. *J. Comp. Neurol.* **278**, 344–52.

Williams, R.W. and **Rakic, P.** (1988*b*). Erratum and addendum to 'Three-dimensional counting: an accurate and direct method to estimate numbers of cells in sections material'. *J. Comp. Neurol.* **281**, 335.

CHAPTER 4

THE USE OF FLUORESCENT PROBES IN CELL-COUNTING PROCEDURES

DANIEL A. PETERSON

1 Introduction

The experimental testing of hypotheses that is at the heart of the scientific method requires objective analysis of data to establish or reject the hypothesis. Quantitative estimates of biological parameters play a central role in this process, because data in numerical form can be subjected to statistical analysis establishing whether the results are due to chance or support the hypothesis. Cell number appears to be tightly regulated in most tissues, and changes in cell number are frequently associated with disease or compromise of tissue function. For experimental studies, the ability to produce reliable estimates of cell number is crucial for evaluating the effect of the experimental condition upon cell proliferation, cell death, or the prevention of cell death. The ability to identify cells or cellular activity through the expression of intracellular fluorescent probes and the increasing need to identify cells through the simultaneous detection of multiple phenotypic markers by use of multiple immunofluorescence microscopy has made the need to count fluorescent probes a common analytical challenge.

Estimating cells in a tissue from sections through the tissue requires careful sampling to ensure an accurate result. In addition to accuracy, researchers are understandably concerned about making these estimates rapidly and efficiently to make the best use of limited resources. After a brief discussion of counting methodology, this chapter describes the use of fluorescent probes in cell-counting procedures. The theoretical and practical considerations for using fluorescent probes in quantitative analysis are presented along with a worked example using the optical fractionator procedure. This section demonstrates the advantages of fluorescent probes for optical sectioning and the preservation of section thickness that results from the preparative methodology making staining with fluorescent probes particularly useful for the implementation of the optical fractionator procedure. The reader is also provided with a description of the limitations of fluorescent probes and cautions about their analysis. The chapter concludes with a summary of the advantages and disadvantages of using fluorescent probes for quantitative stereology. As this chapter shows, when properly applied, fluorescent probes are ideally suited to the three-dimensional sampling requirements of the optical fractionator procedure. Indeed, fluorescent probes are useful markers for determining a variety of quantitative parameters. The ability to use multiple fluorescent labels increases the yield of data per section and the compatibility of these preparations with confocal microscopic analysis recommends the adoption of fluorescent probes for quantitative as well as qualitative studies.

2 Approaches to determining particle number

A comprehensive approach to the theoretical background to determining number in a structure from microtomic sections is beyond the scope of this chapter. The reader should consult other chapters in this volume or other sources (Peterson *et al.* 1996, 1998; Peterson 1999). Some brief comments necessary for introducing the usefulness of fluorescent probes to quantitative procedures follow.

With our current level of instrumentation, the determination of cell number must be made by examining sections produced from the structure of interest in a microscope in order to enlarge the image of the particle or cell for identification and counting. This necessity is so well understood that we often forget that, when we count the number of cells, what we really want to know is how many cells there are in the original three-dimensional structure. Too often, it is the number of cells in the section that is reported and from which conclusions are drawn.

Another common pitfall of counting studies compounds this problem by ignoring the fact that a histological section has a real thickness. Instead, the section is assumed to be a two-dimensional structure and cells are counted at a single focal plane as if these were the only cells present. Together these approaches result in a dimensional reduction, from a three-dimensional object to a (presumed) two-dimensional section.

Yet another potential difficulty arises from the fact that an optical objective lens does not sample a flat plane when focused within the tissue section, but rather has an optical depth of field as a function of its design (discussed at length below; see Fig. 4.1). As a result of this, objects outside of the theoretical plane of focus but within the optical depth of field will appear in the projected image, a phenomenon called overprojection. The opposite problem occurs when cells truly within the theoretical plane of focus, but sampled so tangentially as to fail to be resolved, do not contribute to the projected image, a phenomenon known as truncation.

Many solutions have been proposed to correct for the problems of overprojection and truncation. Collectively, these approaches have been termed model-based approaches since they rely upon a geometric model of the cell for their prediction. Cells are generally modeled as spheres. To the degree that the cell in question does not deviate from the model, these mathematical corrections may be useful for estimating the true number of cells in the optical plane. However, model-based approaches also assume the homogeneity of cells within the tissue. These assumptions for homogeneity in cell shape, size, and distribution are often not met in biological tissue and argue against relying upon model-based approaches.

Even when all assumptions required by model-based approaches are met, these procedures still produce numerical data in the form of a density ratio. This means that the results are expressed as a number per unit area (N_A) or a number per unit volume (N_V). Numerical data should not be expressed as a density for two reasons. The first reason is that this form of data is incomplete since the objective is to determine the total number (N) in the original structure. The second reason is that data left in density form is a ratio of two parameters. Comparing a change in the volumetric numerical density (N_V) between a control and experimental condition does not resolve whether the changes observed result from number change or volume

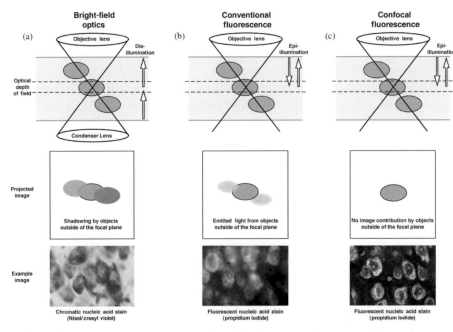

Fig. 4.1 Image production is a function of the optical elements and the staining of the specimen. (a) In conventional bright-field optics, the illumination light passes through the specimen (dia-illumination) and is detected on the other side. The illumination light is focused by a condenser lens that is aligned to coincide with the focal plane of the objective lens. Chromatic staining of the specimen produces light absorption so that stained structures imaged by the objective lens appear dark. The objective lens has a certain optical depth of field that is a function of its numerical aperture (NA); stained structures within this optical depth of field will appear with dark contrast in the projected image. Unfortunately, stained structures above and below the focal plane will also contribute to the projected image by absorbing the illumination light and casting shadows. As seen in the example image, the shadowing by structures outside of the focal plane can make it difficult to discriminate which structures are truly in the plane of focus. (b) In conventional fluorescence optics, a specific wavelength of illumination light is directed down the objective lens to the specimen (epi-illumination). Fluorescent probes excited by the illuminating wavelengths of light respond by emitting a longer wavelength of light that is detected by the objective lens and appears with bright contrast in the projected image. Stained structures outside of the optical depth of field will be exposed to the excitation wavelength and their emitted light can contribute unfocused light concentrations to the projected image reducing the contrast of structures truly in the focal plane. (c) The confocal microscope works on essentially the same principles as conventional fluorescence optics, but adds an aperture directly before the image detector to prevent the contribution of emitted light outside of the focal plane (see Peterson 1999 for details). As a result the image obtained with the confocal microscope is unambiguous in regard to resolving structures in the plane of focus.

change. In other words, if the N_v of the experimental group is higher than the N_v of the control group, did the number of cells go up or was the volume reduced in the experimental condition (Peterson *et al.* 1997, 1998)? Because of this uncertainty, numerical values should always be expressed in reference to the total structure.

Proper sampling is also essential for producing reliable quantitative data. Studies that report results taken from a few sections through the middle of a structure are, by definition, incomplete as they do not provide a representative sample. Few biological tissues are entirely homogeneous in their distribution of cells and the nervous system in particular has few examples of completely homogeneous cell distributions, even within nuclei. The sampling of fields for cell counting should be done in a way that ensures that all regions of the structure being counted have an equal probability in three-dimensions of being sampled (see Fig. 4.2).

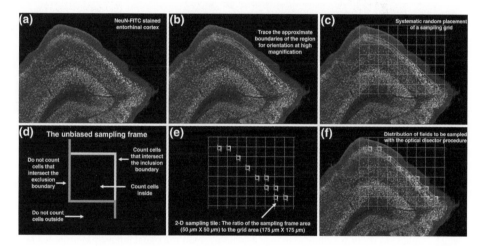

Fig. 4.2 (see also **Plate 1**) Sampling procedures for the optical fractionator using a fluorescent probe does not differ from those described for chromatic stains (West 1993). (a) A section stained with the neuronal marker NeuN detected with the fluorophore FITC (see text for details) is brought into the field of view using a low power objective (4×). (b) The structure to be counted (entorhinal cortex layer II) is outlined with the help of commercial stereology software (StereoInvestigator; MicroBrightField, Inc). The entire region to be sampled does not need to appear in the field of view. If a higher magnification is necessary to define the structure, the software will reposition the stage interactively to define regions outside of the initial, reduced field of view. (c) A regular grid array is placed over the region in a systematic uniform fashion between subsequent samples to ensure that the experimenter does not impose a biased distribution of the sampling. (d) The design of the unbiased sampling frame. Use of the frame ensures that cells visualized in two-dimensions are not over- or undersampled. (e) Grid intersects that overlie the region of interest serve as a positional index for the unbiased sampling frames to be sampled at high magnification (see Fig. 4.3). It is the ratio of these measurements that supplies the value for area sampling fraction (asf) used in the calculation of the optical fractionator result (see Fig. 4.6(b)). (f) All of the sampling elements brought together over the region of interest.

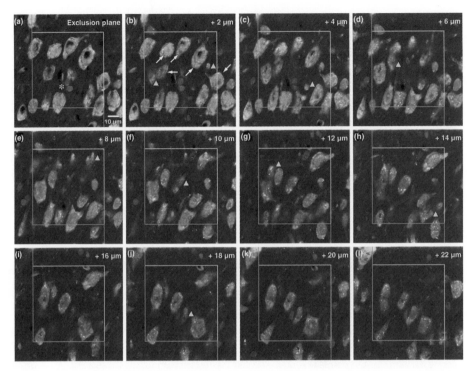

Fig. 4.3 (see also **Plate 2**) The optical disector counting procedure is applied at each sampling site indicated in Fig. 4.2(f). After changing to a suitable high NA immersion objective lens, the software positions the stage at each of the sampling sites in turn. Within each counting frame, sections are sampled with equal probability without bias for cell size, shape, or orientation. All cells were stained with propidium iodide (red) and entorhinal neurons by immunofluorescent labeling against the neuronal marker NeuN (FITC, green) as described in the text. Non-neuronal cells are colored only red (asterisk in panel A) while neurons are labeled with both red and green, which combine to show as yellow to orange depending upon the signal intensity contributed by each primary color. A series of focal planes collected at a sampling site was viewed in sequence and is represented here at discrete 2 μm intervals; in practice, the change of focal planes while counting is a continuous process. The first focal plane (0 μm; panel (a)) imaged after the guard region from the top of the section was used to define an exclusion plane. Any neuron that intersects this plane was excluded from being counted even if it meets the counting criteria in subsequent optical planes (arrows). Subsequent optical planes are viewed and neurons are counted (blue triangles) as they come into view provided they meet the sampling frame requirements (see Fig. 4.2(d)). For example, in (e) a new neuron is first observed and a blue triangle is placed just to the right of it. This neuron is observed in subsequent focal planes (panels (f)–(k)), but it is only the first time it appears that a mark is placed to avoid counting the cell more than once. In this figure, the last focal plane in which a newly visible cell would be counted is at +20 μm (panel (k)). An additional focal plane (l) is shown to illustrate that there should be a guard region at the bottom of the optical disector to evaluate the identity of any new cell that appears in the bottom plane and to verify that it meets the counting rules. Using these counting rules, one counts real cells in a volumetric subsample of the entire tissue. This raw count of cells is used to estimate total neuronal number (see Fig. 4.6).

The difficulty of designing a study to meet the assumptions required for model-based approaches was solved with the introduction of the disector procedure (Sterio 1984), which made the important advance of applying the principle of the unbiased counting frame to three dimensions. The unbiased counting frame (Gundersen 1977, 1978) simply places a sample frame over the image to use in selecting cells with equal probability (see Fig. 4.2(d)). By applying this unbiased frame on two adjacent sections a known distance apart, one could directly sample a three-dimensional volume. With one of these image planes designated for exclusion of cells that intersected it and the subsequent image plane for inclusion of intersecting cells, the disector (so-called because of the two (di-) sectors used for counting; note spelling of disector with a single 's') could directly sample cells without a methodological bias in regard to cell size, shape, distribution, or orientation. Because the procedure for counting is based upon the design of the counting probe rather than upon a model of the cell, the disector is called a design-based approach.

In practice, the physical disector was cumbersome to implement because it relied upon the tedious registration of the same location in adjacent sections. It soon became apparent that the same principle could be applied to cells within a section by focusing through the tissue. Since the cells were already in registration when this 'optical' sectioning was performed, the procedure was highly efficient (see Fig. 4.3). The optical disector, as this procedure came to be called, requires only a section of adequate thickness to focus through and a high resolution lens (discussed below). In conjunction with the fractionator sampling procedure (West *et al.* 1991; West, 1993), this design-based approach can efficiently and rapidly produce estimates of total number in the absence of methodological bias.

3 Axial resolution and optical sectioning

While the optical disector procedure represents a tremendous advance in the efficiency of (methodologically) unbiased three-dimensional sampling, there are certain optical considerations necessary to correctly implement this procedure. The optical disector sampling is essentially a continuous optical sectioning through the sample thickness. As such, it is necessary to achieve optimal axial resolution to distinguish the appearance and boundaries of the particles as the plane of focus moves through the specimen. This high axial resolution produces a narrow optical depth of field. The axial resolution of an objective lens is defined by its numerical aperture, which is in turn determined by its design and manufacture. In addition to the intrinsic design of the lens, the numerical aperture is based upon the refractive index of the immersion medium that is required between the front surface of the lens and the coverslip. For 'dry' lenses, where only air exists between these surfaces, the theoretical maximum numerical aperture (NA) is 1.0. Lenses using oil immersion can achieve a theoretical maximum NA of 1.55 for standard immersion oils. The common alternative immersion media of water and glycerol have inter-mediate NAs as a result of their lower refractive indices. Therefore, for the optical disector procedure, high NA lenses using high refractive index immersion media should be used to obtain the narrow optical depth of field necessary to accurately distinguish cells in the tissue. The cost of this high axial resolution is a shortened working distance of the objective lens. Typically this means that section thickness must be kept below 100 µm. However, advances in lens design and microscopy techniques, such as multiphoton excitation, may increase the useful working distance of high NA lenses.

The method of specimen illumination and detection contributes substantially to the ability to discriminate the appearance and boundaries of cells as they are imaged within the section. In bright-field microscopy, the specimen is trans-illuminated (dia-illumination; Fig. 4.1(a)) with light focused through a condenser lens. When the optical axis is properly aligned, the plane of focus of the transmitted light coincides with the plane of focus of the objective lens. Since the NAs of the objective and condenser lenses seldom match, a complete theoretical alignment is rarely achieved. Furthermore, the chromatic staining applied to cells will cause light absorption by cells lying above or below the focal plane of the objective lens. Therefore, the use of transmitted light combined with chromatic staining produces a shadowing in the projected image (Fig. 4.1(a)) that may make correct resolution of cell edges in the focal plane difficult.

The use of fluorescent probes reduces these resolution concerns. In conventional fluorescence microscopy, the illumination light passes through the objective lens and is focused at precisely the same point in the tissue as the image focal plane (Fig. 4.1(b)). In this epi-illumination, the objective lens serves as both condenser and objective lenses, ensuring perfect optical alignment of illumination and detec-tion. Furthermore, the light returning to contribute to the image is not reflected light that may cause shadowing, but represents specific emitted signal (see below). With conventional fluorescence, the only artifact to contribute to the detected focal plane is the emitted light from cells outside the focal plane that are exposed to the illumination light (Fig. 4.1(b)). Typically, the image produced by conventional fluorescence microscopy presents less difficulty for the identification of cells within the focal plane than does bright-field microscopy. Image resolution of fluorescent probes can be further enhanced by the use of confocal microscopy (Fig. 4.1(c)), where the optical design of the instrument excludes emitted light outside of the plane of focus producing an entirely unambiguous image (see Peterson 1999 for details).

4 Using fluorescent probes for cell counting

The utility of fluorescent probes arises from their ability to respond to illumination by specific wavelengths of light with the specific emission of longer wavelengths of

light. For each fluorescent probe or fluorophore there is a spectral profile that characterizes its absorption of light (excitation) and another profile that characterizes
its emission of light (Fig. 4.4(a)). The distance between the peaks of the excitation
and emission profiles, called the Stokes shift, determines how useful the fluorophore is for microscopic detection. The greater the Stokes shift, the clearer the
detection of the fluorophore emission without interference from the excitation
light. The situation is slightly different for multiphoton microscopy where pulsed
long-wavelength excitation produces emission of fluorophores at lower
wavelengths.

4.1 Types of fluorescent probes

Fluorophores may be vital dyes, such as nucleic acid stains, that can be reacted
directly with the tissue to detect the cellular label of interest. Nucleic acid stains are
particularly valuable for quantitative purposes in order to identify all cells present
in the tissue (Table 4.1). Many of the historical histochemical stains also emit
fluorescent profiles. In fact, the use of fluorescence microscopy in neuroscience
was promoted in the 1960s by the ability to identify neuronal populations expressing biogenic amines in the central nervous system (CNS) by their characteristic
fluorescent profile following appropriate preparation (Falck *et al.* 1982). As yet
unexplored, the modification of historical histochemical stains to specifically limit
their excitation and emission spectra for use in combination with other
fluorophores could considerably expand the methodological approaches available
to current investigators.

The most common use of fluorophores is in conjugation with immunoglobulins
for detection of specific antigenic sites (Table 4.1). Immunocytochemical detection
protocols are in widespread use and are easily adapted to the use of fluorophores
as an alternative to chromogens, such as diaminobenzidine (DAB; see Appendix 1
for an example). Immunofluorescence protocols are also simpler since a
fluorophore conjugated to a secondary antibody is usually adequate to produce a
detectable signal without an amplification step. Studies of neurogenesis have made
widespread use of multiple immunofluorescence staining to simultaneously identify multiple phenotypic labels with precise three-dimensional co-localization
(Eriksson *et al.* 1998).

Another recent use of fluorophore labeling is the introduction of transgenes
encoding for fluorescent proteins derived from jellyfish (Moriyoshi *et al.* 1996;
Kafri *et al.* 1997). While this was originally available only as a green fluorescent
protein (GFP), variants have been developed that fluoresce at a variety of wavelengths (Haseloff 1999). This approach has been further enhanced by the creation
of destabilized fluorescent proteins that can be used to monitor the onset and
duration of gene expression (Li *et al.* 1998). These marker genes can be introduced into individual cell lineages or into whole organisms by using homologous

(a)

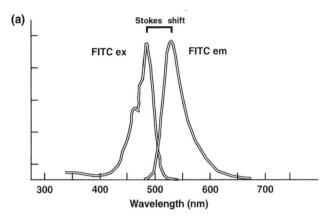

(b)

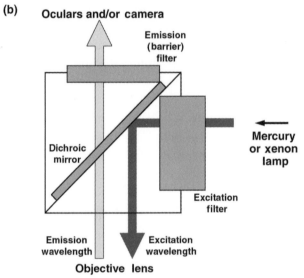

(c)

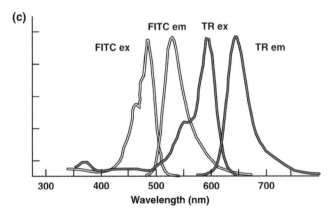

Fig. 4.4 Principles of fluorescence microscopy. (a) A fluorescent probe, or fluorophore, is defined by the wavelengths of light that cause it to absorb energy, or become excited, and the wavelengths at which it releases energy in the form of emitted light. Both the excitation and emission wavelengths have characteristic spectral curves for each fluorophore that are commonly listed by their peak values for excitation (ex) and emission (em; see Table 4.1). This example shows the spectral profiles for the commonly used fluorophore FITC. The distance between the peak values is called the Stokes shift. An adequate Stokes shift is necessary to design a filter set that can fully separate the excitation light from the emission light to provide an image that has a good signal-to-noise ratio. (b) A filter cube is a set of coated glass filters designed to separate the excitation wavelengths from the emission wavelengths. The illuminating light is of high intensity and broad spectrum generated from a mercury or xenon arc lamp that is focused and directed into the filter cube. This light passes first through a glass filter (excitation filter) coated to pass light of only the specific wavelengths that will excite the fluorophore. This filtered excitation light then is reflected off of another coated glass (dichroic) mirror into the objective lens where it is focused within the section. The emitted light from the fluorescent probes is collected by the objective lens and passes into the filter cube where it encounters the dichroic mirror. Although the excitation wavelength was reflected off of the dichroic mirror surface, the optical coating is designed to allow this longer wavelength of emitted light to pass through. Before forming the projected image in the oculars or at a camera, the emitted light passes through another filter (an emission or barrier filter) that screens out all but the specific wavelengths of light emitted by the fluorescent probe. (c) If the spectral profiles of different fluorophores have enough separation, as in this example of the fluorophores FITC and Texas Red (TR), it is possible to label a tissue section with two or more fluorescent probes and detect more than one structure simultaneously.

recombination. Quantitative stereology is an ideal tool for the assessment of endogenous tissue development within an organism or evaluating the fate of transplanted cells expressing the fluorescent marker gene.

In addition to these uses of fluorescent probes for quantitative stereology, there are a wide variety of fluorophores that can be used to monitor cellular physiology and reflect changes in pH, calcium concentration, membrane potential, oxygen radical concentration, etc. (see, for example, the Molecular Probes web site listed in Table 4.2). While undertaking quantitative stereology on living tissue presents a variety of practical challenges, the ability to quantify cells based upon physiological state is at least a potential consideration when using fluorescent probes.

4.2 Detection of fluorescent probes

Detection of fluorophores is accomplished by focusing high-intensity, broad-spectrum light from a mercury or xenon arc lamp through a filter cube (Fig. 4.4(b)). Single or mixed gas lasers emitting specific wavelengths of light are used as the light source for standard scanning confocal microscopy (Peterson 1999). The coated glass elements in the filter cube are designed specifically for the fluorophore in use. It is possible to design filter cubes that permit two or more fluorophores to be detected simultaneously, but this is usually at the cost of reduced light efficiency. Since different fluorophores seldom generate equivalent emission intensities, it is more practical to use single fluorophore filter cubes and examine the tissue for each signal in sequence.

4.3 Use of multiple fluorescent probes in the same section

Because of the specific spectra of excitation and emission, it is possible to label a tissue with more than one fluorescent label at a time (Fig. 4.4(c)). By selecting fluorophores that do not overlap (see Table 4.1), it is easily possible to triple-label a section for counting of three different markers in the same section (Peterson

Table 4.1 Common fluorescent probes*

Fluorescent label (fluorophore)	Maximum excitation (nm)	Maximum emission (nm)
Nucleic acid stains		
Hoechst 33342	350	461
Hoechst 33258 (bisbenzimide)	352	461
DAPI	358	461
SYTO 16	488	518
Acridine Orange	500	526
Ethidium bromide	518	605
Propidium iodide	535	617
ToPro3	642	661
Conjugated fluorophores†		
AMCA	349	448
Marina Blue	365	460
Cy2	490	510
FITC	494	518
Alexa 488	495	519
BODIPY FL	505	513
Cy3	550	570
Tetramethyl Rhodamine	555	580
Lissamine Rhodamine B	570	590
Alexa 568	578	603
Texas Red	595	615
Cy5	650	670

* The number of fluorophores is vast and ever-increasing. This table lists the most commonly used fluorophores that are suitable for both conventional fluorescence and confocal microscopy when matched with the appropriate laser lines.
† These fluorophores are commercially available conjugated to a wide variety of immunoglobulins, anti-digoxigenin, or various amplification compounds such as biotin–streptavidin or tyramide signal amplification.

Table 4.2 Vendor contact information for fluorescence microscopy*

Company	Address	Contact information
Dyes and fluorophores		
Molecular Probes, Inc	PO Box 22010, Eugene, OR 97402–0469, USA	Phone: 800–438–2209 Fax: 541–335–0504 Internet: www.probes.com
Jackson Immunoresearch Laboratories, Inc	872 W. Baltimore Pike, PO Box 9, West Grove, PA 19390, USA	Toll free: 800–367–5296 Fax: 610–869–0171 Email: cuserv@jacksonimmuno.com Internet: www.jacksonimmuno.com
Sigma Aldrich Research	3050 Spruce Street, St. Louis, MO 63103, USA	Toll free: 800–521–8956 Phone: 314–771–5765 Internet: www.sigma-aldrich.com
Perkin Elmer Life and Analytical Sciences	549 Albany Street, Boston, MA 02118, USA	Toll free: 800–762–4000 Internet: www.perkinelmer.com
Filters		
Chroma Technology Corp.	10 Imtec Ln. POBox 489 Brattleboro, VT 05101, USA	Toll free: 800–824–7662 Phone: 802–428–2500 Fax: 802–428–2525 Email: orders@chroma.com Internet: www.chroma.com
Omega Optical	210 Main St., Brattleboro, VT 05301, USA	Phone: 802–254–2690 Fax: 802–254–3937 Internet: www.omegafilters.com
Video cameras		
Dage-MTI, Inc	701 N. Roeska Avenue, Michigan City, IN 46360, USA	Phone: 219–872–5514 Fax: 219–872–5559 Email: info@dagemti.com Internet: www.dagemti.com
Optronics	175 Cremona Drive, Goleta, CA 93117, USA	Toll Free: 800–796–8909 Fax: 805–968–0933 Internet: www.optronics.com
Hamamatsu Photonic Systems	360 Foothill Road, Bridgewater, NJ 08807, USA	Toll Free: 800–524–0504 Phone: 908–231–1116 Fax: 908–231–0852 Email: usa@hamamatsu.com Internet: www.hamamatsu.com
Sony Electronics, Inc	1 Sony Drive, Park Ridge, NJ 07656, USA	Phone: 201–358–4190 Fax: 201–930–4752 Internet: www.sony.com/medical
Fuji Medical Systems USA, Inc	Science Systems Group, 419 West Ave. Stamford, CT 06902, USA	Toll free: 800–431–1850 Fax: 203–327–6485 Internet: www.fujimed.com

* This list represents major suppliers for these materials. Your microscope representative will be able to supply much of this material.

et al. 1996). With certain conditions and with the appropriate equipment, this could be extended to four markers. The ability to quantify multiple labels within a section provides the opportunity to extract more information than would be possible by counting adjacent sections stained with one marker each. Distribution of cell labels and co-expression of cell labels can be determined in the course of obtaining the individual fluorophore counts. If one of the fluorophores is a general cell counterstain, such as a nucleic acid marker, then it also becomes possible to have an internal control for the assessment of cell death. The change in total cell number can be compared against the loss of a phenotypic marker to help determine if the cell has truly died or if a loss of phenotype expression can account for the marker loss. Finally, even in the absence of assessing co-expression of cellular markers, the use of multiple fluorophores can preserve precious tissue sections by providing more information per section. This can be especially important in the use of human tissue where access to many series of sections may be limited (Eriksson *et al.* 1998).

4.4 Fluorescent preparation preserves section thickness

Another important advantage in using fluorescent probes for counting cells is the preservation of section thickness that results from the preparation methods. Most protocols for brightfield microscopy require the air drying of tissue sections on to glass slides and the subsequent treatment of these sections with alcohols and solvents prior to coverslipping in a medium containing organic solvents. However, substantial tissue shrinkage can occur as a result of such processing. This is especially true of paraffin processing where the embedding medium must be removed after sectioning. In contrast, many fluorophores should not be exposed to alcohol or organic solvents and most fluorophores do well with aqueous mounting media. Therefore tissue distortion is minimized as a result of fluorescent processing. This point is illustrated in Table 4.3, where a variety of preparative conditions were examined for their impact upon tissue section thickness. In addition to the use of solvents discussed above, it is clear from Table 4.3 that air drying of the section on to the glass slide should be for the minimum time necessary to keep it in place to avoid section thickness shrinkage.

5 Practical issues in using fluorescent probes

5.1 Tissue preparation for fluorescent probes

Most immunocytochemical protocols and many chromatic stain procedures can be modified to use fluorescent probes (Appendix 1). In most cases, the protocols will become shorter, simpler, and less expensive as a result of the elimination of solvent requirements. It may be necessary to reconsider tissue sectioning approaches. The main advantage of paraffin sectioning and cryostat sectioning to bright-field

Table 4.3 The effect of preparation technique on tissue section thickness

Method of preparing histological sections for microscopy*	Measured thickness (μm) of section cut at 50 μm	Percentage of axial shrinkage
Mounted on slide from microtome knife immediately after sectioning†	50	0
Mounted on a glass slide from a buffer bath‡	48.8	2
Mounted from buffer rinse after storage overnight in cryoprotectant§	45.9	8
Mounted from buffer rinse after storage for 2 years in cryoprotectant	45.2	10
Mounted from buffer bath and air dried for 30 minutes¶	30.1	40
Mounted from buffer bath and air dried overnight	28.9	42
Mounted from buffer bath, air dried overnight, and processed for a bright-field nucleic acid stain‖	10.9	78

* Young adult Fisher 344 rats received transcardial perfusion under deep anesthesia with a solution of 4% paraformaldehyde and 0.1% glutaraldehyde. Brains were removed, postfixed overnight, equilibrated in 30% sucrose, and then sectioned on a freezing sliding microtome. The coefficient of error in the amount of specimen advance on the microtome was less than 1% as measured using a digital encoder. The section thickness of cerebral cortex in at least 9 sections from at least 3 brains was measured for each condition using a microscope equipped with a digital encoder.

† Sections were transferred to a glass slide by a brush and then spread with a drop of 0.1M phosphate buffer. A coverslip was placed over the drop of buffer and measurements were made immediately using both differential interference and phase contrast optics.

‡ Sections were rinsed for 5 min in a solution of the fluorescent nucleic acid stain, propidium iodide (Molecular Probes; diluted 1:1000 in 0.1M phosphate buffer) and then transferred to a bath of 0.1M phosphate buffer for mounting on to chrom-gelatin coated glass slides. Excess buffer was drained off and, within 30 seconds, 75 μl of the PVA–DABCO mounting medium (see Appendix 2) was placed on the slide and a coverslip applied. Slides were left overnight at 4°C to allow the mounting medium to harden before measurement. Section thickness was determined as described above using a combination of fluorescence and contrast optics.

§ Cryoprotectant treatment was individual section storage per well in a 96-well plate with 200 μl of cryoprotectant (solution of glycerol, ethylene glycol, and phosphate buffer) per well. Plates were stored at –20°C.

¶ After mounting, sections dried on the slides for the stated interval before the placement of the mounting medium and the coverslip.

‖ Air-dried slides were defatted by rinsing for 30 min in a 1:1 solution of ethanol and chloroform then rehydrated in a descending ethanol series and rinsed in water. Sections were stained for 15 sec in a 0.5% thionine solution (Nissl stain) at 37°C, then water rinsed and dehydrated through an ascending ethanol series. Following 3 xylene rinses, slides were coverslipped using DPX and dried for 72 h to allow the mounting medium to harden before measurement.

microscopy has been their ability to produce very thin sections that reduce the axial resolution problems discussed above. However, the lack of section height in paraffin and cryostat sectioning makes these techniques poorly suited to the optical disector. With the use of fluorescent probes, only emitted light outside the focal plane contributes to the projected image making it possible to image through much thicker sections with adequate axial resolution. The ability to image through thicker sections coincides nicely with the objectives of the optical disector procedure. The use of fluorescent probes permits the use of thick freezing microtome or vibratome sections and provides an alternative to the traditional methacrylate embedding used in stereology.

One of the historical concerns regarding the use of fluorescent probes has been the impermanence of the preparations. In recent years, substantial improvements have been made to the quality of the fluorophores themselves, the anti-fading agents available, and to mounting media. For greatest longevity, sections should be coverslipped immediately upon mounting to glass slides using one of the aqueous or glycerol based mounting media commercially available. I have successfully used an inexpensive medium made in the lab that polymerizes at the edges so that no nail polish seal is required (see Appendix 2). Once the mounting medium has hardened, these slides can be re-examined for years and the sections are easily stored in standard slide boxes. As with all fluorescent material, one must protect the sections from normal light and storage at 4°C will extend useful life.

5.2 Penetration of fluorescence probes

Despite the retention of section thickness in fluorescent preparations, one must still be cautious about the ability of the probe to penetrate throughout the full section thickness. Even if fully penetrant, the fluorescent probe may demonstrate an attenuation of signal toward the section center that might contribute to error in counting. For this reason, all probes used should be evaluated for section penetration prior to quantitative analysis. As shown in Fig. 4.5, fluorophores conjugated to antibodies may differ widely in their penetration. A probe need not be excluded on the basis of poor penetration (Fig. 4.5(c)), but the limitations should be recognized and the optical disector height adjusted to sample only within the region of signal. Figure 4.5(c) also illustrates the danger associated with tissue section shrinkage. The central, non-stained region of the section shown in Fig. 4.5(c) would be collapsed following processing for typical bright-field microscopy and the assumption could be made that all cells present had been stained, leading to an underestimation of total number even if shrinkage correction calculations were properly performed.

5.3 Equipment for fluorescence microscopy

Most research grade microscopes, both upright and inverted, can accommodate the additional equipment for fluorescent optics. In general terms, one will need to

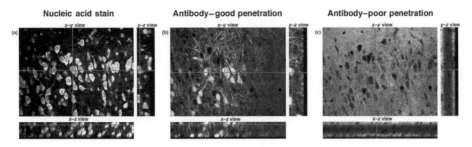

Fig. 4.5 Fluorescent probes vary in their ability to penetrate through the entire section thickness. (a) Nucleic acid stains, such as propidium iodide shown here, typically penetrate completely into a thick section (50 μm) when incubated 'free-floating' prior to mounting since the probe can penetrate from all section surfaces. In the $x - y$ view, the stained cells are seen at a given focal plane within the tissue. Overlying the $x - y$ view are two lines crossing just to the left of center. The horizontal (x-axis) line represents the position within the section from where the $x - z$ view (an edge-on view) below is generated. The $x - z$ view shows the labeling through the section thickness (z-axis) revealing complete penetration of the stain throughout the section thickness. The line running across the $x - z$ view indicates the focal depth of the $x - y$ view. Similarly, the $y - z$ view represents an image generated through the section at the position of the vertical (y-axis) line in the $x - y$ view. (b) When fluorophores are conjugated to antibodies, their penetration can vary as a function of a variety of factors including steric hindrance, avidity, and extent of section pre-treatment with solubilizing factors, such as detergents. Treatment of this section with an antibody against the calcium-binding protein calbindin (rabbit anti-calbindin; 1:1000; Swant) detected with Cy5-conjugated secondary antibody (donkey anti-rabbit-Cy5; 1:250; Jackson Immunoresearch Labs.) revealed good penetration of this antibody with some loss of signal intensity in the middle of the section. (c) In contrast, this combination of an antibody against the excitatory amino acid glutamate (mouse anti-glutamate; 1:1000; Incstar) detected with FITC (donkey anti-mouse-FITC; 1:250; Jackson Immunoresearch Labs.) showed signal penetration of only a few microns into each surface. It may still be possible to sample such labeling with the optical disector; however, the height of the disector will need to be adjusted so that it does not exceed the signal depth at either surface. It should be clear from this example that one must have an understanding of how well the label penetrates before commencing with three-dimensional sampling.

acquire a lamp house and power supply for the arc lamp along with the collector lens and the housing to direct the light to the microscope nosepiece. The filter block will need to housed at the nosepiece at right angles from the arc lamp to the objective lens. A filter block will be needed for each type of fluorophore to be observed (Table 4.2). If excitation wavelengths in the ultraviolet (UV) range of the spectrum will be required (below about 420 nm), then special fluorite objective lenses may be required; otherwise standard, good quality research objective lenses are adequate. Of course, for stereological sampling, a high NA immersion lens will be required as discussed above.

Detection of the image can be done through the oculars for all except the infrared emission of the fluorophore Cy5 (Table 4.1). For stereological analysis, it is more convenient to generate a video image of the focal plane on to a video monitor or directly on to the computer screen when used with one of the commercial stereological systems. To do this, one must have a research grade video camera with adequate low-light sensitivity (Table 4.2). To ensure optimal sensitivity, a video camera should have the ability to integrate or 'build up' the image over time. In this case, the ability to cool the camera will reduce the amount of random noise produced in the image with longer integration periods. A clever alternative to the video camera is a product called Lucivid from MicroBrightField Inc (www.micro-brightfield.com). Rather than taking the video image into the computer, this product delivers the computer video to the oculars so the observer sees the microscope image directly with the computer screen superimposed upon it.

Finally, it is possible to utilize the confocal microscope for stereological quantitation of fluorescent probes. As none of the confocal microscope manufacturers currently include this capability, the end user must make adaptations to the system. For a discussion of this topic, see Peterson (1999). MicroBrightfield, Inc has developed a module of their stereology software suitable for use on most confocal microscopes.

6 The optical fractionator procedure using a fluorescent probe—a worked example

6.1 The region of interest

The use of fluorescent probes does not change the way in which the optical fractionator procedure is implemented. In the example presented here, the number of neurons in layer II of the entorhinal cortex (ECL2) will be determined using the optical fractionator. The entorhinal cortex serves as a major relay between cortical projections to the hippocampus and for hippocampal output to the neocortex, therefore playing an important role in the processing of learning and memory. The output of ECL2 neurons forms the perforant pathway and makes up the principal excitatory projection to the hippocampus. Layer II (ECL2) neurons are one of the earliest neuronal populations to exhibit neurodegenerative pathology in Alzheimer's disease (Gomez-Isla *et al.* 1996). Because of the importance of this neuronal population, I have developed a lesion model to study the vulnerability of ECL2 neurons and their potential rescue through delivery of trophic factors (Peterson *et al.* 1996).

6.2 Tissue preparation

The entorhinal cortex occupies the posterior pole of the rat neocortex and is most clearly discerned in horizontal sections. Layer II (ECL2) neurons form a prominent lamina whose anatomical boundaries have been clearly defined (Fig. 4.2; Peterson *et al.* 1994, 1996). Serial horizontal sections were cut from the frozen brain of a perfusion-fixed, young adult Fisher 344 rat at a thickness of 50 µm on a freezing sliding microtome (see notes to Table 4.3). A one-in-six series of sections was immunostained using a monoclonal antibody against the neuronal marker NeuN (Chemicon; 1:2500). A donkey anti-mouse secondary antibody conjugated with the fluorophore fluorescein isothiocyanate (FITC) was used to detect the location of the primary antibodies (Jackson Immunoresearch Labs; 1:250). Sections were rinsed and incubated for 5 minutes in the nucleic acid stain, propidium iodide (PI; Molecular Probes; 1:1000) before mounting on to glass slides and coverslipping with the PVA–DABCO (polyvinyl acetate–1,4-diazabicyclo[2.2.2]octane) mounting medium (see Appendix 2).

After allowing the mounting medium to polymerize, the sections were examined on an Olympus BX50 fluorescent microscope equipped with a 100 W mercury arc lamp and appropriate filter sets for detecting FITC and propidium iodide (Table 4.1). The combination of these two fluorophores was chosen to show both specific staining of neurons (FITC, green) along with the staining of all cells (propidium iodide, red). This combination is useful for training students on the stereology system. By switching filter sets, the student can compare the morphology of neurons revealed by the propidium iodide staining with the phenotypic marker NeuN. After an appropriate level of training is reached, the propidium iodide can be used for ascertaining the total number of neurons based upon their morphology in conjunction with estimating the co-expression of specific phenotypic markers labeled with other fluorophores to identify subpopulations of neurons.

6.3 Equipment used for the fluorescent optical fractionator

In the example presented below, certain equipment and software are described in sufficient detail to permit this work to be replicated. Alternative equipment and software may be suitable for performing the optical fractionator procedure using fluorescent probes. For additional information about alternative equipment selection, please consult Table 4.2 and Peterson (1999). To begin the optical fractionator procedure, sections were placed on an Olympus BX50 fluorescent microscope (as described above) equipped with a Ludl Biopoint X-Y-Z motorized stage coupled with a Heidenhain digital encoder to provide a closed loop for measurement of axial movement. The motorized stage was interfaced with a PC running the StereoInvestigator software from MicroBrightField, Inc. (www.microbrightfield.com) to control specimen position. The microscope was equipped with a cooled, intensified video camera (300T-IFG; Dage-MTI, Inc) to allow signal amplification of the emission from the fluorescent probes and the video signal was sent to a frame capture board in the PC for use by the StereoInvestigator software.

6.4 Implementation of the fluorescent optical fractionator

Once the proper location is found and a suitable image obtained using a low power lens (4 ×; Fig. 4.2), the operator uses the software to delineate the region to be sampled and to designate the sampling offset for each field to be subsequently

(a)

Data Collection
Sample 97.46.31

		c2	c8	d2	d8	e2	e8	f2	f8	g2	g8	h2
1	Section	c2	c8	d2	d8	e2	e8	f2	f8	g2	g8	h2
2	Region	ECL2	ECL2	ECL2	ECL2	ECL2	ECL2	ECL2	ECL2	ECL2	ECL2	ECL2
3	Number Of Sampling Sites	7	6	8	9	8	10	10	11	12	10	11
4	Contour Area μm²	170160	147979	199938	239465	219592	229870	194115	232146	232839	195578	230918
6	Counting Frame Area (XY) μm²	1225	1225	1225	1225	1225	1225	1225	1225	1225	1225	1225
7	Counting Frame Thickness (Z) μm	16	16	16	16	16	16	16	16	16	16	16
8	Counting Frame Volume (XYZ) μm³	19600	19600	19600	19600	19600	19600	19600	19600	19600	19600	19600
9	Counting Frame Width (X) μm	35	35	35	35	35	35	35	35	35	35	35
10	Counting Frame Height (Y) μm	35	35	35	35	35	35	35	35	35	35	35
11	Sampling Grid (X) μm	175	175	175	175	175	175	175	175	175	175	175
12	Sampling Grid (Y) μm	175	175	175	175	175	175	175	175	175	175	175
13	Sampling Grid Area (XY) μm²	30625	30625	30625	30625	30625	30625	30625	30625	30625	30625	30625
14	Section Thickness μm (measured)	31	29	30.6	29.6	27.8	32.8	32.2	27.6	32.3	31	27.3
15	Neurons counted	28	9	14	15	21	18	19	22	23	25	24

(b)

Calculation of Optical Fractionator Results

Sum of Neurons Counted   Q ˙   218
line 15

	Calculation	Result
Section Sampling Fraction ssf	=1/6	= 0.17

series of sections sampled in line 1

Area Sampling Fraction   asf   = 1225/30625   = 0.04
Counting Frame Area (line 6) / Sampling Grid Area (line 13)

Mean Section Thickness μm   t   30.11
line 14

Counting Frame Height μm   h   16
line 10

Thickness Sampling Fraction   t/h   = 30.11/16   = 1.88

Fraction Sampled = (t/h) · (1/ssf) · (1/asf)
= (1.88) · (1/0.17) · (1/0.04)
= 282.27

N = Q ˙ · Fraction Sampled
= 218 · 282.27
= 61535.5

Conclusion
The optical fractionator procedure estimates a total of 61,500 neurons in layer 2 of the Entorhinal Cortex of this animal

Fig. 4.6 Calculation of the optical fractionator result using one animal as an example. (a) The data collected from 11 sections sampled as illustrated in Figs 4.2 and 4.2 is collected by the software. For each section (c2, c8, d2, etc.) the area of layer II is outlined (line 4; see Fig. 4.2(b)), the number of sites sampled (line 3; see Fig. 4.2(e)), and the sum of neurons counted in that section (line 15) are recorded. Note also that a measurement of the actual section thickness was made for every section (line 14; compare with Table 4.3). The other parameters are the sampling values that were entered into the software at the start of the study. These data are from an actual sampling and the values differ from the illustration in Figs 4.2 and 4.3. (b) The calculation of the optical fractionator procedure essentially involves two steps. In the first step, the fraction of the tissue that was sampled is calculated. In the second step, the sum of the cells counted is multiplied by that fraction to yield the estimate of the total number of cells. On the left-hand side are summarized the values that are required for the calculation and the symbols used in the calculation equation. The results of these fraction calculations are most easily expressed in decimal form. The fraction that was sampled is calculated on the right-hand side and this calculation reveals that, with equally probability in three dimensions, 1/280th of the entorhinal cortex was sampled. By taking the sum of the neurons counted (ΣQ^-) and multiplying it by the fraction sampled, this procedure estimates the number of layer II entorhinal neurons as 61,500. Note that the result would have been different if the section thickness on the microtome setting (50 μm) had been used for the value 't' in the calculation of fraction sampled instead of the mean of the measured section thickness (30.11 μm).

counted at high power using the optical disector procedure (see Fig. 4.3). At the completion of this stage, all of the sampling parameters for the optical fractionator procedure have been determined and entered into the software. The section sampling fraction (ssf) is determined by the interval of sections to be measured. In this example, every sixth section was stained for sampling, producing between 10 and 12 sections for each rat brain, which encompassed the entire extent of the entorhinal cortex. At the next level of sampling, the area sampling fraction (asf) is entered into the software (Fig. 4.2(e)). The asf is defined as the area of the counting frame divided by the area of the sampling tiling and, as its name suggests, represents the percentage of the area within the region of interest (ECL2) that will be sampled. The parameters used in the worked example (see Fig. 4.6) typically produce 8 to 10 sample sites per section for a total of some 100 sample sites per animal. A graphic overlay of the sampling distribution may be generated to visually confirm the settings. The counting frame height is also specified before proceeding with the actual counting of cells at each sample site.

Having established the sampling parameters, the operator now changes to the high NA oil immersion lens (60 × Plan Apochromat oil, 1.4 NA) and the software repositions the specimen so that each of the locations shown in Fig. 4.2(f) will be sampled in turn. At each of these optical disector sites, the software changes the focal plane so that the section is slowly focused through while the operator interactively marks cells that meet the counting criteria (Fig. 4.3). Fluorophore-labeled cells can be counted sequentially for each fluorophore using the appropriate filter set or, with the necessary equipment, they can be counted simultaneously. The software allows for running through the focal sequence as often as needed to confirm the markings and different markers can be used to indicate different fluorophores. The software automatically keeps track of the numbers of marked cells (Fig. 4.6(a)). Calculation of the estimated number of cells is performed at the end of the sampling session (Fig. 4.6(b)) according to the equation for the optical fractionator procedure (West *et al.* 1991; West 1993; Mulders *et al.* 1997).

How one determines the sampling distribution to produce the lowest variance with the optimal efficiency is a complex subject (Gundersen and Jensen 1987; Gundersen *et al.* 1999) and the reader should consult other chapters in this volume dealing with sampling design. On a practical basis, it may be easiest to start with the optical disector dimensions and adjust these so that only 2 or 3 cells are counted per optical disector (see Fig. 4.6(a)). Counting few cells means that each optical disector site can be sampled more quickly and with greater accuracy than the rather cumbersome example shown in Fig. 4.3. By being able to count each site quickly, it becomes practical to count more sites and the asf can be adjusted so that each animal can have on the order of 100 optical disector sites distributed with equal probability throughout the structure of interest. The sampling distribution shown in the worked example (Fig. 4.6(a)) should take a trained operator no more than 2 hours to complete.

7 Cautions and concerns when using fluorescent probes

Fluorescent probes provide a valuable tool for both qualitative study and quantitative assessment of cell number. However, there are some limitations to their use that should be considered before designing a study using fluorescent probes.

7.1 Fading and label longevity

As stated above, proper sample preparation and storage can ensure that fluorescent labels in tissue sections remain detectable for years after staining. However, the mere process of observing a fluorophore can diminish its signal, a process known as fading. There are two phenomena that contribute to fading—quenching and bleaching. Quenching is a reversible loss of signal that results from the local transfer of energy according to the principle of resonance energy transfer. A fluorophore will recover from quenching after removal from the excitation wavelength and quenching can be minimized by using the minimum required excitation light intensity and by reducing the dwell time. Excitation light can be attenuated by the introduction of neutral density filters in front of the lamp house or by reducing the current through the arc lamp (Zeiss Attoarc; Carl Zeiss, Inc). The dwell time is the interval for which a fluorophore is exposed to the excitation wavelength. For example, scanning confocal microscopy produces less quenching than conventional fluorescence, despite using a more intense laser light source, because the scanning of the illumination point across the specimen results in less dwell time per fluorophore. Bleaching is the irreversible loss of fluorophore response to excitation and primarily results from the presence of molecular oxygen. It can be reduced by including compounds that scavenge oxygen radicals (see Appendix 2) and by reducing the dwell time. Therefore, the use of fluorescent probes for quantitative stereology will be suitable when counting a given field of view can be accomplished in a short time (several minutes).

7.2 Interpretation of signal-to-noise ratio and false-positives

The difference in the intensity of emitted light detected from an excited fluorophore (signal) relative to adjacent areas of the focal plane that do not contain a fluorophore (noise or background) is referred to as the signal-to-noise ratio. In addition, light emitted from fluorophores that lie outside of the plane of focus may contribute to the noise (Fig. 4.1(b)). Too much noise will prevent the accurate discrimination of the signal and can lead to errors and loss of efficiency in counting.

There are several ways to improve the signal-to-noise ratio. The first of these considerations lies at the level of sample preparation. Reduction of background in the staining is approached in the same fashion as for bright-field staining. A dilution curve of the fluorescent dye or the dilution of the primary and fluorophore-conjugated secondary antibodies is necessary to determine the optimal dilution required to produce the best signal. In addition, proper controls should be observed including a negative control (typically omission of the primary antibody or a buffer rinse without the fluorescent stain). The parallel staining of a known good structure as a positive control is a good idea for several reasons. A positive control will verify that tissue fixation was adequate, that the staining protocol was working even if the other probe didn't work, and it will provide a working reference for assessment of the signal-to-noise ratio.

At a second level of control in the microscope, the excitation light can be attenuated to the optimal intensity by introducing neutral density filters as described above. One should also consider if the filters being used are appropriate. If the filter allows light to pass that lies too far from the peak value (Fig. 4.4(a)), a substantial amount of background will result. This is referred to as a broad bandpass filter. Standard filter sets sold with fluorescent microscopes are often broad filters to provide the most flexibility in use. If you will be using a particular set of fluorophores, it is worthwhile to invest in filter sets that are narrow and specifically designed for that fluorophore (Table 4.2).

A third level of control for the signal-to-noise ratio is at the level of detection. If using a video camera for detection in conjunction with a commercial stereology system, choose a model that provides for control of gain (signal amplification) and black level (signal contrast). Adjustment of these controls will help to produce an image with higher definition. Some high end cameras allow for real time deconvolution of the image, which is a way to mathematically subtract away noise contributed from outside the focal plane. Real time (on-line) deconvolution is more practical for quantitative stereology than off-line computational deconvolution. The use of a confocal microscope will greatly improve the ability to detect an image with the elimination of most noise (Fig. 4.1(c); see Peterson 1999). However, specimen preparation protocols and the optical considerations discussed above should first be thoroughly investigated to see if the investment in expensive video cameras or a confocal microscope is really necessary. In many cases, very high end equipment is not necessary.

7.3 Quantitative measures of fluorescence

The combination of image acquisition by electronic detectors and digital conversion of this signal by computers provides the opportunity to perform image analysis of the fluorescent signal itself. There is a vast selection of image analysis programs available to help the researcher perform measurements of fluorescent signal, or microfluorimetry. This concept extends beyond the simple pixel counts popular with two-dimensional cell counting in that the image intensity is not converted to a binary value based upon a user-defined threshold. Here, the pixels are counted, but their intensity value is maintained as part of the analysis. Micro-fluorimetry is very popular, both because of the promise of additional data held within the image and the fact that the analysis is a capability in most imaging programs.

While it would be useful to use microfluorimetry to ask questions about levels of fluorescent-tagged protein expression or as an easy way to define the number of small particles (membrane-bound receptors, for instance), there are a number of cautions that must be observed. In order to make direct comparisons between the expression levels of two samples, it must be assumed that the emission that is detected is equivalent and linear or that the deviation from linearity is predictable and can be adjusted for in the analysis. This assumption essentially precludes the use of different fluorophores for use in comparison as their spectral properties would not be comparable, nor would the transmissive efficiency of their respective filter sets be comparable.

Even if the same fluorophore emissions are to be compared, there are a number of important controls that must be observed. At the biological level, it must be established if the stain reaction can be correlated in a linear fashion with the signal intensity. In other words, for an immunocytochemical label, does twice the level of protein being detected produce twice the signal intensity? In the absence of a demonstrable relationship between the protein concentration and the signal inten-sity, careful effort with all of the instrumentation controls described below is futile. Another control at the biological level is the need to perform all tissue preparation at the same time (from the point of tissue fixation onward) using the same solu-tions for both the control and experimental groups. Differences in antibody concentration, duration of incubation, and other procedural manipulations be-tween staining runs can have measurable effects upon detection levels.

There are a number of instrumentation controls that must be observed for microfluorimetry. These controls can be divided into optical instrumentation and electronics. The optical elements in a measurement system, either bright-field or fluorescent, do not remain static between sessions. Changes in arc lamp output or laser intensity (even if all of the electronic settings are the same) will vary over time as will the optical alignment of the elements in the light path. A calibrated light meter should be used to monitor actual output and a non-biological imaging standard (such as uranyl glass) should be used to establish uniformity of illumina-tion across the region of interest. In addition, there may be changes in filter trans-mission and uniformity with time. Potential light scattering from dust in the light path cannot be eliminated as a variable. The electronic elements may also exhibit variability between sessions with changes in the detector and/or amplifier stability or linearity. Furthermore, the detector may exhibit a non-uniform pixel response across the image leading to variance that is unrelated to an actual difference in signal intensity. Using the same settings (i.e. the control dial at the same setting) for image collection between sessions is likewise invalid unless those settings are calibrated against a non-biological standard to produce an equivalent histogram.

In addition to the above concerns, which apply equally to bright-field and fluorescent preparations, there are two additional variables that must be accounted for when using fluorescent microscopy. One must account for the loss (usually nonlinear) in signal intensity from focal planes at the surface of the tissue compared to deeper focal planes. The degree of this depth-dependent signal attenuation will be greater for some fluorophore–substrate combinations than for others. Finally, one must account for the effects of fading with observation discussed above.

The successful use of microfluorimetry is a complex task. Furthermore, the interpretation of these data, even when proper controls are undertaken, is by definition relative to the circumstance in which the data was collected. Quantitative data is most useful when the procedure used estimates the absolute value of the parameter rather than the relative value. If estimates produce absolute data, these can be compared between experiments and by different investigators. Quantitation of fluorescent intensity should only be undertaken when the question to be answered cannot be addressed by more objective means and when the importance of the answer warrants the effort required for proper controls.

7.4 Combining stereology with neuroanatomical tracing

One of the major uses of fluorescence microscopy in neuroanatomical research has been the detection of tracer substances. Many of these tract tracers are either fluorescent or biotinylated, enabling them to be labeled with a streptavidin-conjugated fluorophore. The use of multiple fluorophores within a specimen has enabled the relationship of complex neuronal projections to be unraveled. In addition to the qualitative data available from these preparations, it would be clearly advantageous to derive quantitative estimates of such parameters as the number of neurons projecting to a certain terminal field or the number of contacts made by a labeled projection. There is no difficulty in combining fluorescence with stereological estimators as described earlier in this chapter. While there is no mechanical problem with using fluorescent stereology in this context, a caution must be raised about the interpretation of these data. In most circumstances, there are a number of variables that may undermine the apparent result. These include the efficiency of label uptake and transport by the cell. Furthermore, it may not be possible to ensure that the entire population is equally labeled, resulting in cells that are excluded from the estimate based upon failure of uptake, slower uptake kinetics relative to their neighbors, or uptake at levels that are below detection. Therefore, conclusions based upon studies using tracer substances should be expressed in terms of labeled population without assuming that the data are true of the entire population unless this assumption is validated.

8 Conclusions

There are a few considerations against using fluorescent probes for use with cell counting procedures (Table 4.4). Depending upon the specimen preparation and the time required for counting cells in any given field of view, there may be appreciable fading of the signal with observation. A pilot study should be undertaken to assess this concern before committing resources to a full study. In addition, the need to purchase additional equipment for fluorescence microscopy may add too much expense for the investigation. Finally, obtaining good qualitative images from fluorescent preparations may be difficult using traditional film-based photography. The combination of low light levels, fading with observation, and correct adjustment for reciprocity failure make fluorescence microphotography a challenging field. However, access to a confocal microscope will provide a means to obtain high quality images from fluorescent preparations.

There are a number of advantages to fluorescent probes that recommend them for use with quantitative stereology (Table 4.4). The ability to optically section through thick tissue sections without the shadowing and light scattering of bright-field microscopy is ideally suited to the implementation of the optical disector procedure. Similarly, the detection of specific emission with fluorescent epi-illumination reduces the contribution of artifacts outside of the focal plane to the projected image. The preparation of fluorescently labeled specimens helps to preserve section thickness making this approach optimal for the use of the optical disector. By selecting fluorophores whose excitation and emission profiles do not overlap, it is possible to broaden the analysis by using multiple labels within a single section. Sections stained in this fashion may also be compatible for analysis by confocal microscopy. Finally, many indicators of physiological state use fluorescent probes, raising the possibility that quantitative stereology may be applied in functional preparations. The use of fluorescent probes may be the methodology of choice for a wide variety of quantitative procedures in addition to cell counting.

Table 4.4 Considerations in using fluorescent probes for stereology

Advantages
Improved optical microtomy over bright-field microscopy
Reduction of artifacts outside of the focal plane
Preparation preserves the retention of section thickness
Co-registration of multiple labels allows for broader analysis
Preparation is compatible with confocal microscopy
Potential for using a variety of physiological indicators in counting
Disadvantages
The fluorescent signal can fade with observation
Extra microscope equipment is required
Difficult to obtain good qualitative images if relying on traditional film

9 Appendix 1. Example protocol for multiple immunofluorescence with pooled incubation

Solutions

TBS ++	TBS+
100 μl TX-100	100 μl TX-100
38 ml TBS	40 ml TBS
2 ml donkey serum (5%)	

Where TBS is 0.1 M Tris buffered saline

Procedure: (*replace indicated lines with actual antibody information)

Day 1

1. Rinse 2 × 10 minutes in TBS
2. Block with TBS++ at room temperature for minimum of 1 hour, up to 3 hours
3. Incubate in pooled primary (1°) antibodies diluted in TBS+ overnight at 4°C

* First 1° antibody	supplier	dilution used
* Second 1° antibody	supplier	dilution used
* Third 1° antibody	supplier	dilution used

example: mouse anti-NeuN   Chemicon   1:2500

Day 2

4. Rinse/block 2 × 1 hour in TBS++
5. Incubate with fluorophore conjugated secondary (2°) antibodies in TBS+ for 2 hours at room temperature in dark

* First 2° antibody	supplier	dilution used
* Second 2° antibody	supplier	dilution used
* Third 2° antibody	supplier	dilution used

example: donkey anti-mouse-FITC   Jackson Immunores.   1:500

6. Rinse 2 × 15 min in TBS in dark
7. Optional Nuclear Counterstain (use only for unoccupied wavelengths)
    DAPI (1:30,000 in TBS for 5 minutes)
    Sytox Green (1:30,000 in TBS for 5 minutes)
    Propidium Iodide (1:2500 in TBS for 5 minutes)
    ToPro3 (1:5000 in TBS for 5 minutes)
8. Mount and coverslip with PVA/DABCO (see Appendix 2)

10 Appendix 2. PVA-DABCO coverslipping solution for immunofluorescence

This solution is a glycerol based mounting medium containing an anti-fading reagent for use with immunofluorescence preparations. The same solution may be used with or without DABCO for 'aqueous' mounts of material for brightfield microscopy. The advantage of this solution is that the medium at the perimeter of the coverslip polymerizes upon contact with air forming a strong seal. There is no need to seal with nail polish. The coverslips can be removed by soaking the slide in buffer.

Mount stained sections on glass slides and let briefly air dry (less than one minute) or if using Lab-Tek chamber slides, remove the chamber and tip to remove excess buffer. Apply 75 µl of PVA-DABCO solution for use with a 50 mm coverslip. Use caution with pipetting as air bubbles are easily introduced if the solution is aspirated in any way. I suggest using a repeating pipetteman.

Lay the coverslipped slides flat overnight (covered from light at room temperature) until polymerization is complete. Afterwards, the coverslips may be cleaned with solvents to remove immersion oil and slides can be stored upright in normal slide storage boxes. Store slides in slide boxes at 4°C or –20°C and preparations should remain usable for at least ten years, the longest time point evaluated to date.

For 50 ml of 2.5% PVA-DABCO
1. Using a 50 ml culture tube, weigh 12 gm of glycerol
2. Add 4.8 gm of PVA
3. Mix well by tilting tube upside down until PVA is 'coated' with glycerol
4. Add 12 ml of distilled water
5. Mix overnight using a rotator at room temperature
6. Add 24 ml of 0.2 M Tris-HCl at pH 8–8.5
7. Heat to 50°C in a water bath with mixing (approximately 30 min)
8. Add 1.25 gm of DABCO and mix well
9. Centrifuge at 5000g for 15 min
10. Remove supernatant, aliquot (suggest 1 ml), and store at -20°C

The solution may be kept up to 6 months at -20°C and up to one week at 4°C before becoming milky. Do not refreeze.

PVA-polyvinyl alcohol   Sigma no. P8136
DABCO-1,4 diazabicyclo[2.2.2]octane   Sigma no. D2522

Acknowledgments

I thank my colleagues, including Drs G.R. Chalmers, F.H. Gage, H.J. Karten, G. Kempermann, H.G. Kuhn, and T. Palmer, for many helpful discussions on these topics over the years. I also thank Linda Kitabayashi, Letia Peterson, John Leppert, and Dawn Erickson for valuable assistance. This work was supported by a TLL Temple Foundation grant from the Alzheimer's Association and a grant from the American Federation for Aging Research.

References

Eriksson, P.S., Perfilieva, E., Bjork-Eriksson, T., Alborn, A.M., Nordborg, C., Peterson, D.A., and Gage, F.H. (1998). Neurogenesis in the adult human hippocampus. *Nat. Med.* **4**, 1313–17.

Falck, B., Hillarp, N.A., Thieme, G., and Torp, A. (1982). Fluorescence of catechol amines and related compounds condensed with formaldehyde. *Brain Res. Bull.* **9**, 11–14.

Gomez-Isla, T., Price, J.L., McKeel, D.W. Jr, Morris, J.C., Growdon, J.H., and Hyman, B.T. (1996). Profound loss of layer II entorhinal cortex neurons occurs in very mild Alzheimer's disease. *J. Neurosci.* **16**, 4491–500.

Gundersen, H.J.G. (1977). Notes on the estimation of numerical density of arbitrary profiles: the edge effect. *J. Microsc.* **111**, 219–23.

Gundersen, H.J.G. (1978). Estimators of the number of objects per area unbiased by edge effects. *Microscopica Acta* **81**, 107–17.

Gundersen, H.J.G. and Jensen, E.B. (1987). The efficiency of systematic sampling in stereology and its prediction. *J. Microsc.* **147**, 229–63.

Gundersen, H.J.G., Jensen, E.B.V., Kieu, K., and Nielsen, J. (1999). The efficiency of systematic sampling in stereology—reconsidered. *J. Microsc.* **193**, 199–211.

Haseloff, J. (1999). GFP variants for multispectral imaging of living cells. *Methods Cell Biol.* **58**, 139–51.

Kafri, T., Blömer, U., Peterson, D.A., Gage, F.H., and Verma, I.M. (1997). Sustained expression of genes delivered directly into liver and muscle by lentiviral vectors. *Nature Genet.* **17**, 314–17.

Li, X., Zhao, X., Fang, Y., Jiang, X., Duong, T., Fan, C., Huang, C.C., and Kain, S.R. (1998). Generation of destabilized green fluorescent protein as a transcription reporter. *J. Biol. Chem.* **273**, 34970–5.

Moriyoshi, K., Richards, L.J., Akazawa, C., O'Leary, D.D., and Nakanishi, S. (1996). Labeling neural cells using adenoviral gene transfer of membrane-targeted GFP. *Neuron* **16**, 255–60.

Mulders, W.H., West, M.J., and Slomianka, L. (1997). Neuron numbers in the presubiculum, parasubiculum, and entorhinal area of the rat. *J. Comp. Neurol.* **385**, 83–94.

Peterson, D.A. (1999). Quantitative histology using confocal microscopy: implementation of unbiased stereology procedures. *Methods: A Companion to Methods in Enzymology* **18**, 493–507.

Peterson, D.A., Lucidi-Phillipi, C.A., Murphy, D.P., Ray, J., and **Gage, F.H.** (1996). Fibroblast growth factor-2 protects entorhinal layer II glutamatergic neurons from axotomy-induced death. *J. Neurosci.* **16**, 886–98.

Peterson, D.A., Leppert, J.T., Lee, K.F., and **Gage, F.H.** (1997). Basal forebrain neuronal loss in mice lacking neurotrophin receptor p75. *Science* **277**, 837–8.

Peterson, D.A., Dickinson-Anson, H.A., Leppert, J.T., Lee, K-F., and **Gage, F.H.** (1998). Central neuronal loss and behavioral impairment in mice lacking neurotrophin receptor p75. *J. Comp. Neurol.* **404** (1), 1–20.

Sterio, D.C. (1984). The unbiased estimation of number and sizes of arbitrary particles using the disector. *J. Microsc.* **134** (2), 127–36.

West, M.J. (1993). New stereological methods for counting neurons. *Neurobiol. Aging* **14**, 275–85.

West, M.J., Slomianka, L., and **Gundersen, H.J.G.** (1991). Unbiased stereological estimation of the total number of neurons in the subdivisions of the rat hippocampus using the optical fractionator. *Anat. Rec.* **231**, 482–97.

CHAPTER 5

COUNTING IN SITU HYBRIDIZED NEURONS

BENTE FINSEN, RIKKE GREGERSEN, ELIN
LEHRMANN, STEEN LOVMAND, AND MARK J. WEST

1 Introduction

1.1 General

Since the introduction of hybridization in 1969 by Gall and Pardue, *in situ* hybridization (ISH) histochemistry has become an extremely powerful tool for studying gene expression in individual neurons. It supplements immunohistochemistry in that it provides a means to visualize the site of biosynthesis, rather than the location of a specific peptide or protein, and has been widely used in studies of gene expression in the central nervous system during development and after experimental manipulation (Conn 1992; Wilkinson 1992a; Emson 1993; Valentino *et al.* 1987; Wisden and Morris 1993*a*). The first probes to be developed were radiolabeled cloned cDNA probes, which were followed by cRNA and synthetic cDNA probes (Gall and Pardue 1969). These were ideal for quantitative studies of gene expression in brain nuclei and brain regions (O'Shea and Gundlach 1993; Wisden and Morris 1993*b*), but not optimal for stereological quantification of cell numbers due to limited cellular resolution. The development of sensitive, high-resolution, non-radiolabeled probes in the past decade (Kiyama *et al.* 1989; Augood *et al.* 1991; Finsen *et al.* 1992; Larsson and Hougaard 1993; Lewis *et al.* 1993) has reduced this problem and one can now utilize modern design-based stereological techniques to make unbiased estimates of the total number of *in situ* hybridized neurons in defined regions of the brain (Dalby *et al.* 1996; *West et al.* 1996; Andreassen *et al.* 1999).

1.2 Use of design-based stereological techniques to estimate the number of *in situ* hybridized cells in a defined region

Estimates of the total or true number of neurons are usually the most useful morphological parameter for making evaluations of neuron loss or gain. Data of this type are intuitively easy to understand, can be obtained with stereological techniques that are based on unbiased statistical sampling techniques, and are readily amenable to parametric statistical analyses. The technique described here for obtaining these estimates is referred to as the optical fractionator technique (West *et al.* 1991). With this technique, estimates of the total number of *in situ* hybridized neurons can be estimated from direct counts of labeled cells (neurons) made in a known fraction of the structure of interest. This is accomplished by counting the labeled cells in a known fraction of the section thickness, under a known fraction of the sectional area of the structure, on a known fraction of the sections that pass through the structure of interest. It requires the use of relatively thick sections (optimally 20 μm or more after mounting) and the unequivocal identification of labeled cells throughout the section thickness. When counting *in situ* hybridized cells, these technical requirements can only be fulfilled by utilizing non-radioactive ISH techniques.

Of the two most common stereological techniques designed to ensure unbiased estimates of the total number of neurons in a defined region (Gundersen 1986, West 1993, 1999), the 'optical fractionator' technique was preferred to the $N_V \times V$(ref) technique when dealing with *in situ* hybridized cells. [With the $N_V \times V$(ref) method, one multiplies an estimate of the numerical density, N_V (obtained with multiple optical disector sampling), by the volume of the entire structure, that is, the reference volume, V(ref).] This is because the optical fractionator technique is not affected by tissue shrinkage as long as the tissue shrinkage in the z-axis (which can occur after the section has been cut) is uniform or homogeneously non-uniform. The optical fractionator technique was also chosen because it avoids biases related to the Holmes effect (overprojection). The optical fractionator technique uses the ratio of the height, h, of the optical disector probe to the thickness, t, of the section, t/h = tsf, where tsf is the thickness sampling fraction. If both h and t are measured in the same way, this ratio will be comparable from preparation to preparation and laboratory to laboratory. If the estimates of N_V and V(ref) are made on different sections or tissue slices one must monitor any changes in the volume of the tissue that takes place. This situation does not arise with the optical fractionator technique with the assumption that tissue deformation in the z-axis is uniform or homogeneously non-uniform.

1.3 *In situ* hybridization using alkaline phosphatase (AP)-conjugated oligonucleotide probes

Relatively small (26–30 bases) oligonucleotide probes are conjugated at the 5′ end to an alkaline phosphatase (AP) reporter molecule (Fig. 5.1). Because of their size, they penetrate deeply into lightly fixed brain tissue, provide excellent single cell rendition, and result in preparations that are suitable for optical fractionator analysis (Dalby *et al.* 1996; West *et al.* 1996; Andreassen *et al.* 1999). The tailor-made, commercially synthesized probes and the protocol are extremely user-friendly. The probes can be applied singly, to detect cells with a large number of copies of mRNA (Kiyama *et al.* 1989; Finsen *et al.* 1992; West *et al.* 1996; Andreassen *et al.* 1999), or can be used in a mixture containing two or more probes that are complementary to different non-overlapping parts of the target mRNA molecule to visualize cells

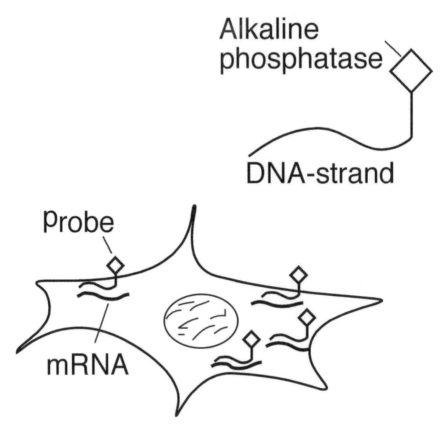

Fig. 5.1 Schematic representation of single neuron *in situ* hybridized with an alkaline phosphatase (AP)-conjugated oligodeoxynucleotide probe.

in which there are only a small number of copies of mRNA (Gregersen *et al.* 2000, Lambertsen *et al.* 2001). They are excellent for the ISH of organotypic slice cultures (Finsen *et al.* 1992, Østergaard *et al.* 1995) and, as radiolabeled probes (O'Shea and Gundlach 1993; Wisden *et al.* 1993), can be used for the semiquantitative analysis of gene expression in individual cells and regions (Kiyama *et al.* 1989; Augood *et al.* 1991, 1993; Andreassen *et al.* 1999). They can also be used in combination with immuno- or enzyme-histochemical detection techniques and in co-expression studies, when combined with a radiolabeled probe (Andreassen *et al.* 2000; Augood *et al.* 1993, 1995). The development of new sensitive fluorescent AP substrates (Paragas *et al.* 1997) makes it also possible to use AP-conjugated probes for fluorescent ISH (FISH).

The protocols and examples provided here were developed for the optical fractionator studies of somatostatinergic interneurons and enkephalinergic projection neurons in the dorsal striatum of normal and neuroleptic-treated rats that were published previously (West *et al.* 1996; Andreassen *et al.* 1999, 2000). The principles, however, apply to any population or subpopulation of neurons or glial cells that express a specific mRNA molecule within a well-defined brain structure or region.

2 Principles of hybridization

Hybridization between a probe and target mRNA takes place by means of hydrogen bonding between bases on single-stranded nucleic acids and their complementary bases and results in the formation of nucleic acid hybrids/duplexes (Fig. 5.1). The stability of oligodeoxynucleotide–mRNA hybrids depends on the factors that affect their melting temperature (T_m). T_m is defined as the temperature at which 50% of the DNA–RNA duplexes dissociate and is related to a number of variables according to the equation (Wilkinson 1992b):

T_m (°C) = 79.8 + 18.5 log(molarity of monovalent cations) + 0.58(% GC content of probe) + 0.12(% GC)^2 – 0.5(% formamide) – 820/(probe length in bases).

The major factors that affect the T_m are the concentrations of Na$^+$-ions and formamide, the guanine–cytosine (GC)-content, and the length of the probe. With a probe of a given composition, the concentrations of Na$^+$ and formamide can be altered to manipulate T_m. In combination with the temperature, which also influences the stability of the hybrids, these factors are referred to as *stringency* factors (Tecott *et al.* 1987). High salt concentration has a stabilizing effect on hybrids, while high temperature and high formamide concentrations have destabilizing effects. The factors that affect the kinetics of hybrid formation and the stability of hybrids have all been characterized for hybrids that form in solution, whereas hybrids that form *in situ* in the tissue have a lower T_m. In practical terms, this means that oligonucleotide probes have an optimal temperature for hybridization when present in 50% formamide at approximately 20–25°C below the calculated T_m value (Tecott *et al.* 1987; Augood *et al.* 1993).

3 AP-conjugated oligonucleotide probes

3.1 Probe design

By using standardized probes with similar lengths and GC-content, hybridization reactions can be directed against different mRNA molecules under similar conditions (see Table 5.1 and also Table 5.4 in Section 8). The AP-labeled probes used in our laboratory typically consist of 26–30 bases (Table 5.1) and are complementary to those parts of the mRNA molecule that are translated into the prepro-, pro-, and the mature protein. Molecular cloning and access to DNA sequences in public databases, such as the National Center for Biotechnology Information (NCBI; http://www.ncbi.nlm.nih.gov/), make it possible to identify probe sequences that are specific for the target mRNA molecules of interest.

Table 5.1 AP-labelled in situ hybridization probes

Messenger RNA*	Probe sequence	No. of bases	%GC	ISH†	Base no.	Species of cDNA	References
Calbindin	5'GATCAAGTTCTGCAGCTCCTTTCCTTCCAG	30	50	V[1], C	91–120	Human	Parmentier et al. 1987
					376–405	Rat	Hunziker and Schrickel 1988
GAPDH	5'CCTGCTTCACCACCTTCTTGATGTCA	26	50	V, C[1,7]	206–235	Mouse	Nordquist et al. 1988
					828–853	Human	Tokunaga et al. 1987
					846–871	Rat	Fort et al. 1985
					752–777	Mouse	Sabath et al. 1990
MBP	5'CTTCTGGGGCAGGGAGGCCATAATGGGTAGT	30	57	C[8,9]	302–331	Mouse	Newman et al. 1987
Pre-prosomatostatin	5'GAATGTCTTCCAGAAGAAGTTCTTGCAGCC	30	46	V[3,4,6,10,11], C, O[5,13]	100–129	Rat	Goodman et al. 1983
					1,076–1,105	Mouse	Fuhrmann et al. 1990
Pre-proenkephalin	5'CGCTAGCAGCCAGATGCAAAGTCTCAGGAA	30	53	V[2], C, O[13]	10–39	Rat	Howells 1986
TNF	5'CTTCTCATCCCTTTGGGGACCGATCACC	28	57	C[7,8]	305–332	Mouse	Pennica et al. 1985
	5'CGTAGTCGGGGCAGCCTTGTCCCTTGAA	28	60	C[7,8]	570–597	Mouse	Pennica et al. 1985
					2,164–2,191	Rat	Shirai et al. 1989
Tyrosine hydroxylase	5'TCAAAGGCTCGGACCTCAGGCTCCTCTGAC	30	60	V[10], C	1,223–1,252	Rat	Grima et al. (1985)

* GAPDH, glyceraldehyde-3-phosphate dehydrogenase; MBP, myelin basic protein; TNF, tumour necrosis factor.
† Indicates excellent ISH results on: V, vibratome sections; C, cryostat sections; and/or O, organotypic slice cultures. The superscripts refer to the following publications: 1, this chapter; 2, Andreassen et al. 1999; 3, Dalby et al. 1996; 4, Dalby et al. 1998; 5, Finsen et al. 1992; 6, Garrett et al. 1994; 7, Gregersen et al. 2000; 8, Jensen et al. 2000a; 9, Jensen et al. 2000b; 10, Khorooshi et al. 1999; 11, Tønder et al. 1994; 12, West et al. 1996; 13, Østergaard et al. 1995.

When selecting the ideal sequence one should pay attention to the occurrence of self-complementarity and deoxyguanine stretches and to the formation of secondary structures. The sequence should not contain stretches with more than three consecutive Gs. The formation of double strands, hairpin, and other secondary structures with a ΔG (change in Gibbs free energy) below -2 kcal/mol or melting temperatures above 45°C should be avoided. In particular, this applies to double-stranded and secondary structures that involve the 5' end. A commonly used program for probe design is Oligo®, which can be purchased from National Biosciences, Inc, Plymouth, MN.

3.2 Probe synthesis and labeling

The probe is synthesized with a modified 5' amino end. The chemical synthesis of the 5' amino modified DNA oligonucleotide is performed by a β-cyanoethyl phosphoramidite reaction with a Perkin Elmer ABI DNA synthesizer. After the synthesis, the 5' amino modified DNA oligonucleotide is purified by Ion-Exchange HPLC and labeled with the AP molecule via the aminoreactive linker, and disuccinimidyl suberate (DSS) in a two-step procedure. In the first step, the amine reactive linker is coupled to the 5' amino modified DNA oligonucleotide. The second step involves the conjugation of the AP enzyme to the linker-oligonucleotide to produce the AP-conjugated oligonucleotide complex (according to an improved, unpublished protocol based on Jablonski *et al.* (1986), Goodchild (1990), and Farmer and Castaneda (1991)). Following conjugation, the AP–oligonucleotide monomeric complex is separated from non-conjugated AP, non-conjugated oligonucleotide, and multimer complexes by Ion-Exchange high-performance liquid chromatography (HPLC) to produce a high quality product with a high specific activity (Beaucage, S.L. and Caruthers, M.H., United States Patent #4.668.777). The AP-conjugated oligonucleotide can be stored, without loss of enzymatic activity, for more than a year in saline buffer containing 0.1% Na-azide, 1 mg/ml bovine serum albumin, 20 mM Tris-HCl, pH 7.5, and 500 mM NaCl at 4°C.

4 Visualization of single cells with AP-conjugated oligonucleotide probes

The cellular resolution of AP-conjugated probes is comparable to that of other non-radiolabeled probes (Augood *et al.* 1993; Larsson and Hougaard 1993; Lewis *et al.* 1993), and exceeds that of radiolabeled probes (compare Fig. 5.2(a)–(e) with 5.2(f); O'Shea and Gundlach 1993; Wisden and Morris 1993b; Lehrmann *et al.* 1998). Most neuronal types, such as somatostatin and enkephalin mRNA-expressing neurons of the striatum (Fig. 5.2(a), (b); West *et al.* 1996; Andreassen *et al.* 1999, 2000), tyrosine hydroxylase mRNA-expressing ventral mesencephalic dopamine

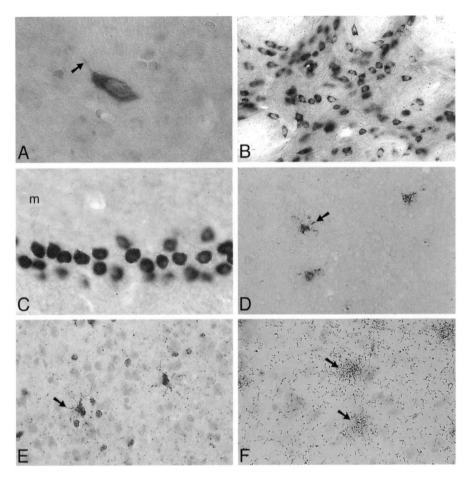

Fig. 5.2 (see also **Plate 3**) High- and medium-power micrographs of ISH reactions performed with (a)–(e) AP-labeled and (f) radiolabeled (^{35}S) probes. (a) Striatal somatostatin (SS) mRNA expressing neuron in adult rat striatum. Counterstained with Neutral red. The arrow marks a SS mRNA containing dendrite. (b) Enkephalin mRNA expressing neurons in adult rat striatum. (c) Calbindin mRNA expressing cerebellar Purkinje cells. (d) Tumor necrosis factor (TNF) mRNA expressing process-bearing (arrow) microglia–macrophages located at the infarct border in mouse subjected to permanent medial cerebral artery occlusion (MCAO). (e) Myelin basic protein (MBP) mRNA expressing oligodendrocytes in the striatum of adult mouse. Note translocation of MBP mRNA out into cellular processes (arrow). (f) Combined ISH for transforming growth factor beta-1 mRNA (black silver grains) combined with immunohistochemistry for microglial-macrophage complement type 3 receptor (red stain). Arrows point to two double-labeled cells located at infarct border in an adult rat subjected to transient MCAO (for details, see Lehrmann *et al.* 1998). The hybridizations shown in (a)–(c) were carried out on 50 μm thick vibratome sections, and those shown in (d)–(f) were carried out on 30 μm (d), (e) or 15 μm (f) thick cryostat sections. m, Molecular layer. Magnification: (a) 1300 ×; (b) 400 ×; (c) 500 ×; (d)–(f) 600 ×.

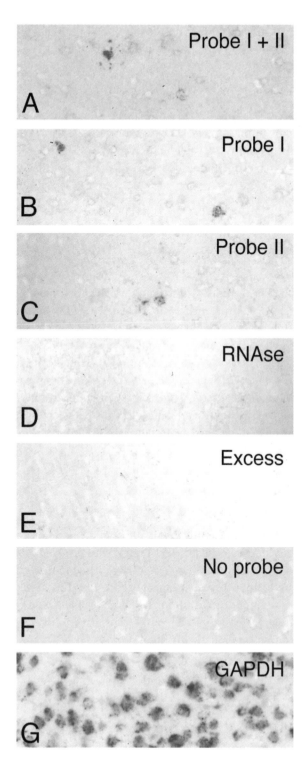

Fig. 5.3 (see also **Plate 4**) Control reactions for specificity of TNF mRNA ISH signal. The controls were performed on parallel sections from a mouse subjected to permanent MCAO. (a) Sections *in situ* hybridized with a probe mixture, consisting of two probes (probe I and II) directed against two different, non-overlapping stretches of TNF mRNA, showed a stronger ISH signal than (b), (c) sections hybridized with each individual probe. (d) Section pretreated with RNAse A prior to ISH shows no signal. Neither do sections hybridized with (e) an excess of unlabeled probe I and II or (f) no probe. (g) The GADPH control demonstrates that mRNA is abundant in the tissue section. Magnification, 400 ×.

neurons (not shown; Kiyama *et al.* 1989, Khorooshi *et al.* 1999), and the calbindin mRNA-containing Purkinje cells of the cerebellum (Fig. 5.2(c)), display homogeneous cytoplasmic labeling, with the hybridization signal occasionally extending into the proximal parts of the dendrites (Fig. 5.2(a)). The high resolution of the AP-conjugated probes also enables one to visualize process-bearing glial cells such as oligodendrocytes (Fig. 5.2(e); Jensen *et al.* 2000*a,b*). Other non-neuronal cells, including astrocytes and activated microglia and brain macrophages, can also be visualized as single cells. Glial fibrillary acidic protein (GFAP) mRNA is used as a marker for astrocytes (data not shown) and specific cytokine mRNAs are used as markers for activated microglia and brain macrophages (Fig. 5.2(d), (f) and Fig. 5.3; Lehrmann *et al.* 1998; Gregersen *et al.* 2000; Jensen *et al.* 2000*a*).

5 Sensitivity of AP-labeled oligonucleotide probes

In agreement with the findings of Augood *et al.* (1993) and Emson (1993), the intensity of the ISH signal obtained with the AP-conjugated probes is comparable to that obtained with radiolabeled probes. Furthermore, the ISH signal can be enhanced by using a probe-mixture that contains two or more AP-labeled probes. This is clearly illustrated in our own studies, in which we have been able to detect cytokine mRNA by using a mixture of two AP-conjugated oligonucleotide probes (Figs 5.2(d) and 5.3; Gregersen *et al.* 2000; Jensen *et al.* 2000*a*, Lambertsen *et al.* 2001). The numbers of copies of cytokine and growth factor transcripts are often small in individual cells and traditionally have been detected with radioactive ISH (Fig. 5.2(f); Kiefer *et al.* 1996; Guthrie *et al.* 1997; Lehrmann *et al.* 1998).

6 Tissue processing

6.1 Perfusion and immersion fixation

Rats were anesthetized with a mixture of pentobarbital (40 mg/kg) and chloral hydrate (180 mg/kg) and perfused transcardially for 4 min with ice-cold 4% paraformaldehyde (PFA) in 0.15M phosphate buffer (PB), pH 7.3. The brains were postfixed for 2 hours in the same fixative and maintained for 22–24 hours at 4°C in RNAse-free, 0.15M PB that contained 0.1% PFA for additional stabilization of the tissue prior to sectioning. The brains were subsequently transferred to ice-cold 0.15M PB and stored until sectioning (maximally 6–8 hours).

6.2 Sectioning

Prior to sectioning, the brains were marked with a cut in the left hemisphere, in order to be able to distinguish between the two hemispheres during the stereological analysis.

Table 5.2 Buffers and solutions*

DEPC–H$_2$O	1% diethylpyrocarbonate (DEPC; Sigma D-5758)	Stir overnight. Autoclave. Store at 4°C
PFA 4%	Heat 1000 ml 0.15M phosphate buffer (PB) to boiling. Add 40 g PFA (Fluka 76240)	Stir until PFA solubilizes. Add DEPC–H$_2$O to 1000 ml. Adjust to pH 7.3
PB 0.6M	Stock solution A, 27.4 ml; stock solution B, 72.6 ml	Adjust to pH 7.3
Stock solution A	20.4 g KH$_2$PO$_4$ in 250 ml DEPC–H$_2$O	
Stock solution B	106.8 g Na$_2$HPO$_4$·2H$_2$O in 1000 ml DEPC–H$_2$O	
20 × SSC	8.82 g sodium citrate (Merck 6448); 17.52 g NaCl (Merck 6404)	Add DEPC–H$_2$O up to 100 ml. Store at 4°C
de Olmos solution	NaPO$_4$-buffer, 700 ml; ethyleneglycol (Merck K2595252), 300 ml; sucrose, 300 g; polyvinylpyrrolidone (PVP, Sigma P528), 10 g	Stir. Store in the freezer at −14°C for 12–15 hours before use
NaPO$_4$-buffer	Na$_2$HPO$_4$·2H$_2$O, 9.50 g; NaH$_2$PO$_4$·H$_2$O, 2.12 g; DEPC–H$_2$O, 700 ml	Adjust to pH 7.3
Hybridization buffer	Formamide, 5000 ml; 20 × SSC, 2000 ml; 40% Denhardt, 250 ml; 50% dextran sulfate (Pharmacia 27–4564–01), 2000 ml; striatal somatostatin salmon DNA (Sigma D9156), 500 mg; DEPC–H$_2$O, 750 ml	Ready-made hybridization buffer can be stored at −20°C for many months
Denhardt 40%	0.04 g Ficoll (Sigma F-4375); 0.04 g PVP; 0.04 g bovine serum albumin (Sigma A4503); 5 ml DEPC–H$_2$O	Add reagents in the indicated order.
Dextran sulfate 50%	5 g dextran sulfate; 10 ml DEPC–H$_2$O	Stir at RT
Formamide (deionized)	100 ml formamide (BDH 44254); 5 g Amberlite (Sigma MB-1A)	Stir for 1 h at RT in the dark. Filter (0.45 mm) and store at −20°C
Tris-HCl, pH 9.5	10 mM tris (Sigma T-1378); 0.15M NaCl	Adjust with HCl to pH 9.5. Store at 4°C
AP-developer	1 ml Tris–HCl, pH 9.5, MgCl$_2$; 4.5 μl NBT stock solution; 3.5 μl BCIP stock solution	Freshly made up. Keep in the dark
Tris–HCl, pH 9.5, MgCl$_2$	0.1M Tris; 0.1M NaCl; 0.05M MgCl$_2$ (Merck A617833)	Adjust with HCl to pH 9.5. Store at 4°C
NBT stock solution	70 mg NBT (Sigma N6639); 1 ml 70% dimethylformamide	Store in the dark at 4°C
BCIP stock solution	50 mg BCIP (Sigma B1026); 1 ml 100% dimethylformamide	Store in the dark at 4°C

Table 5.2 Cont'd

TE (tris–EDTA)-buffer	Tris 2M, pH 7.4, 0.5 ml; EDTA (ethylene diaminetetraacetic acid) 0.5 M, pH 8.0, 20 μl	Add DEPC–H_2O up to 100 ml. Adjust to pH 7.4
Tris 2 M, pH 7.4	Tris, 24.22 g	Add DEPC–H_2O up to 100 ml. Add concentrated HCl to pH 7.4
EDTA 0.5M, pH 8.0	EDTA (Sigma E5134), 4.65 g	Add DEPC–H_2O up to 100 ml. Add 10 N NaOH to pH 8.0
Neutral red solution	1 g Neutral red (Merck 1369); 100 ml 0.05M acetate buffer, pH 4.8	Store at RT
Xylene–phenol–creosote	250 μl creosote (Allchem, UK); 250 g phenol in 2 l xylene	Store in ventilated safe at RT

* Supplier and catalogue number are mentioned when chemical/substance is first encountered in the table.

6.2.1 Experimental brains

The brains were sectioned with a calibrated vibratome in the frontal plane into 10 parallel series of free-floating sections. The sections had a mean thickness of 50 μm and were collected in two microliter plates placed on ice. The first series sampled (a) is a reference series for which each section is placed in serial order, one section per well, in a 24-well microliter plate that contained 0.15M PB. The remaining nine series (b–j) were collected in a 12-well microliter plate, with one series of sections per well. Series (b) and possibly (c), which were *in situ* hybridized immediately after sectioning, are collected in RNAse-free 2 × standard sodium citrate (2 × SSC), while the remaining series (d–j) were collected in RNAse-free de Olmos cryoprotective solution (Table 5.2).

6.2.2 Internal standard sections

Brains from normal animals, which were subsequently used as internal standards for the ISH reaction were cut into sets of 15–20 midstriatal sections. These sections were collected either in 2 × SSC or in the cryoprotective solution (see Section 6.3).

6.3 Handling and storage of sections

The sections in the reference series (a) were mounted in sequence on gelatinized microscope slides, air-dried, and stained with toluidine blue. The reference series was used as a guide when numbering the *in situ* hybridized sections (Fig. 5.4(a), (b)). The sections to be hybridized can be stored for up to 24 hours in 2 × SSC at 4°C. Alternatively, sections can be stored for up to 1 month at −14°C in the cryoprotective solution, without noticeable loss of hybridization signal.

7 *In situ* hybridization

7.1 Hybridization of free-floating vibratome sections

Freshly cut sections along with an internal standard section were transferred to Eppendorf tubes that contained 1.100–1.200 μl hybridization medium (Table 5.3(a)). Prior to hybridization, sections stored in the cryoprotective solution were rinsed twice, for a total of 60 min, in 2 × SSC at room temperature (RT), along with the internal standard sections. The probe concentration was in the range of 2–3 pmol/ml hybridization medium. Care was taken to ensure that all sections were unfolded and fully immersed in the hybridization medium.

7.2 Post-hybridization rinsing

The post-hybridization rinsing removed the non-specifically bound probe. This involved adjusting the stringency factors, that is, the salt and formamide concen-

(a)   (b)

(c)   (d)

(e)   (f)

(g)   (h)

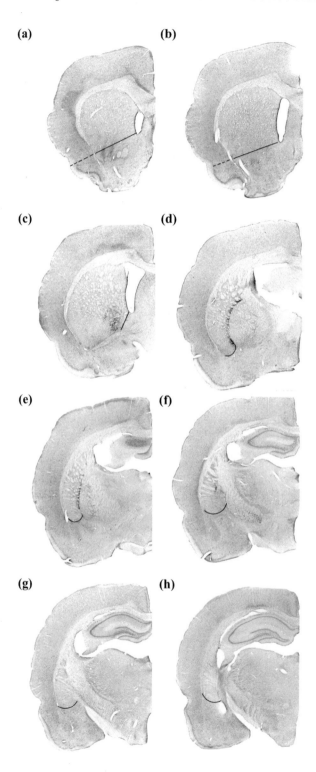

Fig. 5.4 (a)–(h) Low power micrographs of 50 μm thick counterstained somatostatin mRNA *in situ* hybridized sections, taken from eight positions along the rostrocaudal extent of the striatum of a rat, show the ventral borders of the striatum (solid lines) used in this example. Scale bar, 1 mm. (Reproduced with permission from West *et al.* (1996).)

tration and the temperature, as well as the pH of the rinsing buffer. Practically, this was achieved by rinsing the sections as described in Table 5.3(b), at relatively high temperature and salt concentrations, and using no formamide. A protocol that involves medium temperatures and decreasing salt concentrations is equally efficient (Table 5.3(b)).

7.3 Colour development

Colour development was carried out in the dark at RT, using nitroblue tetrazoleum (NBT) and 5-bromo-4-chloro-3-indolyl phosphate (BCIP) as chromogens (Altman 1976) to produce a bluish-purple hybridization signal (Fig. 5.2; Tables 5.2 and 5.3(c)). The free-floating sections were gently shaken in order to ensure that the sections developed evenly. The volume of developer was adjusted to the level of message, because chromogens may be exhausted if the message is abundant. The sections can be developed for up to 72 hours, at which time development is terminated by rinsing in tris ethylenediaminetetraacetic acid (TE)-buffer. The hybridization procedure inactivates endogenous AP activity and thereby eliminates the need to add levamisole to the developer (Lambertsen *et al.* 2001).

7.4 Mounting sections

Sections were transferred individually to distilled water and mounted on gelatin-coated microscope glass slides (Table 5.3(d)). The sections are best mounted in sequence, with the aid of the reference series, to facilitate the subsequent stereological analysis.

7.5 Counterstaining, dehydration, and coverslipping

To expedite the delineation of the region of interest and the actual counting of labeled cells, the *in situ* hybridized sections were counterstained with Neutral Red (Fig. 5.3(a), Tables 5.2 and 5.3(e)), dehydrated in graded acetone (which is less harsh to the AP-reaction product than ethanol), softened in xylene–phenol– creosote, cleared in xylene, and coverslipped with Eukitt (Tables 5.2 and 5.3(f)).

8 Specificity of the hybridization reaction

Controling the specificity of the ISH reaction requires more than one procedure. Although probe specificity can be tested by either Northern blot or RNAse protection assays, these provide no guarantee that the probe hybridizes specifically in the tissue section. However, the specificity can be verified by carrying out a number of

Table 5.3 *In situ* hybridization procedure

(a) *In situ* hybridization

1. Freshly cut vibratome sections (or sections stored in cryoprotective solution) are rinsed as free-floating sections for 2×60 min in $2 \times$ SSC at RT

2. Hybridization medium is made up from hybridization buffer and AP-labeled oligonucleotide probe. Probe and ice-cold hybridization buffer are mixed in a plastic tube by pipetting, followed by vortexing and 5 sec centrifugation at 2000 r.p.m. The hybridization medium is distributed in 1.100–1.200 µl aliquots in Eppendorf tubes (one for each animal) placed in a rack on ice

3. Sections are transferred from the rinsing bath into the Eppendorf tubes containing the hybridization medium. One internal standard section is added to each tube. Care is taken not to dilute the hybridization medium with excess SSC during transfer of sections. The tubes are closed and turned upside-down a few times in order to ensure that sections are completely unfolded and immersed in the hybridization medium

4. Sections are hybridized for 20–24 hours at 37°C. The tubes with the sections are turned upside down a few times two or three times during hybridization to ensure an even hybridization of all sections

(b) Post-hybridization rinsing

1. Hybridization solution with sections are transferred into preheated glasses and quickly rinsed in 55°C warm $1 \times$ SSC, pH 7.8

2. Rinsing 3 times 30 min in $1 \times$ SSC, pH 7.8 at 55°C, or 30 min in $2 \times$ SSC, $1 \times$ SSC, $0.5 \times$ SSC, and $0.1 \times$ SSC at 37°C under gentle shaking

(c) AP colour development

1. Rinsing 2 times 15 min in tris–HCl buffer, pH 9.5 at RT

2. Application of 1.5–2.0 ml freshly made AP-developer to each glass. Signal development takes place in the dark at RT under gentle shaking. Sections may be inspected the next day or occasionally after a few hours. Development is arrested after 48–72 hours

3. Arrestment of AP-development by rinsing 2 time 15 min in TE-buffer at RT, and left in TE-buffer at 4°C until mounting (maximally 24 hours)

(d) Mounting

1. Sections are transferred from TE-buffer to water and mounted on to gelatinized microscope slides and air-dried

(e) Counterstaining

1. Mounted and air-dried sections are immersed in Neutral Red solution (Merck 1369; 1 g in 100 ml 0.05M acetate buffer, pH 4.8). The period of immersion is 8–12 min, but should be optimized in each case. The sections are rinsed in distilled water after staining

(f) Dehydration and coverslipping

1. Graded acetone solutions (50, 70, 90, and 100%) 2 min each, followed by softening in xylene–phenol–creosote (250 ml creosote (Alchem, UK) and 250 g phenol in 2 l xylene). The sections are cleared in xylene and coverslipped with Eukitt.

control reactions. Table 5.4 describes the types and the rationale of the most commonly used specificity controls. The results of some of these controls are shown in Fig. 5.3.

9 Screening of internal standard sections

Prior to the stereological analysis, the internal standard sections, which had been hybridized together with the experimental material (Table 5.3(a)), were carefully analyzed for variations in the intensity of the ISH signal. In cases in which the ISH signal of an internal standard section deviated from the internal standard sections that were hybridized along with sections from the other animals, either the animal from which the section came was eliminated or another series of sections from the same animal was included in the stereological analysis.

10 Stereological quantification

10.1 Defining the region to be analyzed

To make unbiased estimates of the total number of labeled cells in a particular brain structure it is necessary to define the borders of the structure, that is, the sections and the regions of the sections that contain the cells of interest. In most cases, such as the dorsal striatum, this is most readily accomplished by rigorously defining the anatomical borders within which the sampling and counting is to be performed (Fig. 5.4). This can be done roughly with a fine-tip marker on the back of the slide. The drawn outlines can be trimmed with a scalpel under a dissection microscope for the precision needed. For other structures, in which borders are not well defined or may overlap with those of other structures, one could take advantage of some unique morphological characteristics of the cells of interest and ensure that the areas of the sections containing these cells are within the region of the section marked for sampling. In the latter case, the boundaries of the region analyzed would not have to be defined rigorously but, as a minimum, would have to include all the cells of interest.

In the example provided here, the boundaries of the dorsal striatum, were operationally defined in the following manner. The rostral border between dorsal and ventral striatum is defined by a line extending from the ventral part of the ventricle to the dorsal edge of the piriform cortex. Both of these features could be defined readily in Neutral Red counterstained sections (Fig. 5.4). This particular operational definition excluded the most ventral part of the dorsal striatum and a small portion of the core of the nucleus accumbens (West *et al.* 1996). The anterior commissure and the globus pallidus were used to define the more caudal parts of the anterior border (Fig. 5.4). The posterior and ventral borders of the striatum were delineated on the basis of architectonic differences with the adjacent amygdala. The remaining borders are natural borders and readily defined by the internal capsule and the ventricle.

Table 5.4 Control reactions*

Probe specificity	Northern blot RNAse protection assay RT-PCR	Ensures that the probe is specifically directed against the mRNA of interest
Probe specificity and specificity of ISH reaction	ISH with two probes directed against different, non-overlapping parts of the mRNA molecule	The statistical probability that two different 28–30' mer probes hybridize unspecifically and display identical unspecific ISH signal is infinitesimal. Like the above-mentioned assays also this control ensures that both probes are specific for the mRNA of interest, and it may substitute for the other control reactions. Sections hybridized with either probe alone should yield identical regional and cellular location of the ISH signal. Sections hybridized with both probes together as a cocktail display an increased ISH signal as compared to sections hybridized with either probe alone
	ISH with irrelevant probe	An additional control for probe specificity. ISH is performed with probes with similar length and CG-ratio, but specific for mRNA molecules with a different, regional and cellular distribution
	RNAse A treatment of sections prior to ISH	RNAse A digests all types of RNA. Hybridized sections should be blank. If signal is observed the probe binds unspecifically to non-mRNA molecules in the tissue section
	ISH with an excess of unlabelled probe	Ensures that the AP-complex doesn't stick to molecules in the tissue. Unlabelled probe added in an excess of ×100 competes out the AP-labelled probe. Hybridized sections should be blank
	ISH without probe	Test for presence of unspecific enzyme activity in tissue. Sections should be blank
Detection of peptide or protein	Immunohistochemistry for relevant peptide/protein	In many, but not all cases, immunocytochemical detection of the translational product is confirmatory to the ISH reaction
Test of tissue quality	ISH with probe directed against widely expressed mRNA species	ISH of parallel sections with a probe directed against a 'housekeeping' protein like GAPDH may in some cases be advantageous by documenting the overall suitability of the material for in situ hybridization

* RT-PCR, reverse transcriptase polymerase chain reaction.

10.2 Definition of a somatostatin mRNA-expressing neuron

Neuronal nuclei with evidence of the AP reaction product in the perinuclear region of the cell body (Fig. 5.2(a)) were counted (see Section 10.3.4).

10.3 Sampling scheme for estimating the total number of neurons with the optical fractionator technique

10.3.1 Sections

A systematic random sample (Gundersen and Jensen 1987) of an average of 12 sections from all of the sections that passed through the striatum was used in the analysis of each animal. In this case, in which serial sections of the entire brain were cut at 50 μm and the total number of sections passing through the striatum numbered approximately 120, the sections were selected at 10 section intervals after selecting the first section at random from within the first 10 sections of the series. (The selection of approximately 12 sections was based on considerations detailed in Chapter 2.) This gives a section sampling fraction, ssf, of 1/10 (Table 5.5, eqn 1).

10.3.2 Sectional area

A systematic random sample of the area of the sections containing the striatum was achieved by positioning an unbiased counting frame (Gundersen 1977) of known area, a(frame), 4144 μm^2, at the coordinates of a virtual rectangular lattice that was superimposed on the sections (Fig. 5.5(b)) with the aid of computer-controled stepping motors attached to the x- and y-axes of the microscope stage. The distance between the coordinates was 200 μm along each axis, resulting in an area of $(200)^2$ μm^2 (40 000 μm^2) associated with a step from one point in the lattice to the next. This area is referred to as a(step). The fraction of the area of the sections sampled in this manner is the ratio a(frame)/a(step) or the area sampling fraction, asf. In this example asf = 4144/40 000 = 0.103.

10.3.3 Section thickness

The thickness of each section was measured at 3–5 randomly selected positions in the sampling lattice, with a microcator (Heidenhein, Germany) mounted on the stage of the microscope. The mean of the means of the measures made on each section (n = 12) was used as an expression of the mean thickness, t, of the sections from the series from that individual (Fig. 5.5).

A known fraction of the section thickness was sampled with optical disectors at each position in the lattice. This was accomplished by first moving the counting frame, superimposed on a high magnification of a video image of the region of the section being sampled, through a distance, h, of 15 μm of the section thickness. The

Table 5.5   Estimating the total number of striatal somatostatin (SS) messenger RNA expressing neurons and the coefficient of error CE of the estimate in an individual

I   Data used to estimate N, the total number of neurons expressing SS mRNA in the striatum of one adult female Sprague–Dawley rat, and CE(N), the coefficient of error of that estimate

Section	Q^-
1	4
2	16
3	26
4	25
5	26
6	21
7	15
8	8
9	3
10	13
11	2
12	6
ΣQ^-	165

II   Equations used for the calculations
(1) $N = \Sigma Q^- \times 1/\text{ssf} \times 1/\text{asf} \times 1/\text{tsf} = 165 \times 10 \times 9.65 \times 1.33 = 21{,}200$
(2) $\text{CE}(\Sigma Q^-) = 1/(\Sigma Q^-)^{1/2} = 0.8$
(3) $\text{CE}(t) = 0.02$
(4) $\text{CE}(N) = \{(\text{CE}(\Sigma Q^-))^2 + (\text{CE}(t))^2\}^{1/2} = 0.82$

III   Notes on the calculations
(a)   Twelve sections were used in the analysis. These sections represented a systematic random sample (see Chapter 2) of sections taken at 10 section intervals, after a random start within the first 10 section interval. ΣQ^- is the number of neurons counted in the optical disector samples made on each of the 12 sections and is 165. Because every tenth section was used in the analysis, the section sampling fraction, ssf, was 1/10 and 1/ssf = 10 in equation (1).

(b)   The disector samples were made at regular 200 μm intervals in the x and y dimensions of each section, as shown in Fig. 5.6. The x,y step area was consequently 40,000 μm². The area of the counting frame used to construct the optical disectors was 4,144 μm². The sampling fraction, asf, was consequently 4,144/40,000 = 0.1036 and 1/asf = 9.65 in equation (1).

(c)   The height of the disectors, h, was 15 μm and the mean of the measured section thicknesses, t, was measured to be 20 μm. As a consequence, the thickness sampling fraction, tsf, was 0.75 and the reciprocal of the tsf used in equation (1) is 1.33.

(d)   The estimate of N is obtained by multiplying the total number of neurons counted ($\Sigma Q^- = 165$) by the product of the reciprocals of the fractions sampled (see eqn (1) above), i.e. the reciprocal of the fraction of the volume of the striatum sampled (1/ssf × 1/asf × 1/ tsf = 1/128) 128.

(e)   The coefficient of error of the estimate, CE(N), is the square root of the sum of the squares of the relative variances related to the sampling of the sections, CE(ΣQ^-) (eqn 2), and the relative variance of the measures of section thickness, CE(t) (eqn 3). In both cases the CE is calculated as it would be for independent samples.

(f) Considering the sampling to be independent in all three dimensions, as is done here, represents a notable change from the approach used in the original paper describing this method (West *et al.* 1996). In that paper, the total variance of the estimate included a component that stemmed from the variations in counts from section to section (the systematic random sampling, 'SRS' variance, along the axis of sectioning). The calculation of the SRS variance involved a special formula, the 'quadratic approximation formula' (Gundersen and Jensen 1987) in an attempt to obtain a more accurate estimate of CE(N). The change adopted here has been prompted by a subsequent paper published by Gundersen *et al.* (1999). The latter demonstrates that, when more than 5–6 sections, selected in a systematic random manner, are used in the analysis, the variance related to intersectional variations (i.e. the SRS variance) is insignificant compared to the variance stemming from the intrasectional sampling and that only the latter need be considered. The approach used here represents a simplification and improvement in the method for calculating the variance of the estimate of N that can be used when more than 5–6 sections are used in the analysis (see also Chapter 2).

(g) The coefficient of error of an estimate, e.g. CE(N), has no biological significance, but is useful when designing and optimizing the sampling scheme, as described in detail in Chapter 2.

fraction of the section thickness sampled, h/t, that is, the thickness sampling fraction, tsf, was on average 0.75 (Table 5.5, eqn 1).

10.3.4 Counting

Optical disector counting rules (Gundersen *et al.* 1988; West and Gundersen 1990) were used to count the number of neurons, $\bar{Q}$, expressing striatal somatostatin mRNA in the optical disectors. Accordingly, neurons were counted as they came into focus within the counting frame as the latter was focused through the section (West *et al.* 1996). The nucleus of the counterstained labeled cells is used as the counting unit, as it is easier to determine when this comes into focus than to determine when the perikaryal reaction product around the nucleus first comes into focus (Fig. 5.6).

10.3.5 Estimates of total number

The total number of labeled neurons in each individual, N, is estimated by multiplying the number of neurons counted in all optical disectors, $\Sigma\bar{Q}$, by the product of the reciprocals of the fractions of the structure sampled.

$$\text{Estimated } N = \Sigma\bar{Q} \times 1/\text{ssf} \times 1/\text{asf} \times 1/\text{tsf}.$$

10.3.6 The precision of the estimation procedure

The precision of an individual estimate, expressed as the coefficient of error, CE_N, was calculated as the reciprocal of the square root of the number of objects counted

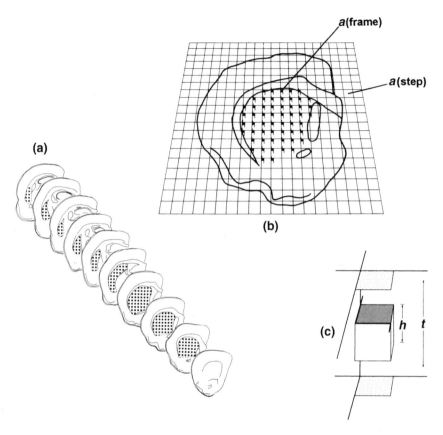

Fig. 5.5 Diagrammatic summary of the sampling scheme used to estimate the total number of neurons expressing SS mRNA in the striatum of the rat. (a) A systematic random sample of 10–13 sections that spans the entire length of the striatum is selected for analysis. The sections are selected at equal intervals in the series (in this case, every tenth section, after a random start within the first 10 sections; see Fig. 5.6) and therefore constitute a known fraction of the section series, the section sampling fraction, ssf. (b) The labeled neurons are counted under a known fraction of the section area (the area sampling fraction, asf). This fraction corresponds to the ratio of the area of the disector counting frames, a(frame) (shown here as small black rectangles) to the area associated with each step movement of the slide, a(step) (shown here as large white rectangles); asf = a(step)/a(frame). (c) The neurons are counted in optical disectors positioned in the central part of the section thickness. The height of the optical disector, h, constitutes a known fraction of the section thickness, t. The ratio h/t is the thickness sampling fraction, tsf. The area of the counting frame, a(frame), is shaded. The sum of the number of neurons in the disectors, ΣQ^-, along with the ssf, asf, and tsf are used to estimate the total number of neurons, N, as shown in the equation for N in the text (see also Section 10.3.5). The distance between the disector samples (large open squares in (b)), the areas of the disector counting frame (small dark squares in (b)), and the height of the disector (h in (c)) are dimensioned so that approximately 150 neurons are counted when the set of sections is sampled. (Reproduced with permission from West *et al.* (1996).)

and the variance of the measures of section thickness. An example of how this can be calculated is shown in Table 5.5, eqn 4, along with an example of the primary data used to make an estimate of the total number of striatal somatostatin mRNA expressing cells in one individual. The manner in which the precision of an individual estimate was used to evaluate the appropriateness of the sampling scheme described here has been described in detail elsewhere (Gundersen and Jensen 1987; Gundersen *et al.* 1999).

11 Discussion

11.1 Methodological considerations

When applying the optical fractionator method to *in situ* hybridized tissue the optimal thickness of the tissue sections is influenced by two opposing factors:

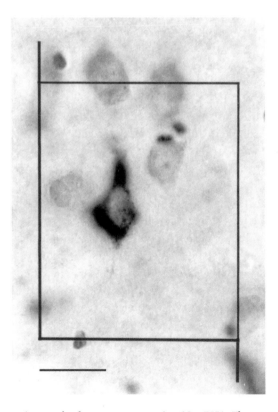

Fig. 5.6 High power micrograph of a neuron expressing SS mRNA. The area of the unbiased counting frame a(frame) is 4144 μm^2, the same area as that used with the optical disectors. Scale bar, 20 μm.

(1) the sections should be relatively thick so that optical disectors can readily be positioned within the sections; (2) relatively thin sections are needed to ensure that the probe penetrates the entire thickness of the section and that all cells in the sections are labeled. The protocol described here, in which a short synthetic AP-conjugated oligonucleotide probe is applied to 50 μm thick vibratome sections, has generally been successful in determining the appropriate section thickness for this type of material.

There are four major criteria for optimal fixation: The fixative should: (1) stabilize the tissue for sectioning; (2) preserve the tissue histologically; (3) retain mRNA within the cells; and (4) allow the probe to penetrate deep into the tissue section. Even though formaldehyde is a relatively weak fixative (e.g. it involves the formation of considerably fewer cross-links than glutaraldehyde), a slight prolongation of either the perfusion time (e.g. from 4 to 7–8 min) or the immersion fixation time in 4% PFA (e.g. from 2 to 3–4 hours) significantly reduces penetration of the probe. We have found that the present protocol stabilizes the tissue to a degree that makes it possible to obtain the uniform, high quality vibratome sections required for the stereological analysis of the particular type of tissue used here.

The time required to apply the method may be of major importance if the technique is to be used for diagnostic purposes. The sections are routinely hybridized for 20–24 hours. However, the ISH signal for abundant messages, such as somatostatin and enkephalin mRNA, can be obtained with the same technique after 3–4 hours of hybridization. We have not determined the minimal hybridization time required for complete probe penetration and we have not attempted to enhance the penetration of the probe into thicker sections by prolonging the hybridization time or by using proteolytic treatment or detergents. Potential practical problems could arise in both cases because the reporter AP enzyme is sensitive to prolonged exposure to high formamide concentrations and because proteolytic treatment makes the lightly fixed, free-floating, vibratome sections more fragile.

In addition to the destabilizing effect of the stringency factors, the pH also has a significant impact on the stability of the DNA–RNA hybrids. The hybridization signal is significantly reduced if the pH lies outside the interval of 7.5–8.0 and it is almost non-existent below pH 5 and above pH 9.5. The sensitivity of the hybrids to pH is also temperature-dependent, making it necessary to redefine the optimum pH of the rinsing solution when changing the temperature of the post-hybridization rinses.

In some experimental paradigms, the NBT-substrate in the AP-developer may be reduced to formazan by hybridization-resistant (most likely mitochondrial) enzymes (Altman 1976), and may produce a homogeneous bluish color in specific cortical laminae. This problem has been encountered in acutely lesioned and electroconvulsed rats (Dalby *et al.* 1998). The induction of reducing enzymes may result in interpretative problems in that NBT is removed from the developer, resulting in a suboptimal ISH signal. The problem can be overcome, without loss of hybridization signal, by subjecting the free-floating sections to HCl treatment prior to ISH in order to inactivate the enzymes.

11.2 Interpretive caveats

Unique problems arise in the interpretation of stereological data of the type presented here. These originate from the criteria used to define the counting unit, and are not directly related to the general stereological methodology. The only requirement for applying the stereology method employed here is that all cells containing appropriate levels message are labeled. In this protocol for counting neurons expressing striatal somatostatin mRNA, the counting unit is a nucleus associated with a labeled message. Because it is not known how many labeled molecules of mRNA must be present before expressing cells can be identified microscopically, the counting unit used in this study is to some extent arbitrary. This emphasizes the necessity of carrying out the ISH procedure under extremely standardized conditions. In the material used to exemplify this protocol (West *et al.* 1996) and in the study by Andreassen *et al.* (1999), there is an unusually high degree of rendition of the labeled neurons and a very low degree of uncertainty when identifying labeled cells. In these studies stable, well-defined biological entities are apparently being studied. This may not be the case when other oligonucleotides are used to visualize the same message or when attempts are made to visualize: (1) less abundant messages; (2) messages that show cyclic variations; or (3) messages that are transiently downregulated as a consequence of experimental manipulations. These problems should be dealt with on an individual basis and should take into account the normal dynamics of the gene of interest. A further discussion of the practical problems and interpretive caveats associated with stereological counting of striatal somatostatin and enkephalin mRNA expressing neurons can be found in West *et al.* (1996) and Andreassen *et al.* (1999).

11.3 Use of supplementary series of parallel sections

Because no amplification steps are involved and the AP histochemical reaction displays linear kinetics during the first several hours of development, the intensity of the ISH signal is proportional to the amount of hybridized probe and therefore to the amount of mRNA present within individual cells. As with radiolabeled probes (O'Shea and Gundlach 1993; Wisden and Morris 1993*b*), the AP-labeled probes thereby meet the criteria required for densitometric analysis of either regional or single cell gene expression (Augood *et al.* 1991, 1993). Densitometric analyses of *in situ* hybridized sections has tradionally been applied to thin cryostat or paraffin sections (Augood *et al.* 1991, 1993). As discussed here they can also be performed on thicker vibratome sections. Data of this type can supplement data obtained from the stereological analysis (Andreassen *et al.* 1999, 2000).

Other supplementary methods include dual ISH (Augood *et al.* 1993, 1995) and ISH combined with conventional enzyme- or immunohistochemistry (Andreassen *et al.* 2000; Augood *et al.* 1993; Gregersen *et al.* 2000; Jensen *et al.* 2000*a*), all of which can be carried out successfully with freshly cut or cryoprotected vibratome sections. We have successfully applied the protocol of Trembleau *et al.* (1993), which combines ISH for either somatostatin mRNA, calbindin mRNA, biocytin, or TNF mRNA with the immunohistochemical staining of neuropeptides and astroglial GFAP (Gregersen *et al.* 2000; Jensen *et al.* 2000*a*). Accordingly, tissue sections are hybridized, rinsed, and then subjected to the full immunohistochemical procedure, including the histochemical diaminobenzidine development. The sections are then returned to the *in situ* hybrization protocol and subjected to AP-development (Gregersen *et al.* 2000). The denaturation that occurs during the initial step of hybridization may affect some antigens and necessitate modifications in the procedure. One approach is to perform the primary antibody-incubation prior to visualization of the mRNA molecule (Kiefer *et al.* 1996; Lehrmann *et al.* 1998). Dual ISH has traditionally been carried out on thin cryostat or paraffin sections (Augood *et al.* 1993, 1995) and has permitted the use of a non-radiolabeled and a radiolabeled probe simultaneously. Dual ISH on vibratome sections can be performed by simultaneously hybridizing vibratome sections with two non-radiolabeled probes and by applying a digoxigenin-labeled (Lewis *et al.* 1993) or a biotinylated probe (Larson and Hougaaard 1993) along with the AP labeled probe.

12 Conclusion

A protocol for obtaining stereological estimates of the total number of *in situ* hybridized neurons has been described. Unbiased estimation of the total number of neurons or glia expressing a certain gene adds an extremely powerful analytic tool to the techniques that are routinely used to quantify gene expression, such as solution hybridization and densitometric assessment of gene expression. Together, these methods provide unique opportunities to obtain information about the regulation of gene expression within specific neuronal or glial subpopulations of anatomically well-defined brain regions in normal and pathological states.

Acknowledgements

We acknowledge the excellent technical assistance of Dorte Lyholmer in particular, and of Lene Jørgensen and Grethe Jensen for their contributions to the presented ISH protocol, and the expert photographical assistance of Albert Meyer, Institute of

Anatomy, Aarhus University, and Neurophotography, Montreal Neurological Institute, McGill University, Montreal. This work was supported by the Danish Multiple Sclerosis Society, The Novo Nordic Foundation, the Danish Medical Research Council, the Danish State PharmaBiotec Research Program, the University of Aarhus, and the University of Southern Denmark/Odense University.

References

Altman, F.P. (1976). Tetrazoleum salts and formazans. *Progr. Histochem. Cytochem.* **9**, 1–51.

Andreassen, O.A., Finsen B., Østergaard, K., Sørensen, J.C., West, M.J., and **Jørgensen, H.A.** (1999). The relationship between oral dyskinesias produced by long-term haloperidol treatment, the density of striatal preproenkephalin messenger RNA and enkephalin peptide, and the number of striatal neurons expressing preproenkephalin messenger RNA in rats. *Neuroscience* **1**, 27–35.

Andreassen, O., Finsen, B., Østergaard, K., West, M., and **Jørgensen, H.** (2000). Reduced number of striatal neurons express preprosomatostatin mRNA in rats with oral dyskinesias after long-term haloperidol treatment. *Neurosci. Lett.* **279**, 21–24.

Augood, S. J., Kiyama, H., Faull, R.L., and **Emson, P.C.** (1991). Dopaminergic D1 and D2 receptor antagonists decrease prosomatostatin mRNA expression in rat striatum. *Neuroscience* **44**, 35–44.

Augood, S.J., McGowan, E.M., Finsen, B.R., Heppleman, B, and **Emson, P.C.** (1993). Non-radioactive ISH using alkaline phosphatase-labeled oligonucleotides. In *In situ hybridization* (ed. W. Wisden and B.J. Morris), pp. 81–97, Biological Techniques Series. Academic Press, New York.

Augood, S.J., Herbison, A.E., and **Emson, P.C.** (1995). Localization of GAT-1 GABA transporter mRNA in rat striatum: cellular coexpression with GAD_{67} mRNA, GAD_{67} immunoreactivity, and parvalbumin mRNA. *J. Neurosci.* **15**, 865–74.

Conn, M.P. (ed.) (1992). *Gene expression in neural tissues*, Methods in neurosciences, Vol. 9. Academic Press Inc, New York.

Dalby, N. O., Tønder, N., Voldby, D., Finsen, B., West, M. J., and **Bolwig, T. G.** (1996). No loss of hippocampal somatostatinergic neurons after electroconvulsive shocks. A stereological and *in situ* hybridization study. *Biol. Psych.* **40**, 54–60.

Dalby, N.O., West, M.J., and **Finsen, B.** (1998). Hilar somatostatin-mRNA containing neurons are preserved after perforant path kindling in the rat. *Neurosci. Lett.* **255**, 45–8.

Emson, P.C. (1993). In-situ hybridization as a methodological tool for the neuroscientist. *Trends Neurosci.* **16**, 9–16.

Farmer, J.G. and **Castaneda, M.** (1991). An improved preparation and purification of oligonucleotide–alkaline phosphatase conjugates. *Biotechniques* **11**, 588–9.

Finsen, B.R., Tønder, N., Augood, S., and Zimmer, J. (1992). Somatostatin and neuropeptide Y in organotypic slice cultures of the rat hippocampus. An immunocytochemical and *in situ* hybridization study. *Neuroscience* **47**, 104–13.

Fort, P., Marty, L., Piechaczyk, M., el Sabrouty, S., Dani, C., Jeanteur, P., and Blanchard, J.M. (1985). Various rat adult tissues express only one major mRNA species from the glyceraldehyde-3-phosphate-dehydrogenase multigenic family. *Nucleic Acids Res.* **13**, 1431–42.

Fuhrmann, G., Heilig, R., Kempf, J., and Ebel, A. (1990). Nucleotide sequence of the mouse preprosomatostatin gene. *Nucleic Acids Res.* **18**, 1287.

Gall, J.G. and Pardue, M.L. (1969). Formation and detection of RNA–DNA hybrid molecules in cytological preparations. *Proc. Natl Acad. Sci., USA* **63**, 378–83.

Garrett, B., Finsen, B., and Wree, A. (1994). Parcellation of cortical areas by *in situ* hybridization for somatostatin mRNA in the adult rat: frontal, parietal, occipital and temporal regions. *Anat. Embryol.* **190**, 389–98.

Goodchild, J. (1990). Conjugates of oligonucleotides and modified oligonucleotides: a review of their synthesis and properties. *Bioconjugate Chem.* **1**, 165–87.

Goodman, R.H., Aron, D.C., and Ross, B.A. (1983). Rat pre-prosomatostatin. Structure and processing by microsomal membranes. *J. Biol. Chem.* **258**, 5570–3.

Gregersen, R., Lambertsen, K., and Finsen, B. (2000). Microglia/macrophages are the major source of tumor necrosis factor in permanent middle cerebral occlusion in mice. *J. Cereb. Blood Flow Metab.* **20**, 53–65.

Grima, B., Lamouroux, A., Blanot, F., Biguet, N.V., and Mallet, J. (1985). Complete coding sequence of rat tyrosine hydroxylase mRNA. *Proc. Natl Acad. Sci., USA* **82**, 617–21.

Gundersen, H.J. (1977). Notes on the estimation of the numerical density of particles: The edge effect. *J. Microsc.* **111**, 219–23.

Gundersen, H.J. and Jensen E.B. (1987). The efficiency of systematic sampling in stereology and its predictions. *J. Microsc.* **147**, 229–63.

Gundersen, H.J., Bagger, P., Bentsen, T., Evans, S.M., Korbo, L., Marcussen, N., Møller, A., Nielsen, K., Nyengaard, J.R., Pakkenberg, B., Sørensen, F.B., Vesterby, A., and West M.J. (1988). The new stereological tools: disector, fractionator, nucleator and point sampled intercepts and their use in pathological reasearch and diagnosis. *Acta Pathol. Microbiol. Immunol. Scand.* **96**, 857–81.

Gundersen, H.J., Jensen E.B., Kieu, K., and Nielsen, J. (1999). The efficiency of systematic sampling in stereology—reconsidered. *J. Microsc.* **193**, 199–211.

Guthrie, K.M., Woods, A.G., Nguyen, T., and Gall, C.M. (1997). Astroglial ciliary neurotrophic factor mRNA expression is increased in fields of axonal sprouting in deafferented hippocampus. *J. Comp. Neurol.* **386**, 137–48.

Howard, C.V. and Reed, M.G. (eds.) (1998). *Unbiased stereology: three-dimensional measurement in microscopy*. Bios Scientific Publishers, Oxford.

Howells, R.D. (1989). Proenkephalin biosynthesis in the rat. *NIDA Res. Monogr.* **70**, 43–65.

Hunziker, W. and **Schrickel, S.** (1988). Rat brain calbindin D28: six domain structure and extensive amino acid homology with chicken calbindin D28. *Mol. Endocrinol.* **2**, 465–73.

Jablonski, E., Moomaw, E.W., Tullis, RH., and **Ruth, J.L.** (1986). Preparation of oligodeoxynucleotide–alkaline phosphatase conjugates and their use as hybridization probes. *Nucleic Acids Res.* **14**, 6115–29.

Jensen, M.B., Hegelund, I.V., Lomholt, N.D., Finsen B., and **Owens, T.** (2000*a*). IFNγ enhances microglial reactions to hippocampal axonal degeneration. *J. Neurosci.* **20**, 3612–21.

Jensen, M.B., Poulsen, F.R., and **Finsen, B.** (2000*b*). Axonal sprouting regulates myelin basic protein gene expression in denervated mouse hippocampus. *Int. J. Dev. Neurosci.* **18**, 221–35.

Kiefer, R., Funa, K., Schweitzer, T., Jung, S., Bourde, O., Toyka, K.V., and **Hartung, H.-P.** (1996). Transforming growth factor-β1 in experimental autoimmune neuritis: cellular localization and time course. *Am. J. Pathol.* **148**, 211–23.

Khorooshi, M. H., Meyer, M., Pedersen, E.B., and **Finsen, B.** (1999). Lack of effect of short-term depletion of plasma complement C3 on the survival of syngeneic dopaminergic neurons following grafting into the intact rat striatum. *Cell. Transplant.* **8**, 489–99.

Kiyama, H., Emson, P.C., and **Ruth, J.L.** (1989). Distribution of tyrosine hydroxylase mRNA in the rat central nervous system visualized by alkaline phosphatase in situ hybridization chemistry. *Eur. J. Neurosci.* **2**, 512–24.

Lambertsen, K., Gregersen, R., Lomholt, N.D., Owens T. and **Finsen, B.** (2001). A specific and sensitive protocol for detection of tumor nectosis factor in the murine central nervous system. *Brain. Res. Prot.* **7**, 175–91.

Larsson, L.I. and **Hougaard, D.M.** (1993). Sensitive detection of rat gastrin mRNA by *in situ* hybridization with chemically biotinylated oligonucleotides: validation, quantitation, and double-staining studies. *J. Histochem. Cytochem.* **41**, 157–63.

Lehrmann, E., Kiefer R., Christensen T., Toyka K.V., Zimmer J., Diemer, N.H., Hartung, and **Finsen, B.** (1998). Microglia and macrophages are major sources of locally produced transforming growth factor-β1 after transient middle cerebral artery occlusion in rats. *Glia* **24**, 437–48.

Lewis, E.M., Robbins, E., and **Baldino, F.** (1993). *In situ* hybridization histochemistry with radioactive and non-radioactive cRNA and DNA probes. In *Molecular imaging in neuroscience. A practical approach* (ed. D. Rickwood and B.D. Hames), pp. 1–22, The Practical Approach Series. IRL Press, Oxford.

Newman, S., Kitamura, K., and **Campagnoni, A.T.** (1987). Identification of a cDNA coding for a fifth form of myelin basic protein in mouse. *Proc. Natl Acad. Sci., USA* **84**, 886–90.

Nordquist, D.T., Kozak, C.A., and **Orr, H.T.** (1988). cDNA cloning and character-ization of three genes uniquely expressed in cerebellum by Purkinje neurons. *J. Neurosci.* **8**, 4780–9.

O'Shea, R.D. and **Gundlach, A.L.** (1993). Quantitative analysis of *in situ* hybridiza-tion histochemistry. In *In situ hybridization protocols for the brain* (ed. W. Wisden and B.J. Morris), pp. 57–78, Biological Techniques Series. Academic Press, New York.

Østergaard, K., Finsen, B., and **Zimmer, J.** (1995). Somatostatin (SS), neuropeptide Y (NPY), enkephalin (ENK), and NADPH in organotypic slice cultures of the rat striatum: an immuno-cytochemical, histochemical and in situ hybridization study. *Exp. Brain Res.* **103**, 70–84.

Paragas, V.B., Zhang, Y.Z., Haugland, R.P., and **Singer, V.L.** (1997). The ELF-97 alkaline phosphatase substrate provides a bright, photostable, fluorescent signal amplification method for FISH. *J. Histochem. Cytochem.* **45**, 345–57.

Parmentier, M., Lawson, D.E., and **Vassart, G.** (1987). Human 27-kDa calbindin complementary DNA sequence. Evolutionary and functional implications. *Eur. J. Biochem.* **170**, 207–15.

Pennica, D., Hayflick, J.S., Bringmann, T.S., Palladino, M.A., and **Goeddel, D.V.** (1985). Cloning and expression in *Escherichia coli* of the cDNA for murine tumor necrosis factor. *Proc. Natl Acad. Sci., USA* **82**, 6060–4.

Sabath, D.E., Broome, H.E., and **Prystowsky, M.B.** (1990). Mouse glyceraldehyde-3-phosphate dehydrogenase mRNA, and translated products. *Gene* **91**, 185–91.

Shirai, T., Shimizu, N., Horiguchi, S., and **Ito, H.** (1989). Cloning and expression in *Escherichia coli* of the gene for rat tumor necrosis factor. *Agric. Biol. Chem.* **53**, 1733–6.

Tecott, L.H., J.H. Eberwine, J.D. Barchas, and **Valentino, K.L.** (1987). Metholodological considerations in the utilization of *in situ* hybridization. In *In situ hybridization—applications to neurobiology* (ed. K.L. Valentino, J.H. Eberwine, and J.D. Barchas), pp. 3–24. Oxford University Press, Oxford.

Tokunaga, K., Nakamura, Y., Sakata, K., Fujimori, K., Ohkubo, M., Sawada, K., and **Sakiyama, S.** (1987). Enhanced expression of a glyceraldehyde-3-phosphate dehydrogenase gene in human lung cancers. *Cancer Res.* **47**, 5616–19.

Tønder, N., Kragh, J., Finsen, B.R., Bolwig, T.G., and **Zimmer, J.** (1994). Kindling induces transient changes in neuronal expression of somatostatin, neuropeptide Y, and calbindin in adult rat hippocampus and fascia dentata. *Epilepsia* **35**, 1299–308.

Trembleau, A., Roche, D., and **Calas, A.** (1993). Combination of non-radioactive and radioactive *in situ* hybridization with immunohistochemistry: a new method allowing the simultaneous detection of two mRNAs and one antigen in the same brain tissue sections. *J. Histochem. Cytochem.* **41**, 489–98.

Valentino, K.L., Eberwine, J.H., and **Barchas, J.D.** (eds.) (1987). *In situ hybridiza-tion—applications to Neurology.* Oxford University Press.

West, M.J. (1993). New stereological methods for counting neurons. *Neurobiol. Ageing* **44**, 275–85.

West, M.J. (1999). Stereological methods for estimating the total number of synapses: issues of precision and bias. *Trends Neurosci.* **22**, 51–61.

West, M.J. and **Gundersen, H.J.G.** (1990). Unbiased stereological estimation of the number of neurons in the human hippocampus. *J. Comp. Neurol.* **196**, 1–22.

West, M. J., Slomianka, L., and **Gundersen, H. J. G.** (1991). Unbiased stereological estimation of the total number of neurons in the subdivisions of the rat hippocampus using the optical fractionator. *Anat. Rec.* **231**, 482–97.

West, M.J., Østergaard, K., Andreassen, O.A., and **Finsen, B.** (1996). Counting *in situ* hybridized neurons with modern unbiased stereological methods. *J. Comp. Neurol.* **370**, 11–22.

Wilkinson, D.G. (ed.) (1992a). *In situ hybridization, a practical approach.* IRL Press, Oxford.

Wilkinson, D.G. (1992b). The theory and practice of *in situ* hybridization. In *In situ hybridization, a practical approach* (ed. D.G. Wilkinson), pp. 1–13. IRL Press, Oxford.

Wisden, W. and **Morris, B.J.** (eds.) (1993a). *In situ hybridization protocols for the brain,* Biological Techniques Series. Academic Press, New York.

Wisden, W. and **Morris, B.J.** (1993b). *In situ* hybridization with synthetic oligonucleotide probes. In *In situ hybridization protocols for the brain* (ed. W. Wisden and B.J. Morris), pp. 9–34, Biological Techniques Series. Academic Press, New York.

CHAPTER 6

NUMBER IN ELECTRON MICROSCOPY: ESTIMATION OF TOTAL NUMBER OF SYNAPSES IN THE MAIN REGIONS OF HUMAN NEOCORTEX

JENS R. NYENGAARD, KARL-ANTON DORPH-PETERSEN, AND YONG TANG

1 Introduction

The total number of synapses is one of several parameters characterizing the neuronal circuits of the brain. Despite unresolved issues concerning the exact functions and efficacy of synapses of different types and targets, the number of synapses directly represents the functional connectivity of the neurons of a region. Therefore, together with information about the total number of neurons, the total number of synapses of a region of the nervous system may provide a structural correlate for robust quantification of changes in neuronal functional capacity.

The term 'synapse' was first used by Sir Charles Sherrington in 1897 (see Foster and Sherrington 1897). It was derived from the Greek words meaning 'to fasten together' and was used to signify a site where the axon terminal of one neuron comes into functional contact with a second neuron. It follows from the neuron doctrine that a synapse is not a site of cytoplasmic confluence between neurons, but is rather an interface at which they are functionally related.

Each interneuronal chemical synapse consists of a presynaptic element and an opposed postsynaptic element separated by an intercellular space of 10–20 nm called the synaptic cleft. The presynaptic and postsynaptic membranes display densities on their cytoplasmic faces, and these specialized membranes together with the synaptic cleft are defined as the synaptic junction. The presynaptic element contains an accumulation of synaptic vesicles; the pre- and postsynaptic membranes come into opposition with only a narrow interstice between them and with filamentous or granular material condensed in the adjacent cytoplasm. The dense material may be limited to small areas, or may extend for the entire area of the junction (Peters *et al.* 1991).

Significant disagreements exist among previous morphometric studies of synapse number in relation to memory deficits associated with aging and diseases such as Alzheimer's disease. There may be a number of methodological reasons for these conflicting results: lack of unbiased sampling design; use of two-dimensional sampling and model-based methods; reporting of densities; and ignorance of postmortem changes as described later.

Many of the problems mentioned above are overcome by using a design-based stereological approach. The total processing time per brain including the counting of the synapses adds up to approximately 10 days of work, which may be a limiting factor in its application. The method described is a modification of a method previously published (Tang *et al.* 2001); however, here we report the original unchanged data.

2 Material

The material comprised five male brains collected in accordance with Danish laws on autopsied human tissue. The subjects had an average age of 21.8 years (range

19–28 years), average height of 180 cm (175–183 cm), and average weight of 72 kg
(67–77 kg). The average brain weight was 1519 g (1400–1750 g) and the post-
mortem interval (time from death to fixation) was less than 60 hours. None of the
subjects had prior neurological or psychiatric disorders or a medical record of any
diseases (including alcohol or drug abuse) that may affect the central nervous
system. All subjects were previously included in the study by Pakkenberg and
Gundersen (1997).

3 Methods

3.1 Estimation of neocortical volume

The brains were fixed in 0.1M sodium phosphate buffered formaldehyde (pH = 7.2,
4% formaldehyde). Right or left hemisphere was chosen at random. Four neocortical regions, that is, frontal, temporal, parietal, and occipital, were delineated
and painted on the pial surface using different colors (Brændgaard *et al.* 1990;
Pakkenberg and Gundersen 1997). Each hemisphere was embedded in 6% agar
and cut coronally at 7-mm intervals with a random starting point. The neocortical
volume of each region was estimated from the coronal sections according to
Cavalieri's principle (see Fig. 6.1(a)–(d)).

3.2 Uniformly random sampling of tissue blocks from each region of neocortex

From the coronal slabs, every fourth slab was sampled systematically uniformly
random. From the slabs sampled, the slabs from the same region (frontal, parietal,
temporal, and occipital) were identified. Then, two tissue wedges were sampled
uniformly random from each of the four cortical regions as described in
Fig. 6.1(e)–(f). Approximately 2 mm was sliced from the top of each sampled
wedge (Fig. 6.1(g)). Each slice was divided into 3–7 blocks of approximately equal
size and one tissue block was uniformly random sampled among those that
contained neocortex (Fig. 6.1(h)–(i)). The eight tissue blocks sampled in this way
were used for electron microscopic (EM) investigations.

3.3 Tissue preparation for EM and staining of synapses

The sampled tissue blocks were embedded in agar in order to stabilize them for the
subsequent cutting on a vibratome. Each block was cut into 75 mm thick slices
(Fig. 6.1(j)) using a vibratome (Micro-cut H 1200, Bio-Rad, USA). These slices
were stored in 0.1M phosphate buffer (pH = 7.4) for 1 day, and then treated with
a mixture of osmium and maleic acid for half an hour and rinsed three times in the

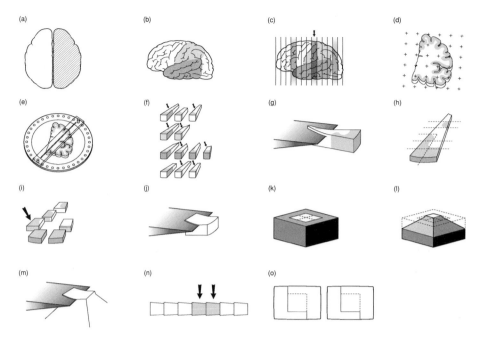

Fig. 6.1 Overview of the multiple steps of sampling and processing for synapse counting in human neocortex. (a) Right or left hemisphere of the brain is sampled randomly. (b) The four regions (frontal, parietal, temporal, and occipital) of the sampled hemisphere are painted in different colors. (c) The hemisphere is cut coronally in 7 mm thick slabs and every second slab is sampled systematically, uniformly randomly, producing a sample of approximately 13 slabs. (d) By point counting on the slabs the volume of the neocortex of the four regions can be estimated using Cavalieri's principle. (e) Using a coronal slice-sampling device with a rotatable knife holder, tissue wedges are systematically, uniformly randomly sampled through the sets of slabs. Two opposed wedges per slab are cut producing approximately 26 wedges in total. The tissue wedges are sorted according to regions and from each region two wedges are systematic, uniformly randomly sampled resulting in a final sample of eight tissue wedges as shown by arrows in (f). (g) A slice is cut from the top of each sampled wedge. (h) From each of the eight tissue slices, 3–7 tissue blocks of approximately equal size and containing neocortex are cut. From the set of tissue blocks shown in (i), one (indicated by the arrow) is randomly sampled and its area estimated by point counting for shrinkage correction. Since one block is chosen from each wedge, a total of eight blocks per brain (two per region) are sampled. (j) From each tissue block a 75 mm thick slice is cut using a vibratome. The slice is processed for electron microscopy and stained with E-PTA. (k) The slab embedded in Epon is shown, and its area is re-estimated using point counting for shrinkage correction. (l) The Epon block is afterwards trimmed so only a trapeze-shaped tissue surface is at the top. (m) From the trimmed Epon block a series of sections of a thickness of 50–90 nm is cut using an ultramicrotome. (n) From the series of sections cut, one pair of consecutive sections is chosen. Thus eight section pairs are generated per brain—two from each region. Electron micrographs at a final magnification of approximately 12 000 × are made in the vicinity of the four corners of each section pair. (o) On all 32 pairs of micrographs a set of unbiased counting frames is superimposed and the synapses are counted using the physical disector principle.

Table 6.1 Tissue preparation and staining of the brain samples for synapse number estimation

Steps in tissue preparation

1   Fix in 4% formaldehyde
2   Put into agar
3   Cut into vibratome slices 75 μm thick
4   Phosphate buffer
5   Fix in osmium and phosphate buffer
6   Dehydrate in graded ethanol solutions
7   Stain in 1% E-PTA at 60°C
8   Propylene oxide
9   Embed in Epon
10   Cut into ultrathin sections ~70 nm thick
11   Sample two consecutive sections
12   Examine in an electron microscope

buffer at room temperature. Subsequently, the slices were dehydrated in a graded series of ethanol solutions (30–50–70–90%), 4 min each at room temperature. The slices were stained for 2.5 hours at 60°C in 1% ethanol phosphotungstic acid (E-PTA) solution containing a trace of water. Then, the slices were treated with water-free acetone (two changes, 10 min each) and put into a mixture of acetone and Epon (1:1) for at least 1 hour at room temperature. In order to embed the slices as flatly as possible, a small tube containing hardened Epon was used as an embedding mold and the slice was placed in the mold and covered with a drop of Epon. The embedded blocks were dried in an oven at 60°C for at least 24 hours (see Table 6.1).

3.4 Preparation of serial ultrathin sections and micrographs and estimation of section thickness

Each Epon block was divided into four parts and one part was sampled uniformly random. The other three parts were trimmed away (Fig. 6.1(k)–(l)). The sampled Epon pyramide was cut into sections (Fig. 6.1(m)) with a thickness of 50–90 nm using an LKB 2088 ultramicrotome (Sweden). Ribbons of six to seven consecutive sections were picked up on a one-hole grid that was covered with a Formvar film. Since the tissue blocks were sampled uniformly random from the neocortex of human autopsy brains, sections could be arbitrarily selected from ribbons under the condition that selection of sections was independent of synapses inside sections. The thicker sections were preferred; however, the criterion for the selection of a disector pair was that at least one fold was present in each of the two sections. The first two consecutive sections both with folds were then chosen and examined in a Philips 300 electron microscope (Fig. 6.1(n)). Since the same position had to be photographed in corresponding parts of consecutive ultrathin sections, the photographs were taken from undamaged places very close to the four corners of the sections. Electron micrographs were enlarged to a final magnification of ~12 000 ×. A magnification standard (grating replica) was photographed and printed together with each series of electron micrographs. At least one fold and up to three folds appearing on each chosen section were photographed and used for the estimation of section thickness. The thickness of the ultrathin sections was estimated with Small's smallest fold method (Small 1968; Williams 1981). The folds were photographed at a magnification of ~50 000 ×. The smallest fold width was measured on the photograph. Since minimal folds stand perpendicular to the section plane and have a width that equals twice that of the section thickness the ultrathin section thickness was half of the smallest fold width obtained.

3.5 Estimation of synaptic density in neocortex using the disector

The physical disector principle (Sterio 1984) was used in order to count synapses. The electron micrograph from the first section was termed the reference plane and the micrograph from the next ultrathin section was called the look-up plane (Figs 6.1(o) and 6.2). The pre- and postsynaptic densities were used as counting units. A transparency with an unbiased counting frame (Gundersen 1977) was superimposed on each of the micrographs. A guard area, which exceeded the largest dimension of the largest pre- and postsynaptic densities, separated the edges of the frame and the micrograph. Synapses were labeled on the reference micrograph if their pre- and postsynaptic density profiles were located entirely or partly within the counting frame and did not intersect the forbidden lines or their extensions. The synapses and 'synaptic bridges' present in the reference plane, and not in the look-up plane, were counted (Fig. 6.2). Then the reference and look-up planes were reversed to double the sample size. The numerical density of synapses in each region, $N_V(\text{syn/reg})$, was calculated using the formula

$$N_V(\text{syn/reg}) := \frac{\sum Q_1^-(\text{syn}) \cdot n_1 + \sum Q_2^-(\text{syn}) \cdot n_2}{\sum v_1(\text{dis}) \cdot n_1 + \sum v_2(\text{dis}) \cdot n_2} \qquad (3.1)$$

where $\sum Q_1^-(\text{syn})$ and $\sum Q_2^-(\text{syn})$ are the total number of synapses minus the total number of 'synaptic bridges' counted in the disectors in the first and the second block, respectively, from the neocortical region. Likewise, $\sum v_1(\text{dis})$ and $\sum v_2(\text{dis})$ are the total volume of the four disectors from the first and the second block, respectively. The total volume of the four disectors in each block is: $\sum v(\text{dis}) = 4a(\text{frame}) \cdot (t_1 + t_2)$, where $a(\text{frame})$ is the frame area and t_1 and t_2 are the section thicknesses of each of the two sections in the disector. n_1 and n_2 are the numbers

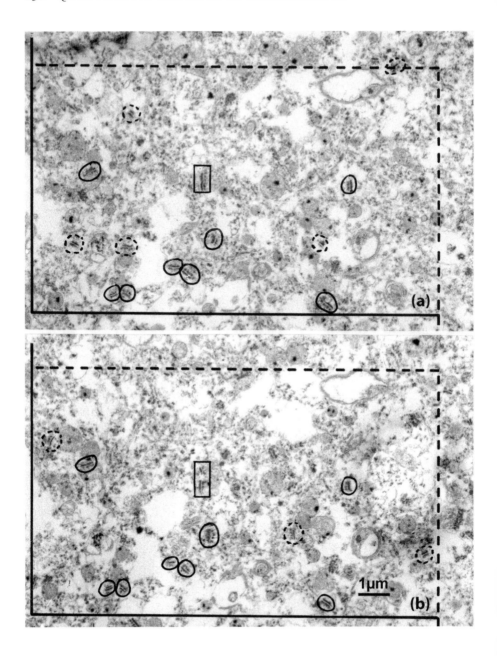

of neocortex-containing blocks cut from each of the two wedges from the region. They are used to provide the proper weights of the synapse counts and volumes, compensating for the fact that only one block is sampled from each wedge independent of its amount of neocortex. The sign ':=' refers to the fact that it is an estimate.

◄ **Fig. 6.2** The electron micrographs of two consecutive ultrathin sections from the neocortex of a human brain are shown to demonstrate the application of the disector method for synapse counting. Each disector consisted of two parallel planes, where the first plane (a) was called the reference plane and the second (b) the look-up plane. Then, the reference and look-up planes were reversed to double the sample size. Unbiased sampling frames were superimposed on the pair of micrographs. The pre- and postsynaptic densities were used as a counting unit. The synapses present in the reference plane and not in the look-up plane were counted (dashed circles which equal Q^-) if their pre- and postsynaptic densities were completely or partly inside the sampling frame without touching the exclusion lines and their extension. The synapse profiles appearing in both planes (solid lines) were not counted. The two separate synapse profiles marked with (full square) in (b) were found to belong to one complex synapse when examining the consecutive section shown in (a). This event is called a bridge. When evaluating the total number of synapses the number of bridges is subtracted from Q^- and the horseshoe-shaped synapses are counted correctly (modified from Tang *et al.* 2001).

3.6 Estimation of the total number of synapses in each neocortical region and the total number of synapses in the whole neocortex

The total number of synapses in each neocortical region, N(syn,reg), was calculated as

$$N(\text{syn, reg}) := N_V(\text{syn/reg}) \cdot V(\text{reg}) \tag{3.2}$$

where V(reg) is the volume of each neocortical region in two hemispheres. However, the numerical density of synapses was first corrected for tissue shrinkage as explained in Section 3.7. The total number of synapses in the whole neocortex of a brain was estimated as the sum of the estimates of total number of synapses in the four regions.

3.7 Estimation of neocortex shrinkage

All eight tissue blocks sampled per brain were used for estimation of shrinkage. Before being processed, the cross-sectional areas of the tissue blocks were measured using point counting (Fig. 6.1(i)). After Epon embedding, the areas of the tissue blocks were measured again using point counting (Fig. 6.1(k)). The processing-induced volume shrinkage for one brain was estimated as

$$\text{Volume shrinkage} := 1 - \left(\frac{\sum A(\text{after})}{\sum A(\text{before})} \right)^{\frac{3}{2}} \tag{3.3}$$

where A(before) is the area of a tissue block before the processing, A(after) is the area of the same tissue block in the Epon. In this implementation of the physical disector, the final section thickness—and not only the block advance—was estimated. Therefore, one could ignore the difference between the cross-sectional area of the embedded tissue and the area of the same tissue on ultrathin sections—the section compression. This is because the section compression does not alter the reference volume, and therefore leaves the numerical density of synapses unchanged. In the present study, the processing-induced volume shrinkage (after fixation) of the brain neocortex was approximately 20%. As the estimates of the reference volume were performed before this shrinkage, the corresponding estimates of the numerical density (N_V) of synapses in the human brain neocortex had to be corrected for tissue shrinkage. The correction was performed by multiplying the numerical density of synapses by (1 – volume shrinkage), as shown in Table 6.3, eqn (b) (in Section 4).

3.8 Statistical analysis

Variability within groups was estimated using the observed coefficient of variation (CV=SD/Mean). The observed overall variation of the estimates among individual brains depends on the extent of sampling error at the various sampling levels and on the inherent biological variation, CV_{bio}, between individual brains. The biological coefficient of variation of the stereological estimates between brains was estimated from the calculated stereological sampling variation (the mean of the coefficient of error observed of the individual estimates, CE^2_{ste}) and the CV

$$CV^2 := CV^2_{bio} + CV^2_{ste}. \tag{3.4}$$

The total number of synapses in neocortex was calculated from the product of density estimates and reference volume estimates, that is, the total volume of neocortex. Therefore, the total intrabrain CE has to be calculated from these two-step analyses.

A systematic, random sample of slabs was used to estimate the neocortical volume. The systematic, random sampling in the present methods increased the efficiency of sampling compared to simple random independent sampling (Gundersen et al. 1999); CE[V(cortex)] = 0.029. The estimation of the numerical density of synapses in the neocortex is a ratio estimator; therefore, the equation in Kroustrup and Gundersen (1983) was used to estimate their CE.

The CE for the estimation of the numerical density of synapses in neocortex, CE[N_V(syn/reg)], was calculated according to the formula

$$CE_n\left[\frac{\sum Y}{\sum X}\right] := \sqrt{\frac{n}{n-1}\left[\frac{\sum X^2}{(\sum X)^2} + \frac{\sum Y^2}{(\sum Y)^2} - \frac{2\sum(XY)}{\sum X\sum Y}\right]} \tag{3.5}$$

where ΣY and ΣX are ΣQ^-(syn) and Σv(dis), respectively. The number of blocks studied is n.

The CE estimated for the volume estimations of neocortex and that estimated for the density estimates of synapses in neocortex were added in the usual way to obtain the total CE (Brændgaard *et al.* 1990; Tang and Nyengaard 1997). The CE for the estimation of the total number of synapses in each neocortical region was obtained using the formula

$$\mathrm{CE}\big[N(\mathrm{syn,\ reg})\big] := \sqrt{\mathrm{CE}^2\big[N_V(\mathrm{syn/reg})\big] + \mathrm{CE}^2\big[V(\mathrm{reg})\big]}. \qquad (3.6)$$

When estimating the total variance for the product of density and reference volume, the variances of both parameters and their co-variance at the level of individuals should be taken into account. In the present small study, the co-variance has been ignored because we do not expect and have no evidence of co-variation between the density and the reference volume.

When estimating the numerical density of synapses in the neocortex, the mean section thickness used was obtained by measuring two adjacent sections and measuring one or more folds in all the sections. CE[t(hemi)] is the uncertainty as to how well the mean thickness obtained by measuring folds reports the real thickness of the section. The coefficient of error for the estimate of the total number of synapses in neocortex per hemisphere, CE[N(syn, hemi)], was calculated according to the formula

$$\mathrm{CE}\big[N(\mathrm{syn,\ hemi})\big] := \sqrt{\mathrm{CE}^2\big[N_V(\mathrm{syn/hemi})\big] + \mathrm{CE}^2\big[V(\mathrm{hemi})\big] + \mathrm{CE}^2\big[t(\mathrm{hemi})\big]}. \qquad (3.7)$$

Again, this relationship assumes no co-variation between estimates of numerical density, reference volume, and section thickness per hemisphere.

As it turned out, the variance contribution from the measurement of section thickness was negligible because section thickness was measured on all sections, and the variance contribution from the volume estimate was also negligible.

4 Results

4.1 Number of synapses

We found that the synaptic densities in the occipital, parietal, temporal, and frontal regions were quite similar, approximately 0.3 synapses per μm^3 (see Tables 6.2 and 6.3). The total number of synapses in the neocortex of young male brains was 164×10^{12} with a coefficient of variation CV = 0.17 (Tang *et al.* 2001).

Table 6.2 The results—number of synapses as a mean of five human brains*

Region	V(neo) (cm³)	CE (%)	ΣQ^-(syn)	ΣQ(syn)	h_{ave}(dis) (nm)	N_V(syn) (μm⁻³)	CE (%)	N(syn) (10¹²)	CE (%)
Occipital	69 (19)	4.9	64		64	0.298 (61)		22.0 (84)	
Parietal	149 (5)	4.1	75		62	0.276 (27)		41.5 (30)	
Temporal	133 (16)	4.6	82		62	0.361 (18)		42.0 (23)	
Frontal	239 (12)	5.6	67		64	0.244 (13)		58.9 (24)	
Total	590 (9)	2.7	288	525	63	0.278 (12)	14	164 (17)	14

* V(neo), total volume of neocortex; ΣQ^-(syn), total number of synapses counted in disectors; ΣQ(syn), total synaptic profiles counted in reference and sampling sections; h_{ave}(dis), average height of disectors; N_V(syn), numerical density of synapses of neocortex; N(syn), total number of synapses in neocortex. The coefficient of error of individual estimates, CE, is shown next to the quantity it refers to. The numerical density of synapses was corrected for the tissue shrinkage induced by tissue processing (not including fixation). Coefficients of variations (in %) are shown in parentheses.

4.2 Analyzing and optimizing the sampling scheme

Stereological sampling of synapses and of brain neocortex involves a hierarchy of sampling levels, including the number of fields of view or the number of disectors, tissue blocks, individual brains, and groups of brains. In order to evaluate the appropriateness of the sampling scheme, it is necessary to estimate the variance contribution from all sampling levels to the total group variation among individual brains. For two adjacent sampling levels, the relative variance observed, OCV_j^2, at a higher, jth, level is the sum of the relative variance expected, CV_j^2, at that level and the mean relative variance observed at the level below, CE_{j-1}^2, that is, $OCV_j^2 = CV_j^2 + CE_{j-1}^2$. In planning the sampling scheme, stereological sampling at a lower level may be sufficient when the computable variance of a stereological estimate at this level, CE_{j-1}^2, is about half or less than half of the total variance at the higher level, OCV_j^2, that is, when CV_j^2 contributes most to the variation observed at the higher level.

After the contribution to the total variance observed is known for each sampling level the most efficient way to improve the precision of stereological estimates is to increase the amount of sampling at the level that contributes most to the overall variance if necessary.

As already stated in the statistical section of this chapter, the calculations of the coefficient of error (CE_{est}) for the estimates were made on the second lowest level of sampling, that is, the number of blocks. Therefore, eqn (3.4) in Section 3.8 is used (Table 6.3). In the present study, an average of 32 disectors were used and 288 synapses were counted per brain with a mean coefficient of error observed (CE_{est}) of 0.14. The coefficient of variation observed (CV) for the estimation of the total number of synapses in the neocortex was 0.17. That is, the stereological sampling

Table 6.3 Estimating the total number of the synapses in human neocortex and the coefficient of error of the estimate in a 21-year-old male individual

Block no.	Q^-(syn)	v(dis) (μm^3)
1	26	138
2	36	174
3	45	99
4	51	159
5	58	170
6	30	84
7	45	108
8	37	136
$n = 8$	ΣQ^-(syn) = 328	Σv(dis) =1068 μm^3

Equations used in calculations

(a) $N'_V\left(\text{syn/neocortex}\right) = \dfrac{\Sigma Q^-\left(\text{syn}\right)}{\Sigma v\left(\text{dis}\right)} = \dfrac{328}{1068\ \mu m^3} = 0.307\ \mu m^{-3}$

(b) N_V(syn/neocortex) = N'_V(syn/neocortex) $\cdot$ (1 − volume shrinkage) = 0.307 μm^{-3} $\cdot$ (1 − 0.22) = 0.239 μm^{-3}

(c) N(syn/neocortex) = N_V(syn/neocortex) $\cdot$ V(neocortex) = 0.239 μm^{-3} $\cdot$ (657 × 10^{12}) μm^3 = 157 × 10^{12}

(d)

$$CE\left[N'_V\left(\text{syn/neocortex}\right)\right] = \sqrt{\frac{n}{n-1}\left[\frac{\Sigma\left(Q^-\right)^2}{\Sigma Q^-\Sigma Q^-} + \frac{\Sigma\left(v\left(\text{dis}\right)\right)^2}{\Sigma v\left(\text{dis}\right)\Sigma v\left(\text{dis}\right)} - \frac{2\Sigma\left(Q^- v\left(\text{dis}\right)\right)}{\Sigma Q^-\Sigma v\left(\text{dis}\right)}\right]}$$

$$= \sqrt{\frac{8}{8-1}\left[\frac{14\ 256}{328\times328} + \frac{150\ 518}{1068\times1068} - 2\cdot\frac{44\ 688}{328\times1068}\right]}$$

$$= 0.10$$

(e) $CE\left[N\left(\text{syn/neocortex}\right)\right] = \sqrt{\left(CE^2\left[N_V\left(\text{syn/neocortex}\right)\right] + CE^2\left[V\left(\text{neocortex}\right)\right] + CE^2\left[t\left(\text{neocortex}\right)\right]\right)}$

$$= \sqrt{\left(0.10\right)^2 + \left(0.021\right)^2 + \left(0.034\right)^2} = 0.11$$

Notes on the equations

(a) N'_V(syn/neocortex) indicates the numerical density of the synapses in neocortex uncorrected for shrinkage. ΣQ^-(syn) is the total number of the synapses counted in the disectors and Σv(dis) is the total volume of the disectors, which is calculated from the area of the counting frame multiplied by the height of the disector. Notice that the number of blocks were not used as weights in this original calculation due to a slightly different sampling design (see details in Tang *et al.* 2001).

(b) N_V(syn/neocortex) is the numerical density of the synapses in neocortex corrected for shrinkage. This is obtained by multiplying the uncorrected numerical density by 1 − volume shrinkage (see eqn (3.3)).

Table 6.3 Cont'd

(c) The total number of the synapses in neocortex, N(syn, neocortex), is obtained by multiplying the numerical density of the synapses in neocortex by the total volume of cerebral neocortex, V(neocortex) (see eqn (3.2)). The value of V(neocortex), 657 cm³, was obtained from another study (Pakkenberg and Gundersen 1997).

(d) CE[N_V(syn/neocortex)] is the coefficient of error for the numerical density estimate of the synapses in neocortex. $n = 8$ is the number of blocks. ΣQ^- is the total number of the synapses counted in all disectors, and Σv(dis) is the total volume of the disectors used in 8 blocks. Since the estimate of the numerical density of the synapses in neocortex is a ratio estimate the one-pass equation from Kroustrup and Gundersen (1983) is used to estimate CE[N_V(syn/neocortex)].

(e) The overall coefficient of error for the estimation of the total number of the synapses in neocortex, CE[N(syn/neocortex)], is estimated by adding all the known variance contributions. CE[t(neocortex)] = SEM(t)/mean(t) indicates the variability of the section thickness estimate. CE[V(neocortex)] is the coefficient of error for the estimate of the neocortical volume, which was obtained from Pakkenberg and Gundersen (1997). Please note that for reasons of simplicity this example calculates the total number of neocortical synapses (and its CE) from the global number density and total volume of neocortex. In reality, as described under eqn (3.2), it was estimated as the sum of the estimates of the total number of synapses in the four regions.

---

variance contributes with $0.14^2/0.17^2 = 66\%$ of the total variance observed for the estimate of the total synapse number of the human brain neocortex. Improvement of the overall precision of the sampling scheme is possible by increasing the number of blocks studied. This will, however, increase the workload considerably as this is by far the most elaborate part of the procedure.

5 Discussion

5.1 Poor sampling design for estimation of the numerical density of synapses

Over the past four decades, numerous attempts have been made to estimate the amount of synapses in various regions of the central nervous system. However, of the previous studies, either two-dimensional density, that is, the number of synaptic profiles per unit area of sections, was estimated, or three-dimensional numerical density was calculated by means of assumption-based stereological methods based on a random section.

There are several inherent problems related to the estimation of two-dimensional synaptic profile density. The synaptic profile density fails to take account of the effects of the size, shape, and orientation of the synapses. For example, larger synapses have a higher probability of being counted on a random section and therefore greater effort in counting will only produce a very precise overestimate of these. If a pre- and postsynaptic density profile curves, a synapse may be intersected more than once by a single plane of section. These multiple intersections of a synapse may appear as apparently unrelated profiles in a single, thin section. On the other hand, it has been found that small or tangentially cut synaptic profiles, which are produced by sectioning of more than 20% of all synapses, cannot be reliably recognized in a random section due to their negligible size (Curcio and Hinds 1983; Geinisman *et al.* 1986*a*). This problem is even more compounded by the presence of the so-called complex perforated synaptic contacts (Calverley and Jones 1987*a, b*, 1990; Geinisman *et al.* 1986*b*). In a random section, a perforated synapse can mistakenly be counted as two or more non-perforated ones or not counted at all if the section passes through the interval between separate pre- and postsynaptic density segments.

The synaptic profile density ignores the effects of section thickness on the density estimate. The common assumption of previous studies is that random sections, from which the synaptic profile density is obtained, are true planes without thickness. Obviously this is not true; even the thickness of an extremely thin section can still be quite large in comparison with the size of synapses. The effect of non-zero section thickness on the estimation is called overprojection or Holmes effect (Cruz-Orive 1983). Finally, the synaptic samples obtained from random sections are also biased by an actual or apparent disappearance of small synaptic profiles from sections due to lost caps or lack of resolution (truncation effects; Brændgaard and Gundersen 1986; De Groot and Bierman 1986; Gundersen 1986). Therefore, the number of synaptic profiles seen in one section has no known relationship to the number of synapses in a three-dimensional reference space of volume.

5.2 Model-based estimation of the numerical density of synapses

Some investigators have used statistical and geometrical methods to transform the number of synaptic profiles per area into the number of synapses per volume. The principles underlying model-based stereological methods have relied on several unverified and unverifiable assumptions regarding the size, shape, and orientation distribution of the particles studied. In general, the synapses have usually been treated by applying a 'best fitting' geometric form, which was generally accepted to be a convex, flat, circular disk (Mayhew 1979; West *et al.* 1972). There are no flaws in the reasoning behind these studies. The difficulty is that the assumptions are approximate and unrealistic (De Groot and Bierman 1986). An increasing amount of data has been reported in the literature indicating that synapses have much more complex shapes than that of a simple disk (Calverley and Jones 1987*a, b*; Calverley *et al.* 1988; De Groot 1985; De Groot and Bierman 1987; Joo *et al.* 1987). Synapses may even exhibit a curvature (Dyson and Jones 1980). The conclusions will be biased to the extent that the approximations diverge from reality, and the effects of these potential biases on the results of synapse count are impossible to evaluate from the data obtained.

5.3 Estimation of the numerical density of synapses with the disector

Since 'number' is a zero-dimensional structural characteristic, the only way it can be sampled and estimated unbiasedly is in three dimensions, that is, using a sampling volume. A simple and unbiased three-dimensional probe is the disector (Sterio 1984). The central idea of the disector is that arbitrarily shaped particles may be counted unambiguously by using at least two consecutive section planes. The probability that a particle is hit by the sampling section plane, but not by the parallel look-up section plane, is the same for both large and small particles. Therefore, the disector method samples isolated particles with a uniform probability in a three-dimensional space. Furthermore, it does not require assumptions about the size, shape, and orientation of the particles counted. The orientation of the section planes can be chosen completely arbitrarily in the disector, and the estimation is independent of truncation and overprojection in addition to the fact that nothing is assumed about the particles counted. By using, alternatively, both sections of the disector, the efficiency of the estimate is doubled, since different particles are being sampled.

The disector has been used to count synapses in the brain of various experimental animals (Calverley and Jones 1987a; Calverley et al. 1988; De Groot and Bierman 1986, 1987; De Groot et al. 1995; Fukui and Bedi 1991; Geinisman et al. 1991, 1992, 1996; Hunter and Stewart 1989; Kleim et al. 1996; Plummer and Behan 1993; Siklos et al. 1990). However, the estimation of the numerical density of synapses using only two adjacent sections may be complicated by an identification problem when complex (perforated) synapses are present. One complex synapse may give rise to two or more profiles in a section plane. It was reported that, in the stratum pyramidale of the hippocampus CA3 area of four rats, 10–20% of synapses appeared to be perforated (De Groot and Bierman 1983). In our study only a few per cent of the sampled synapses split into two profiles on the adjacent section. We have corrected our number estimate for this event in accordance with the explanation in Fig. 6.2.

5.4 The importance of estimating total quantity

Previously, only synaptic densities in various parts of the brain have been reported. As already mentioned, those previous studies on synapses either estimated two-dimensional numerical density, that is, the synaptic profiles per unit area of sections, or calculated three-dimensional numerical density by means of assumption-based stereological methods. It is only recently that the total number of synapses in a region of rabbit hippocampus has been estimated (Geinisman *et al.* 1996). The density estimates serve only as an intermediate observation that allows the estimation of the total quantity. When reporting the density estimates only, the possible age- and disease-related changes in the reference space, that is, the total volume of neuropil, were not taken into account. Shrinkage of the reference volume may happen during tissue processing, and this shrinkage may be unequal between young and aged individuals, as well as between control subjects and diseased individuals. In addition, modifications of preparative procedures could differentially alter the reference space dimensions. Biological conclusions based on density measurements are very difficult to interpret since it will never be known whether any changes in the density are caused by an alteration of the total quantity and/or a change in the reference volume. It is possible that the densities increase due to shrinkage of the reference volume, although the total quantity actually decreases. For example, possible shrinkage of cortical neuropil in patients with Alzheimer's disease has been reported (Hubbard and Anderson 1981, 1985; Mann *et al.* 1985; Miller *et al.* 1980), which may have masked the synaptic density change. It is likely, therefore, that the widespread use of density estimates has contributed to the apparent contradictions in the earlier literature.

One of the more important features of the present methods is that they are designed to provide an estimate of the total quantities. The total number of synapses in the brain neocortex is obtained by multiplying the density estimates with the volume of neocortex. The volume of neocortex is the so-called reference volume estimated with the Cavalieri principle. This methodology eliminates the questionable assumption implicit in comparisons of density data, that is, that the volumes of regions compared are the same. Therefore, our results are the estimates of total quantities of synapses in brain neocortex and can be interpreted unambiguously.

5.5 The post-mortem changes in the numerical density of synapses in the brain neocortex

The present study is based on autopsy materials. In quantitative studies using post-mortem tissue, various difficulties may arise, such as the influence of terminal illnesses, the agonal duration, and variation in post-mortem conditions.

For the quantitative study of synapses in human autopsy brains, a crucial problem is the resistance of the synapses to the post-mortem autolysis. The two-dimensional density of synaptic profiles in the dorsal motor–sensory cortex of the rat during varying periods of post-mortem fixation delay was estimated by means of a model-based method based on the typical structural characteristic of synapses (Petit and LeBoutillier 1990). They found that, in the E-PTA-stained tissue, synaptic profile densities showed an initial non-significant slow steady drop beginning within/after 1 hour, a significant drop after 6 hours, and dropped approximately 30–40% of their initial levels after 10–15 hours post-mortem. In contrast, in the osmium-stained tissue, synaptic counts showed an initial marked and significant drop after 1 hour. Following this initial drop, there was little change in synaptic counts until 15 hours post-mortem at which time there was another marked drop. Therefore, they postulated that the initial drop in osmium tissue may partly be due to the characteristics of the osmium staining technique. Following the initial post-mortem period, the decrease in the number of 'osmium' synapses approximately paralleled the increase in non-vesicular contacts, suggesting that the reduction of the number of synapses may primarily be due to the loss of synaptic vesicles. The results suggested that the requirement of the presence of one or more synaptic vesicles in order to count a synapse (a rule generally applied in quantitative studies of synapses) may not be useful in research on autopsy material. In the present study using E-PTA-stained tissue, the numerical density of synapses in the neocortices of five big animal brains was estimated by means of the disector technique. The mean difference of the synaptic numerical density between 'in life' and 2 days post-mortem fixation delay was 3.9%. From the data obtained, it would appear that the synapses of brain neocortex can be analyzed with a reasonable degree of accuracy in tissue fixed within a few days after death (Tang *et al.* 2001).

5.6 Tissue shrinkage

In order to obtain estimates of the total number of synapses in the brain neocortex, the density estimates have to be multiplied by the volume of the reference space, that is, the total volume of neocortex. In order to verify the assumption that the reference volume is identical for both estimates, any change in tissue volume between the tissue used for the Cavalieri estimation of volume and the density estimation should be estimated. It has been shown that embedding in agar does not affect the brain volume (Pakkenberg and Gundersen 1997; Bente Pakkenberg, personal communication). Paraffin is the embedding medium that induces the largest shrinkage, about 40% by volume (Iwadare *et al.* 1984), while epon and other plastic embedding materials (such as glycolmethacrylate) induce much less shrinkage—0–10% by volume (Gerrits *et al.* 1987; Brændgaard *et al.* 1990).

The total synapse numbers estimated were unaffected by shrinkage due to formalin fixation because both the neocortical volume and the numerical density of synapses in neocortex was estimated from formalin fixed brains. Hence, the only assumption was that no synapses disappeared from the neocortex during formalin fixation. Any change in tissue volume between the tissue used for the Cavalieri estimation of neocortical volume and the estimation of numerical density of synapses in neocortex should be checked. The shrinkage of the reference space due to tissue processing for electron microscopy is known to be unequal in young and aged individuals (Haug 1985) and modifications of preparative procedures could differentially alter the dimensions of the reference space (Geinisman *et al.* 1992). In the present study, we embedded the tissue in epon and stained synapses using the E-PTA method. We found that the mean volume shrinkage induced by the histologaical processing was 22%. This change is not negligible. The numerical density of synapses was corrected for the volume shrinkage of the reference space. In previous studies, the shrinkage of the reference space due to tissue processing for electron microscopy was not taken into account.

5.7 Alternative methods for synapse number estimation

At present, there are other methods available that can be used to estimate the total number of particles, that is, the fractionator, the selector, and the double disector. In general, the fractionator technique (Gundersen 1986) will be the most attractive in order to obtain an unbiased estimate of the number of arbitrarily shaped particles without taking shrinkage into account. However, the fractionator involves the determination of the number of synapses (with disectors) in a known fraction of the regional volume. Consequently, when using fractionator sampling it is important to keep track of all sections cut, in order to determine the fraction of the region quantitated. This approach would be very demanding due to extensive subsampling and exhaustive ultramicrotomy inherent in the application of the fractionator at the EM level. This technique is therefore not the preferred method for estimating the total number of synapses. In principle, the selector (Cruz-Orive 1987) is a disector of unknown thickness. Instead of measuring the volume of the disector, the ratio of the mean particle volume and the reference volume is estimated. In the case of synapses, this technique is very difficult to use since synapses do not have a clear boundary and have a very small dimension in one direction, namely, perpendicular to the pre- and post-synaptic densities, and therefore synaptic volume cannot be reliably estimated. The double disector (Marcussen 1992) is able to obtain estimates of the number of particles inside or associated with other particles without requiring knowledge of the section thickness. More importantly, the estimate is unaffected by tissue shrinkage. The reason why we did not choose the double disector for this study is that we could not make certain that our criteria to distinguish between glia cells and small neurons would be similar at the light microscopical level and at the electron microscopical level.

References

Brændgaard, H. and **Gundersen, H.J.G.** (1986). The impact of recent stereological advances on quantitative studies of the nervous system. *J. Neurosci. Meth.* **18**, 39–78.

Brændgaard, H., Evans, S.M., Howard, C.V., and **Gundersen, H.J.G.** (1990). The total number of neurones in the human neocortex unbiasedly estimated using optical disectors. *J. Microsc.* **157**, 285–304.

Calverley, R.K.S. and **Jones, D.G.** (1987*a*). A serial-section study of perforated synapses in rat neocortex. *Cell Tissue Res.* **247**, 565–72.

Calverley, R.K.S. and **Jones, D.G.** (1987*b*). Determination of the numerical density of perforated synapses in rat neocortex. *Cell Tissue Res.* **248**, 399–407.

Calverley, R.K.S. and **Jones, D.G.** (1990). Contributions of dendritic spines and perforated synapses to synaptic plasticity. *Brain Res.. Rev.* **15**, 215–49.

Calverley, R.K.S., Bedi, K.S., and **Jones, D.G.** (1988). Estimation of the numerical density of synapses in rat neocortex. Comparison of the 'disector' with an 'unfolding' method. *J. Neurosci. Meth.* **23**, 195–205.

Cruz-Orive, L.M. (1983). Distribution-free estimation of sphere size distributions from slabs showing overprojection and truncation, with a review of previous methods. *J. Microsc.* **131**, 265–90.

Cruz-Orive, L.M. (1987). Particle number can be estimated using a disector of unknown thickness: the selector. *J. Microsc.* **145**, 121–42.

Curcio, C.A. and **Hinds, J.W.** (1983). Stability of synaptic density and spine volume in dentate gyrus of aged rats. *Neurobiol. Aging* **4**, 77–87.

De Groot, D.M.G. (1985). Disc-like and complex-shaped synapses: number, size and dense projections. A critical note. *Acta Stereol.* **4**, 147–51.

De Groot, D.M.G. and **Bierman, E.P.B.** (1983). The complex-shaped 'perforated' synapse, a problem in quantitative stereology of brain. *J. Microsc.* **31**, 355–60.

De Groot, D.M.G. and **Bierman, E.P.B.** (1986). A critical evaluation of methods for estimating the numerical density of synapses. *J. Neurosci. Meth.* **18**, 79–101.

De Groot, D.M.G. and **Bierman, E.P.B.** (1987). Numerical changes in rat hippocampal synapses. An effect of 'ageing'? *Acta Stereol.* **6**, 53–8.

De Groot, D.M.G., Bierman, E.P.B., Bruijnzeel, P.L.B., and **Woutersen, R.A.** (1995). The 'disector', a tool for quantitative assessment of synaptic plasticity. An example on hippocampal synapses and synapse-perforations in ageing rat. *Eur. J. Morphol.* **33**, 305–9.

Dyson, S.E. and **Jones, D.G.** (1980). Quantitation of terminal parameters and their inter-relationships in maturing central synapses: a perspective for experimental studies. *Brain Res.* **183**, 43–59.

Foster, M. and **Sherrington, C.S.** (1897). *A text book of physiology*, Part III: *The central nervous system*, 7th edn. Macmillan, London.

Fukui, Y. and **Bedi, K.S.** (1991). Quantitative study of the development of neurones and synapses in rats reared in the dark during early postnatal life. 1. Superior colliculus. *J. Anat.* **174**, 49–60.

Geinisman, Y., de Toledo Morrell, L., and **Morrell, F.** (1986*a*). Loss of perforated synapses in the dentate gyrus: morphological substrate of memory deficit in aged rats. *Proc. Natl Acad. Sci., USA* **83**, 3027–31.

Geinisman, Y., de Toledo Morrell, L., and **Morrell, F.** (1986*b*). Aged rats need a preserved complement of perforated axospinous synapses per hippocampal neuron to maintain good spatial memory. *Brain Res.* **398**, 266–75.

Geinisman, Y., de Toledo Morrell, L., and **Morrell, F.** (1991). Induction of long-term potentiation is associated with an increase in the number of axospinous synapses with segmented postsynaptic densities. *Brain Res.* **566**, 77–88.

Geinisman, Y., deToledo-Morrell, L., Morrell, F., Persina, I., and **Rossi, M.** (1992). Age-related loss of axospinuous synapses formed by two afferent systems in the rat dentate gyrus as revealed by the unbiased stereological disector technique. *Hippocampus* **2** (4), 437–44.

Geinisman, Y., Gundersen, H.J.G., van der Zee, E., and **West, M.J.** (1996). Unbiased stereological estimation of the total number of synapses in a brain region. *J. Neurocytol.* **25**, 805–19.

Gerrits, P.O., Van Leeuwen, M.B., and **Boon, M.E.** (1987). Floating on a water bath and mounting glycolmethacrylate and hydroxypropyl methacrylate sections influence final dimensions. *J. Microsc.* **145**, 107–13.

Gundersen, H.J.G. (1977). Notes on the estimation of the numerical density of arbitrary profiles: the edge effect. *J. Microsc.* **111**, 219–23.

Gundersen, H.J.G. (1986). Stereology of arbitrary particles. A review of unbiased number and size estimator and presentation of some new ones, in memory of William R. Thompson. *J. Microsc.* **143**, 3–45.

Gundersen, H.J.G., Jensen, E.B.V., Kiêu, K., and **Nielsen, J**. (1999). The efficiency of systematic sampling in stereology—reconsidered. *J. Microsc.* **193** (3), 199–211.

Haug, H. (1985). Gibt es Nervenzellverluste waehrend der Aelterung in der menschlicher Hirnrinde? Ein morphometrischen Beitrag zu dieser Frage. *Nervenheilkunde* **4**, 103–9.

Hubbard, B.M. and **Anderson, J.M.** (1981). A quantitative study of cerebral atrophy in old age and senile dementia. *J. Neurol. Sci.* **50**, 135–45.

Hubbard, B.M. and **Anderson, J.M.** (1985). Age-related variations in the neuron content of the cerebral cortex in senile dementia of Alzheimer type. *Neuropathol. Appl. Neurobiol.* **11**, 369–82.

Hunter, A. and **Stewart, M.G.** (1989). A quantitative analysis of the synaptic development of the lobus parolfactorius of the chick (*Gallus domesticus*). *Exp. Brain Res.* **78**, 425–34.

Iwadare, T., Mori, H., Ishiguro, K., *et al.* (1984). Dimensional changes of tissues in the course of processing. *J. Microsc.* **136**, 323–7.

Joo, F., Siklos, L., Dames, W., and **Wolff, J.R.** (1987). Fine-structural changes of synapses in the superior cervical ganglion of adult rats after long-term administration of GABA. A morphometric analysis. *Cell Tissue Res.* **249**, 267–75.

Kleim, J.A., Lussnig, E., Schwarz, E.R., Comery, T.A., and **Greenough, W.T.** (1996). Synaptogenesis and Fos expression in the motor cortex of the adult rat after motor skill learning. *J. Neurosci.* **16**, 4529–35.

Kroustrup, J.P. and **Gundersen, H.J.G.** (1983). Sampling problems in an heterogeneous organ: quantitation of relative and total volume of pancreatic islets by light microscopy. *J. Microsc.* **132**, 43–55.

Mann, D.M.A., Yates, P.O., and **Marcyniuk, B**. (1985). Some morphometric observations on the cerebral cortex and hippocampus in presenile Alzheimer's disease, senile dementia of Alzheimer type and Down's syndrome in middle age. *J. Neurol. Sci.* **69**, 139–59.

Marcussen, N. (1992). The double disector: unbiased stereological estimation of the number of particles inside other particles. *J. Microsc.* **165**, 417–26.

Mayhew, T. (1979). Stereological approach to the study of synapse morphometry with particular regard to estimating number in a volume and on a surface. *J. Neurocytol.* **8**, 121–38.

Miller, A.K.H., Alston, R.L., and **Corsellis, J.A.N.** (1980). Variation with age in the volume of grey and white matter in the cerebral hemispheres of man: measurements with an image analyser. *Neuropathol. Appl. Neurol.* **6**, 119–32.

Pakkenberg, B. and **Gundersen, H.J.G.** (1997). Neocortical neuron number in humans: effect of sex and age. *J. Comp. Neurol.* **384**, 312–320.

Peters, A., Palay, S.L., and **Webster, H.D.** (1991). *The fine structure of the nervous system*, 3rd edn. Oxford University Press, New York.

Petit, T. and **LeBoutillier, J.** (1990). Quantifying synaptic number and structure: effects of stain and post-mortem delay. *Brain Res.* **517**, 269–75.

Plummer, K. and **Behan, M.** (1993). Development of corticotectal synaptic terminals in the cat: a quantitative electron microscopic analysis. *J. Comp. Neurol.* **338**, 458–74.

Siklos, L., Parducz, A., Halasz, N., Rickmann, M., Joo, F., and **Wolff, J.R.** (1990). An unbiased estimation of the total number of synapses in the superior cervical ganglion of adult rats established by the disector method. Lack of change after long-lasting sodium bromide administration. *J. Neurocytol.* **19** (4), 443–54.

Small, J.V. (1968). Measurement of section thickness. Fourth European Conference on Electron Microscopy 1968, pp. 609–10.

Sterio, D.C. (1984).The unbiased estimation of number and sizes of arbitrary particles using the disector. *J. Microsc.* **134**, 127–36.

Tang, Y. and **Nyengaard, J.R.** (1997). An unbiased stereological method for estimating the total length and the size distribution of the myelin fibres in human brain white matter. *J. Neurosci. Meth.* **73**, 193–200.

Tang, Y., Nyengaard, J.R., De Groot, D.M.G., and **Gundersen, H.J.G.** (2001). Total regional and global number of synapses in the human brain neocortex. *Synapse* **41** (3), 258–73.

West, M.J., Coleman, P.D., and **Wyss, U.R.** (1972). A computerised method of determining the number of synaptic contacts in a volume of cerebral cortex. *J. Microsc.* **95**, 277–83.

Williams, M.A. (1981). Section of determined thickness for use in stereological estimations on cells. *Stereol. Jugosl.* **3**, 369–74.

CHAPTER 7

THE NUMBER OF MICROVESSELS ESTIMATED BY AN UNBIASED STEREOLOGICAL METHOD APPLIED IN A BRAIN REGION

ANNEMETTE LØKKEGAARD

1 Introduction

The microvascular network provides the means by which tissues exchange nutrients and are supplied with oxygen. The microvascular network is crucial for the function of tissues and organs, and the structure and density of the microvascular network is closely related to the metabolic demand and function of the tissue.

The microvascular network is influenced by both physiological and pathological changes. Changes in the structure (i.e. length, diameter, surface area, density, or number) of the vascular network may reflect important changes in the tissues, and have relevance for the understanding of physiological and pathological states (Secomb 1995; Tufto and Rofstad 1998; Jain 1999).

Angiogenesis is an important descriptor of changes in tissues, especially in pathological states where pro-angiogenic or anti-angiogenic therapy has become a possibility. Angiogenesis as a determining factor in the development of cancer has prompted a search for efficient anti-angiogenic treatments. A reliable quantification and evaluation of changes in the microvascular network is therefore essential (Vermeulen *et al.* 2002).

A description of microvascular networks can also be important in physiological states. In the human placenta the development of the microvascular network has been described as different for non-smoking and smoking mothers (Bush *et al.* 2000).

A description of the microvascular network often includes the length, surface area, diameter, and 'number density' (i.e. the number of profiles per area in the microscope—a measure that can be used to describe the length of the vessels and not an estimate of the number of microvessels). The number of branches or connections has usually not been estimated. However, an investigation of the type of angiogenesis is important for the understanding of physiological or pathological states, namely, whether the angiogenesis involves branching or elongation of vessels, sprouting or intussusceptive growth. Also, the number of microvessels in a possibly non-expanding network is important. The number of branches in microvascular networks can describe important differences between different types of tissue. Therefore both the length and the number of connections in the network describe the property and function of the microvasculature.

2 A stereological description of the structure of vascular networks

The vascular network is a network of interconnected vessels (tubes), which are much longer than they are wide. The number of microvessels is much larger than the number of larger vessels. The network is extensive in many organs and, in general, one large artery supplies a very large number of microvessels. Properties such as length, number of connections, number of vessels, and the diameter and surface of these vessels can be used to describe the microvascular network. The microvascular network can be considered to be an interconnected structure. It makes no sense to imagine vessels not connected to circulation.

The topology is a description of the structure that does not change with deformation or transformation of the structure (Kroustrup and Gundersen 2001). It describes the interconnectedness of the structure. The Euler number is the number of redundant connections in the topological entity. Or, in the vascular network, the number of connections minus one. The Euler number can be estimated using modern stereological methods, thereby providing an unbiased estimate of the number of vessels in the network.

The Euler number is given by

Euler number = number of objects + number of cavities – number of redundant connections.

In a microvascular network there are no isolated parts in the network that are not connected to the circulation. Also, it is reasonable to assume that there are no holes within microvessels, since a vessel is a tube. That leaves us with

Euler number = 1 + 0 − number of redundant connections.

The number of redundant connections can be estimated from two-dimensional sections in three-dimensional space, either using physical disectors, in which changes from one thin section to an adjacent thin section can be found, or using the optical disector, in which changes are observed while scanning through a thick section. The features observed and counted in the sections are different, but will ultimately give the same stereological estimate of number.

In the physical disector the observed events seen in the network that can be used to describe the Euler number are

Euler number = number of islands + number of holes − number of bridges
(Fig. 7.1).

An unbiased estimator of the number of microvessels has been presented (Nyengaard and Marcussen 1993).

The connectivity of the microvascular network can also be estimated using optical disectors. The changes in the network are not seen in going from one section to the next; the changes are observed while scanning through thick sections of tissue (Fig. 7.1). The changes observed are the branching of microvessels. In accordance with the description of the Euler number as an estimate of the number of redundant connections in the microvascular network using the physical disector, the number of redundant connections can be estimated using the optical disector. The number of redundant connections is the number of branches that can be removed without creating new separate parts. Where a microvessel is branching there is a node. From the number of nodes and branches that meet in the nodes, the number of redundant connections can be estimated, that is, the number of nodes and the number of segments joining in the nodes describe the connectivity of the network. The number of microvessels is thus estimated from the direct counting of the nodes and segments of the network in the optical disector.

3 Method

The number of microvessels can be estimated using the optical disector in either a fractionator design sampling scheme or using a disector/Cavalieri type sampling scheme.

In a fractionator design a systematically random sampled known fraction of the tissue is sampled with disectors, and an unbiased estimate of the number of microvessels for this known fraction of the whole organ or region is estimated. Thereby

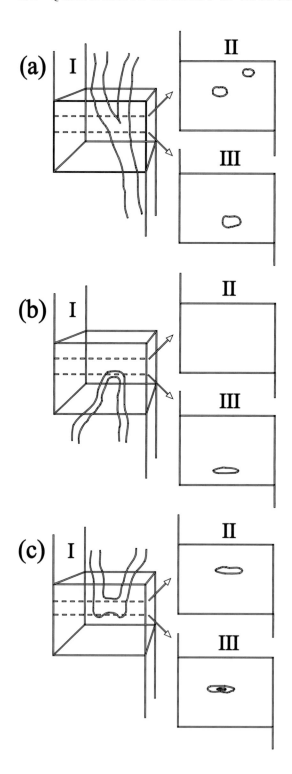

Fig. 7.1 (a) I, The optical disector: counting of one node. II and III, The physical disector: the counting of one bridge from the reference section (II) to the look-up section (III). One connection was counted. (b) I, The optical disector: no nodes are counted. II and III, The physical disector: one island was counted from the reference section (II) to the look-up section (III). (c) I, The optical disector: no nodes are counted. II and III, The physical disector: one hole was counted from the reference section (II) to the look-up section (III).

the estimate for the entire region can be deduced. When using some fractionator designs the region or the organ of interest must be cut exhaustively to provide a known fraction of tissue. This may prohibit the use of the tissue in other settings or protocols. However, a huge advantage is the ability to totally neglect shrinkage or deformation of the tissue, since the estimate is valid for the known fraction of tissue before processing.

In the example presented here, a Cavalieri type sampling scheme was used. The total number, $W(\text{cap})$, of microvessels was therefore estimated by multiplying the estimates of the total volume, $V(\text{ref})$, of the region of interest by the estimates of the numerical density, $W_V(\text{cap/ref})$, of microvessels in the region, $W(\text{cap})$ = $W_V(\text{cap/ref}) \cdot V(\text{ref})$.

3.1 Estimation of volume

The reference volume can be estimated using the Cavalieri principle (Gundersen and Jensen 1987). The analysis must be made on a uniform random sample of the volume of interest. The sections are sampled systematically throughout the entire extent of the region, ensuring that all parts of the tissue have an equal probability of being sampled. On each sampled section a grid with a tessellation of points is placed, and the number of points hitting the region of interest counted (Fig. 7.2

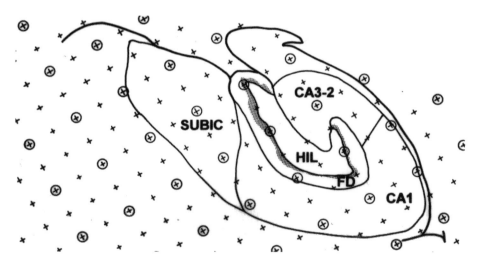

Fig. 7.2 A drawing of a transverse section of the dorsal-medial part of the temporal lobe illustrates the outline of the hippocampal region with the delineations between subdivisions. An example of the grid that was used for the Cavalieri estimation of volume has been superimposed. In practice the grid was placed on the histological section and point counting was performed directly on the section. Distance between points, 1 mm. (Reproduced with permission from Lokkegaard *et al.* (2001).) SUBIC (subiculum), CA1, CA3-2, HIL (Hilus), FD (Fascia deutata)

Table 7.1 Estimation of volume*

Section number	Number of points, P
1	0
2	0
3	10
4	37
5	27
6	20
7	17
8	18
9	63
10	4
Sum	196

* The total volume of interest, the reference volume, V(ref), was estimated with point counting. On each section of the hippocampus with fascia dentata a grid with a tessellation of points was placed, and the number of points hitting the fascia dentata counted. Each point represents a known fixed area, area per point a(p). From the total number of points P hitting the dentate gyrus and the area per point, the area of the fascia dentata on the sections was found, namely, A(ref) = a(p) · P. We used a tessellation of points with a(p) = 0.5 mm², so that A(ref) = 0.5 mm² · 196 = 98 mm².
The volume V(ref) was estimated from the known thickness of the slabs represented by each section. V(ref) = A(ref) · t = 98 mm² · 3.5 mm = 343 mm³.

and Table 7.1). An estimate of the area of the region, A, is obtained by multiplying the number of points, ΣP, with the area associated with each point, a(p). The reference volume V(ref) can then be estimated in accordance with the relationship, V(ref) = t · ΣP · a(p) = t · ΣA, where t represents the distance between sections.

3.2 Estimation of the number of microvessels

When counting the number of microvessels it is necessary to be able to define the beginning and end of each microvessel. The microvascular network can be described as a network with nodes and segments, in which the microvessels meet in nodes. The number of vessels is thus correlated to the number of nodes in which they end or begin, and to the number of joining vessels in each node. This correlation can be used to estimate the number of microvessels in the network. It is important to note that the number of segments between the nodes is not equal to the number of microvessels. The creation of one microvessel can result in the creation of one to three segments. This is shown in Fig. 7.3. The node is described

Microvessels	# Nodes		$\Sigma\left(\frac{n-2}{2}\right)\cdot P_n$	# Microvessels	# Segments
	P_n				
	$n = 3$	$n = 4$		W (cap)	
	0	0	0	1	1
	2	0	1	2	4
	2	1	2	3	6
	4	1	3	4	9

Fig. 7.3 An illustration of the relationship between the number of nodes and the number of microvessels in a network. The first drawing on the left illustrates a single microvessel. Each of the subsequent three drawings represents the formation of one new microvessel. In describing the properties of a microvascular network it is important to distinguish between the number of segments and the number of microvessels. The number of segments is different from the number of microvessels, and varies in relation to how the microvessels are connected. In the last drawing, for example, the number of microvessels increases by one and the number of segments increases by three. (Modified from Nyengaard *et al.* (1988).)

by the number of microvessels joining in the node; this number is called the valence of the node.

The optical disector (Gundersen 1986; Gundersen *et al.* 1988) can be used to esti-mate the number of nodes and segments in the network (Fig. 7.1). The unbiased counting rules can be applied to the counting of nodes, and in each counted node the number of segments can be registered. Every node that comes into focus within the unbiased counting frame (Gundersen 1977) will, as it is moved continuously through the distance h of the section thickness, be sampled (Fig. 7.4).

When using the unbiased counting rule it is important to have an unambiguous definition of when to count the objects of interest. A possible definition is to count the nodes when the cusp of the smallest angle comes into focus within the count-ing frame. A node can be considered to be a part of the microvascular network when at least one vessel joining in the node is a microvessel. The inner diameter is measured in each vascular segment that joins a sampled node and only nodes with at least one microvessel are counted (in the following example that is if one or more of the vessels has a diameter of $\leq 10\ \mu m$). For each counted node the total number of joining segments, that is, the valence, is registered (Table 7.2).

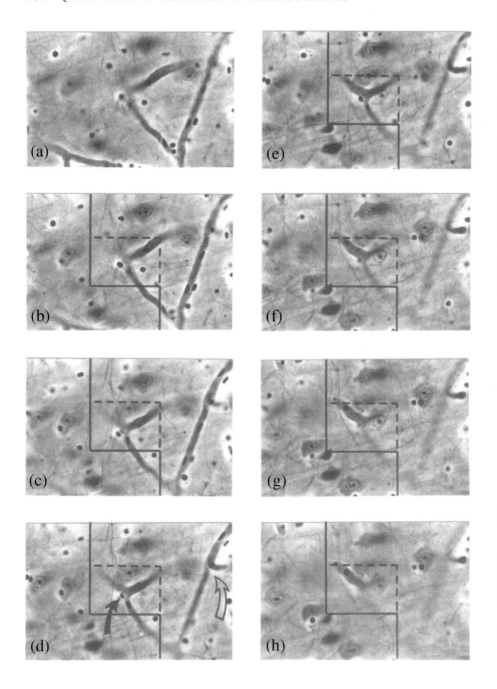

Fig. 7.4 This figure depicts an optical disector containing a capillary node with a valence of 3. (a)–(h) represent consecutive focal planes in the section, with a separation of 4 µm. Superimposed on each plane is an unbiased counting frame of known area, a(frame). Nodes are counted when the cusp of the smallest angle first comes into focus within the counting frame. This optical disector contains one node that is counted in (d) (black arrow) when starting at the first level (a), and proceeding to the last level, (h). One node coming into focus outside the counting frame in (d) is not counted (white arrow). The length of the horizontal line in the counting frame is equal to 100 µm. (Reproduced with permission from Lokkegaard *et al.* (2001).)

Table 7.2 Estimation of number of counted nodes in the fascia dentata*

N	$n = 3$	$n = 4$	$n = 5$	Disectors
1	0	0	0	0
2	0	0	0	0
3	14	0	0	175
4	41	2	0	578
5	26	0	0	397
6	18	0	0	300
7	18	1	0	315
8	26	0	0	293
9	84	3	1	1034
10	3	0	0	45
Sum: P_n	230	6	1	3137
$\frac{n-2}{2} \cdot P_n$	115	6	1.5	
$\Sigma(\frac{n-2}{2} \cdot P_n)+1$	123			

* The number of nodes P_n was counted in each subdivision of the hippocampal region. The valence n of each of the counted nodes was noted. In this table an example of the counted nodes in the fascia dentata is presented. Only nodes with valence 3, 4, and 5 were seen. A valence of three was far more common than the other two. The number of microvessels found in the volume of the disectors was $\Sigma(\frac{n-2}{2} \cdot P_n)+1 = 123$.

The volume of the disectors, v(dis), was estimated from the relationship, Σv(dis) $= \Sigma h \cdot a$(frame), in which h is the height of the disector and a(frame) is the area of the counting frame. $h = 20$ µm and a(frame) $= 0.00181$ mm² so Σv(dis) $= 3137 \cdot 20$ µm $\cdot 1813$ µm² $= 113\ 747\ 620$ µm³.

In order to find the total number of microvessels, W(cap), the number of microvessels in the sampled volume in the disectors was multiplied by the total reference volume, which was estimated using the Cavalieri method (see Table 7.1 and Chapter 8) as V(ref) $= 343$ mm³. The number of microvessels found in the total volume of the fascia dentata was

$$W\ (\text{cap}) = \frac{\Sigma(\frac{n-2}{2}) \cdot P_n + 1}{\Sigma v(\text{dis})} \cdot V(\text{ref}) = 123/0.114 \cdot 343 = 0.37 \text{ million.}$$

From the number of nodes and the valence of the nodes the number of microvessels, W(cap), can be estimated (see Fig. 7.3 and Table 7.2)

$$W_V\left(\text{cap/ref}\right) = \frac{\Sigma\left(\frac{n-2}{2}\right) \cdot P_n + 1}{v(\text{dis})}.$$

P_n is the number of nodes of valence n. Since both ends of the segments are being sampled and counted, we have to divide by the factor of 2.

The nodes are sampled in a volume of disectors. The volume of disectors, v(dis), can be estimated from the relationship

$$\Sigma v(\text{dis}) = \Sigma h \cdot a(\text{frame})$$

where h is the height of the disector and a(frame) is the area of the counting frame.

The total number of microvessels, W(cap), can then be estimated from the relationship

$$W(\text{cap}) = W_v(\text{cap/ref}) \cdot V(\text{ref})$$

where the density of the microvessels is multiplied by the total volume.

3.3 Sampling strategy

The precision of the estimates is of importance for the sampling strategy: the number of individuals to investigate and the amount of sampling necessary in these individuals.

The observed variation, OCV, of the group mean of the estimates can be found from the relationship

$$\text{OCV}^2 = \text{CV}_{\text{biol}}^2 + \text{OCE}^2$$

The mean biological variance between individuals, $\text{CV}_{\text{biol}}^2$, is deduced from the total relative observed variance, OCV^2, and the mean relative observed variance of the individual estimates, OCE^2. OCV^2 is calculated as $(\text{SD/mean})^2$.

The mean relative variance of the individual estimates, OCE^2, is dependent on the sampling strategy. The sampling scheme can be designed to ensure that OCE^2 is less than half the biological variance on each level of sampling. The sufficient amount of sampling can thus be determined so that a lower observed group variance OCV^2 would not have been achieved by additional sampling within individuals. For a description of the estimation of OCE^2 see Section 4 (Tables 7.3 and 7.4) and Gundersen *et al.* (1999).

4 Example

We have estimated the number of microvessels in a brain region. The hipppo-campal region is an extensively examined and well-described region in the human brain. It is well defined anatomically, and its functional importance has been sought on the basis of several brain pathologies in which it is known to be a part of the pathophysiological process. The function of the hippocampal region is known to be important in memory processes; neuronal loss in the hippocampal

Table 7.3 Precision of the estimate of the reference volume on the basis of OCE (Cavalieri)*

Section	P_i	A $P_i \cdot P_i$	B $P_i \cdot P_{i+1}$	C $P_i \cdot P_{i+2}$
1	0	0	0	0
2	0	0	0	0
3	10	100	370	270
4	37	1369	999	740
5	27	729	540	459
6	20	400	340	360
7	17	289	306	1071
8	18	324	1134	72
9	63	3969	252	
10	4	16		
Sum	196	7 196	3 941	2 972

* An evaluation of the precision of the sampling scheme was carried out on the basis of the estimates of OCE² and OCV². The mean OCE² was less than half of the value of the OCV². This indicates that the biological variance contributes more than the stereological variance to the observed group variance, which *a priori* was defined as the level of precision desired in this study.

The CE was estimated from the relationship

$$\text{CE}(\text{Cavalieri}) = \frac{\sqrt{\text{Var}(\text{noise}) + \text{Var}(\text{SURS})}}{\Sigma \text{Pi}}$$

where

$$\text{Var}(\text{noise}) = 0.07 \cdot \sqrt{\frac{b}{a}} \cdot \sqrt{n \cdot \Sigma P} = 0.07 \cdot 9 \cdot \sqrt{10 \cdot 196} = 27.9$$

and

$$\text{Var}(\text{SURS}) = \frac{3(A - \text{Var(noise)}) - 4B + C}{240} = \frac{3(7196 - 27.9) - 4 \cdot 3941 + 2972}{240} = \frac{8712}{240} = 36.3$$

(SURS stands for 'systematic uniform random sampling'.) Therefore,

$$\left(\text{CE}(\text{Cavalieri})\right) = \frac{\sqrt{27.9 + 36.3}}{196} = 0.0$$

Table 7.4 Precision of the estimate of the number of nodes on the basis of OCE (Disector)

N	Q^-	A $Q_i^- \cdot Q_i^-$	B $Q_i^- \cdot Q_{i+1}^-$	C $Q_i^- \cdot Q_{i+2}^-$
1	0	0	0	0
2	0	0	0	0
3	7	49	157.5	91
4	22.5	506.25	292.5	202.5
5	13	169	117	130
6	9	81	90	117
7	10	100	130	465
8	13	169	604.5	19.5
9	46.5	2 162.25	69.75	
10	1.5	2.25		
Sum	122.5	3 238.75	1 461.25	6 721.25

* These calculations were performed as follows.

$$CE(\text{Disector}) = \frac{\sqrt{\text{Var}(\text{noise}) + \text{Var}(\text{SURS})}}{\Sigma Q^-}$$

where $\text{Var}(\text{noise}) = \Sigma Q^- = 123$ and

$$\text{Var}(\text{SURS}) = \frac{3(A - \text{Var}(\text{noise})) - 4B + C}{240} = \frac{3(3239 - 123) - 4 \cdot 1461 + 1025}{240} = \frac{4529}{240}$$

$$= 18.9$$

Therefore,

$$CE(\text{Disector}) = \frac{\sqrt{123 + 43.79}}{123} = 0.10.$$

The combined CE for the total number of microvessels is
$$CE\ (W(\text{cap})) = \sqrt{CE^2(\text{Cavalieri} + CE^2(\text{Disector})} = \sqrt{0.04^2 + 0.10^2} = 0.11.$$

region has been found in, for instance, Alzheimer's disease and temporal lobe epilepsy.

The hippocampal region contains five subdivisions: fascia dentata, hilus, CA3–2, CA1, and subiculum. In this example we have estimated the total number, $W(\text{cap})$, of microvessels in fascia dentata.

4.1 Material

The material used in this analysis was obtained from a 52-year-old male with no known history of neurological disease. The brain was immersion fixed in formaldehyde (4%) after autopsy. The hippocampal region in the left temporal lobe was embedded in 6% agar following free dissection. The entire rostrocaudal extent of

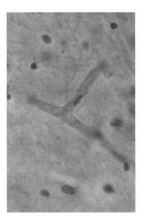

Fig. 7.5 A picture of a Bielchowsky silver-stained section. The vessel is stained unambiguously and the endothelial cells can be seen.

the hippocampal region was cut uniformly random into 3.5 mm slabs. To ensure systematic random sampling, the first cut was placed randomly within the first 3.5 mm interval of the region. This procedure ensured that all parts of the hippocampal region had an equal probability of being sampled. The resulting 10 slabs were dehydrated in graded alcohol solutions of increasing concentration. The slabs were subsequently embedded in glycolmethacrylate.

The sections from the rostral surfaces of each slab were cut with an LKB Historange microtome, at a setting of 40 µm, mounted on glass and stained with the Bielchowsky silver method. This staining procedure ensured evenly stained vessels throughout the entire depth of the sections. The walls of the vessels and the endothelial cells were stained unambiguously (Fig. 7.5).

4.2 Delineation of the sampled area: the hippocampal subdivisions

In order to recognize the borders of the area to be sampled it is important to have appropriately stained sections. It is possible to delineate the subdivisions of the hippocampal region using Bielchowsky silver stained sections. The borders of the subdivisions were delineated based on the definitions by West and Gundersen (1990). The delineations were made unambiguously to avoid sampling bias. The delineated areas were used in the estimation of the reference volume, $V(\text{ref})$, with the Cavalieri method as well as in the estimation of $W(\text{cap})$, the number of microvessels.

The sections were examined with a dissection microscope and the borders were outlined using a marker pen. The delineations were controlled using a Zeiss light-microscope with a lens ×4.

For the estimation of volume, a grid with a tessellation of points was placed on the 10 sections and the number of points hitting the outlined region counted. The design of the sampling strategy ensured that 100–200 points were counted in accordance with rules for efficient sampling (Table 7.1).

4.3 Counting of nodes

Optical disectors (Gundersen *et al.* 1988; West 1993) were used for the direct counting of capillary nodes while scanning through the thick sections. Optical sectioning was performed throughout the depth of the section using a high numerical aperture oil lens. Optical disector counting rules were applied (West and Gundersen 1990). Every node that came into focus within the unbiased counting frame (Gundersen 1977) was sampled as it was moved continuously through the distance *h* of the section thickness. As previously noted, the nodes were considered to be within the counting frame when the cusp of the smallest angle came into focus within the counting frame (Fig. 7.4).

Estimation of the precision of the estimate is shown in Tables 7.3 and 7.4.

5 Methodological considerations

The size of the counting frame was determined in order to take the following points into consideration. The guard area around the counting frame needs to be large enough for an unambiguous decision regarding the joining of vessels in the nodes. Each vessel must be clearly distinguishable, and it must be possible to estimate an inner diameter. This cannot be done reliably very close to the node because of a slight widening or dilation of the vessels just as they join the node. On the other hand, the counting frame must be as large as possible in order to diminish edge effects and to increase the efficiency of the sampling scheme. These different and opposing considerations must be met when choosing the size of the counting frame.

When counting nodes it can be difficult to decide the valence of the nodes. It can be impossible to distinguish between two nodes of valence 3 very close to each other, and one node of valence 4. But it is not a crucial factor, since the number of redundant connections in one node of valence 4 and two nodes of valence 3 is the same. However, when the decision influences whether the investigator counts one or two nodes, it may influence the number estimate. A large counting frame can diminish this problem.

We did not make any corrections for shrinkage in the study design described here because we used glycolmethacrylate. Previous experiences with the use of glycolmethacrylate in the human hippocampus have indicated that shrinkage is minimal (West and Gundersen 1990). Generally, data suggest that large slices in the brain such as those used in this study have very little shrinkage. However, it is not possible to make deductions from large to small pieces of tissue, and from one region to the other.

In theory the thickness of the sections could be even larger than described here. But it is important to be able to see evenly through the staining of sections. The Bielchowsky stained sections had a constant maximum thickness of 40 μm because, if they were thicker, the sections would be stained too intensely. It is also important that the staining is constant throughout the x-, y-, and z-directions of the section. Bielchowsky silver staining unambiguously stains microvessels so that both the wall of the vessels and the endothelial cells can be clearly distinguished. However, staining with specific antibodies gives a unique possibility to distinguish between microvessels with different properties, that is, arteries, venules, and capillaries. Unfortunately, the use of glycolmethacrylate precludes the use of antibodies. Also, the type of tissue used in this study is not optimal for staining with antibodies. The post-mortem human brain is not as preserved as animal tissue, and perfusion techniques are, of course, not an option. The advantage of the Bielchowsky silver staining is the robustness of the parts of the tissue stained with this method as well as the very good discrimination between the vessels and other structures in the brain (Fig. 7.5).

References

Bush, P.G., Mayhew, T.M., Abramovich, D.R., Aggett, P.J., Burke, M.D., and Page, K.R. (2000). A quantitative study on the effects of maternal smoking on placental morphology and cadmium concentration. *Placenta* **21**, 247–56.

Gundersen, H.J. (1986). Stereology of arbitrary particles. A review of unbiased number and size estimators and the presentation of some new ones, in memory of William R. Thompson. *J. Microsc.* **143**, 3–45.

Gundersen, H.J. and Jensen, E.B. (1987). The efficiency of systematic sampling in stereology and its prediction. *J. Microsc.* **147**, 229–63.

Gundersen, H.J.G. (1977). Notes on the estimation of the numerical density of arbitrary profiles: the edge effect. *J. Microsc.* **111**, 219–23.

Gundersen, H.J.G., Bagger, P., Bendtsen, T.F., Evans, S.M., Korbo, L., Marcussen, N., Møller, A., Nielsen, K., Nyengaard, J.R., Pakkenberg, B., Sørensen, F.B., Vesterby, A., and West, M.J. (1988). The new stereological tools: disector, fractionator, nucleator and point sampled intercepts and their use in pathological research and diagnosis. *Acta Pathol. Microbiol. Immunol. Scand.* **96**, 857–81.

Gundersen, H.J.G., Jensen, E.B.V., Kiêu, K., and Nielsen, K. (1999). The efficiency of systematic sampling in stereology—reconsidered. *J. Microsc.* 199–211.

Jain, R.K. (1999). Transport of molecules, particles, and cells in solid tumors. *Annu. Rev. Biomed. Eng.* **1**, 241–63.

Kroustrup, J.P. and **Gundersen, H.J.** (2001). Estimating the number of complex particles using the ConnEulor principle. *J. Microsc.* **203**, 314–20.

Lokkegaard, A., Nyengaard, J.R., and **West, M.J.** (2001). Stereological estimates of number and length of capillaries in subdivisions of the human hippocampal region. *Hippocampus* **11**, 726–40.

Nyengaard, J.R. and **Marcussen, N.** (1993). The number of glomerular capillaries estimated by an unbiased and efficient stereological method. *J. Microsc.* **171**, 27–37.

Nyengaard, J.R., Bendtsen, T.F., and **Gundersen, H.J.** (1988). Stereological estimation of the number of capillaries, exemplified by renal glomeruli. *Acta Pathol. Microbiol. Immunol. Scand.* **4** (suppl.), 92–9.

Secomb, T.W. (1995). Mechanics of blood flow in the microcirculation. *Symp. Soc. Exp. Biol.* **49**, 305–21.

Tufto, I. and **Rofstad, E.K.** (1998). Interstitial fluid pressure, fraction of necrotic tumor tissue, and tumor cell density in human melanoma xenografts. *Acta Oncol.* **37**, 291–7.

Vermeulen, P.B., Gasparini, G., Fox, S.B., Colpaert, C., Marson, L.P., Gion, M., Belien, J.A., de Waal, R.M., Van Marck, E., Magnani, E., Weidner, N., Harris, A.L., and **Dirix, L.Y.** (2002). Second international consensus on the methodology and criteria of evaluation of angiogenesis quantification in solid human tumours. *Eur. J. Cancer* **38**, 1564–79.

West, M.J. (1993). New stereological methods for counting neurons. *Neurobiol. Aging* **14**, 275–85.

West, M.J. and **Gundersen, H.J.** (1990). Unbiased stereological estimation of the number of neurons in the human hippocampus. *J. Comp. Neurol.* **296**, 1–22.

VOLUME

CHAPTER 8

SECTION INTRODUCTION

JENS R. NYENGAARD AND STEPHEN M. EVANS

1 Introduction

Information concerning the size of a brain structure and how this changes with normal development and disease provides vital information on the function of the structure. Modern stereology offers a series of methods for estimating size at both the macroscopical level and the microscopical level. These methods can be employed on physical sections or images derived from imaging modalities, such as X-rays, ultrasound, computerized tomography (CT), magnetic resonance imaging (MRI), and positron emission tomography (PET) scans. For example, modern stereological methods have produced fast and efficient methods for quantifying the loss of brain tissue with dementia, (Regeur 2000) and changes caused by multiple sclerosis in MRI brain images (Gadeberg *et al*. 1999).

In addition to providing information about the pathophysiology of a disease, changes in size can also be used to monitor the progress of the disease and any response to treatment; for example, the TMN (tumor/metastases/nodes) classification for staging of cancer growth and spread relies on a method for estimating tumour size.

The volume of a specific brain region may reflect the activity of that region. On a microscopical level the volume of the neuron may reflect its viability and function. This in turn will influence the function of the area of the brain to which the neurons belong. For example, the size distribution of neocortical neurons is heavily skewed to the right, that is, there is a large number of small neurons compared to large ones (see the example in Gundersen 1988).

The estimation of the mean volume of any brain structure can be influenced by the way in which it is sampled. Most people understand the mean volume of an object to be a number-weighted mean volume, that is, it is obtained from structures that are sampled with equal probability. That is to say, the objects are selected irrespective of their size and shape, and then their volume is estimated. This is by far the most common and most important definition of a mean volume. However, there are three different definitions of mean volumes that can be estimated using stereological sampling principles.

1 Number-weighted mean volume. Each object is sampled with equal statistical weight regardless of its shape and size. The sampling is performed by the disector (Sterio 1984).

2 Volume-weighted mean volume. Each object has a statistical 'weight' according to its volume. The larger the volume the higher the probability of being chosen—the sampling is performed by test points (Gundersen and Jensen 1983, 1985).

3 Star volume. This is defined as the average volume of the object that can be seen unobscured from a random point within the object. Star volume is estimated by the use of point-sampled intercepts (Serra 1982; Gundersen and Jensen 1985).

Once the structures have been disector-sampled, their *number-weighted* mean volume may be estimated by three fundamentally different methods.

1 A global volume estimator uses the principle of Cavalieri (Gundersen and Jensen 1987)

2 The local volume estimators comprise:
   • the nucleator (Gundersen 1988);
   • the planar rotator (Jensen and Gundersen 1993);
   • the optical rotator (Tandrup *et al.* 1997);
   • the selector (Cruz-Orive 1986).

3 A ratio of global estimators generates an indirect estimate of mean number-weighted volume.

2 Global number-weighted volume estimators (the principle of Cavalieri)

An ancient principle first put forward by the Italian mathematician Cavalieri (Gundersen and Jensen 1987) has provided stereology with a powerful tool for

Fig. 8.1 (see also **Plate 5**) The Cavalieri principle is illustrated on the human neocortex in the three parts of this figure. (Left) The four neocortical regions, that is, frontal, temporal, parietal, and occipital, are delineated on the pial surface using different colors (Brændgaard *et al.* 1990; Pakkenberg and Gundersen 1997). The frontal (left) and occipital (right) regions are unstained. The parietal region is dark blue and the temporal region is red. (Middle) One hemisphere is embedded in agar and cut in 7 mm thick slabs. A known fraction (1/4) of the slabs is taken by systematic uniformly random sampling. (Right) On each sampled slab a counting grid with test points is placed at random and test points hitting neocortex are counted. The volume of neocortex equals the multiplication of the distance between the tops of adjacent sampled slabs, the area associated to each test point, and the total number of test points hitting neocortex.

estimating total tissue volume. The Cavalieri principle is quite general and defines how to estimate the total volume of, for example, the brain neocortex, V(neocor); see Fig. 8.1. By cutting the entire object into slices with a known thickness, t, and counting the number of test points, P, associated with a known area, a(p), and hitting neocortex on the cut surfaces of the slices, the total volume of neocortex may be estimated as

$$V(\text{neocor}) = a(\text{p}) \cdot t \cdot \Sigma P(\text{neocor}). \qquad (2.1)$$

If the object were sliced into more than 10–12 slices, a systematic, uniform random sample of the slices could be taken; however, this would depend on the complexity of the structure being studied. When calculating the total volume, the thickness t is then defined as the distance between the top of one slice and the top of the next sampled slice. One of the advantages of the Cavalieri principle is that it does not rely on any assumptions with regard to the object's shape. The use of the Cavalieri principle simply requires that: (1) the position of the first slice hitting the object must be random; (2) the slices are parallel; (3) the thickness of the slices should be constant. The Cavalieri principle has been used extensively in neuroscience and has been used to estimate the volume of a human brain cortex (Brændgaard *et al.* 1990; Pakkenberg and Gundersen 1997) and human brain white matter (Yong *et al.* 1997), to mention just two examples. The volume of a brain or its regions can also be estimated using non-invasive techniques such as CT (Pakkenberg *et al.* 1989) or MRI (Mayhew and Olsen 1991) in combination with the Cavalieri principle.

A three-dimensional structure is represented as an area on a set of two-dimensional plane sections. All point-counting methods are based on the principle that test points 'feel' volume and point counting, as illustrated in Figs 8.1 and 8.2, is still one of the cornerstones in many stereological applications. The old principle of point-counting provides an estimate of volume without any assumptions about the structure and it does so more efficiently than delineating the structure by digitizers or image analyzers (Gundersen *et al.* 1981; Mathieu *et al.* 1981).

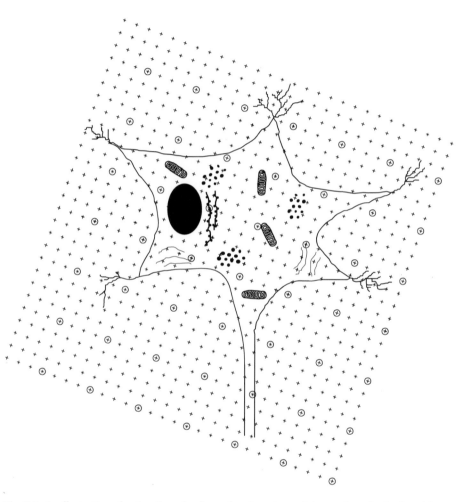

Fig. 8.2 An illustration of estimation of volume density using point-counting of mitochondria in a neuron. The point-counting grid contains 49 units (encircled crosses) each consisting of 16 points (crosses). All crosses including encircled crosses are used to estimate the volume of mitochondria and only encircled crosses are used to estimate the reference volume, which is equal to the volume of the perikaryon. Crosses hit an object if the object can be seen in the upper right corner of the cross. Seven crosses hit mitochondria and nine encircled crosses hit the cell. In this field of view the volume fraction of mitochondria per neuron constitutes $7/(9 \times 16) = 0.05$, because there are 16 times more crosses used for estimating mitochondria volume as volume of perikaryon.

To estimate, for example, the total volume of mitochondria in specific neurons, V(mito, neur), the Cavalieri principle and point-counting can be used to estimate the volume of the neocortex, V(neocor), as discussed in Fig. 8.1. Light microscopy and point counting is used to estimate the volume density of specifically stained neurons in the neocortex, V_V(neur/neocor). Finally, the volume density of mitochondria in the same neurons, V_V(mito/neur), is obtained by electron microscopy and point-counting, as discussed in Fig. 8.2

$$V(\text{mito, neur}) := V(\text{neocor}) \cdot V_V(\text{neur/neocor}) \cdot V_V(\text{mito/neur}) \qquad (2.2)$$

This is the classical hierarchical (Weibel 1979) design for the estimation of total volumes on systematic, uniformly random sampled (SURS) sections. Numerous examples of the use of this principle in neuroscience exist in the literature.

3 Local volume estimators of number-weighted mean volume

The common use of the local volume estimators in neuroscience is to estimate neuron or glia cell volume. The advantage of the local volume estimators is that they may provide the distribution of volume of the individual structures under study. In general, the rotator(s) and the nucleator are equally efficient for estimating individual volume and the selector is less efficient. Both the nucleator and rotator(s) rely on a unique point, for example, a nucleolus, as a reference point from where the volume estimation is performed. This feature makes them ideal for optical sectioning using light microscopy or confocal laser scanning microscopy as shown in Fig. 8.3. The selector does not need a reference point for volume estimation, but requires time-consuming serial sectioning.

Considering the nucleator method, the nucleolus of a neuron or glia cell is disector-sampled and the distance from the nucleolus to the cell boundary is measured, cubed, $\overline{l_n^3}$, and the cell volume estimated as

$$\overline{v_N} = \frac{4\pi}{3} \cdot \overline{l_n^3} \qquad (3.1)$$

The measurements are made in an isotropic direction in three dimensions. The local number-weighted volume estimators have been used extensively in neuroscience (Gundersen *et al.* 1988; Janson and Møller 1993).

4 Indirect estimate of number-weighted mean volume

The indirect estimate of mean volume obtained by the volume density and the number density of the structures, will be the most efficient method for estimating

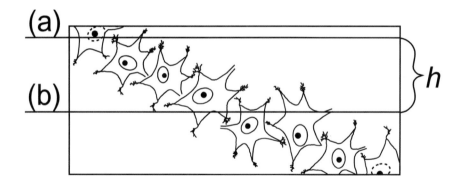

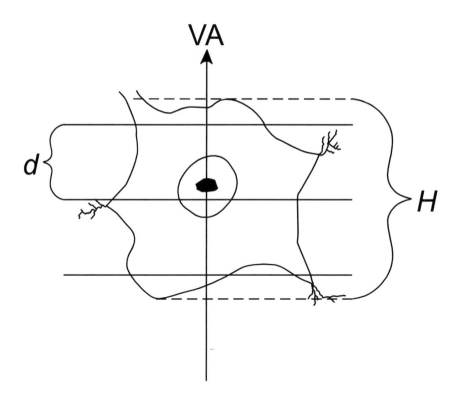

the average volume of brain structures, provided that the volume of individual brain structures is not needed.

$$\bar{v}_N := \frac{V_V}{N_V} \qquad (4.1)$$

◀ **Fig. 8.3** (Upper) The optical disector is illustrated at the top using a thick section (> 25 μm) viewed edge-on. The neuronal nuclei are visualized in an appropriately sampled field of view, and the microscope, equipped with oil objectives of a high numerical aperture, is focused some μm down (a) in order to avoid distortion and unevenness of the section surface. All neuronal nucleoli above or in focus at level (a) are disregarded (one nucleolus). From here, all nucleoli coming into focus when focusing the microscope down to level (b) are counted. Nucleoli in focus at level (b) are also counted. The height of the disector is *h*. If the whole volume between (a) and (b) is used for counting, three neurons will be counted here. The reason why there is a guard area of some μm above and some μm below the optical disector is to avoid ambiguities in the identification of neurons and to avoid the problems of 'lost caps' (nuclei drawn with broken lines). In frozen sections or vibratome sections the guard area almost has the height of a neuron (Andersen and Gundersen 1999). (Lower) For the illustration of the local estimation of the number-weighted mean volume of a neuron, one of the neurons sampled above is magnified at the bottom viewed in the focal plane. In order to use the planar rotator the section is a 'vertical section'. The arrow VA shows the direction of the vertical axis. Uniform, random and parallel test lines (solid lines), a distance *d* apart, are applied to the sampled neuron. The lattice of the test lines must be placed so the vertical axis passes through the sampled nucleolus. For each test line crossing the neuron of height *H*, linear intercepts from VA to the perikaryon border are measured on both sides of VA. The perikaryon volume is estimated by squaring all intercepts at both sides of VA and summing over all test lines. The lowermost test line to the right of VA intersects the perikaryon membrane twice. The length of this test line is calculated by adding up the squared lengths with alternating signs, with the intersection most distant from the VA being positive.

where V_V and N_V are obtained by point-counting and disector-sampling, respectively. A typical example is the estimation of the number-weighted mean volume of specifically marked neurons where N_V is estimated using optical disectors or a physical disector. Point-counting is performed on thin sections. The drawback is that the investigator needs to be able to identify every profile of the brain structures under study. Practical examples of this methodology do, however, not exist in neuroscience.

5 Local estimator of volume-weighted mean volume

The term 'volume-weighted' means that the structures are sampled in proportion to their volume using points such that the greater the volume the more likely they are sampled. The volume-weighted mean volume $\bar{v}_V$ may be estimated with point-sampled linear intercepts (Gundersen and Jensen 1983, 1985)

$$\bar{v}_V = \frac{\pi}{3} \cdot \overline{l_0^3} \tag{5.1}$$

where $\overline{l_0^3}$ is the average of the cubed intercept length across the structure of interest through the sampling point. If both the mean volume-weighted and mean number-weighted volumes of a cell population are estimated, the coefficient of variation in the ordinary number-weighted distribution can be calculated. An advantage of the volume-weighted mean volume is that the measurements can be performed on thin single sections. The drawback is that conclusions from $\bar{v}_V$ are ambiguous with respect to $\bar{v}_N$. An enlarged volume-weighted mean volume can be the result of a greater number-weighted volume, a greater intraindividual variation, or both. The interpretation of the $\bar{l}_0^3$ results can therefore be quite difficult.

6 Local estimator of star volume

Star volume $\bar{v}*$ is related to $\bar{v}_N$ by being a volume-weighted size estimator. Star volume is defined as the mean volume of all parts of an object, which can be seen unobscured in all directions from a particular point (Serra 1982; Gundersen and Jensen 1985). The measurements are performed in the same manner as for point-sampled intercepts and it is a very efficient detector of changes in volume and shape of complex structures. The estimate is taken over a uniform sample of points. If the object is convex the star volume is equal to the mean volume. A straight tubular structure will get a smaller star volume if it bends and a marrow separated by trabeculae will get a greater star volume when holes appear in the trabeculae.

7 Estimation of volume and tissue deformation

It is important to emphasize the difference between the volume of the brain region or neuron *in the section* and in the living brain. It is therefore of no use to apply sophisticated techniques to obtain the volume of a brain region or a neuron in the section if it has suffered from considerable deformation in size (shrinkage). This deformation may either be global or differential—global meaning that all brain compartments change in size to the same degree, whereas different brain compartments change unequally in size during differential tissue deformation. It is very difficult to correct for differential tissue deformation. One possibility is to cut a tissue block in two parts, process tissue from the one 'cut surface' as desired, fast-freeze the other 'cut surface', do the same estimation on the two sets of tissue following staining, and compare the two results. Neither are there any 'smart' unbiased ways to obtain information about global tissue deformation during tissue fixation and processing. However, the area of a piece of tissue before and after fixation/processing may be estimated and the tissue deformation on a volume basis calculated as

$$\text{Volume shrinkage} := 1 - \left(\frac{\text{area after}}{\text{area before}} \right)^{1.5}. \tag{7.1}$$

Table 8.1 The five general estimators of cell volume* (modified from Hans Jørgen G. Gundersen, personal communication)

Advantages	Disadvantages
Nucleator/(rotator), $\overline{v}_N = \dfrac{4\pi}{3} \cdot \overline{l_n^3}$	
(1) Optical disector	(1) Binucleated cells
(2) Over/underprojection of low impact	(2) Very irregular or long cells
(3) May have very low CE	(3) Isotropic test lines
(4) Good approximation to real size distribution	
(5) Useful for *any* cell organelle: nucleus, mitochondria, etc.	
Indirect, $\overline{v}_N = \dfrac{V_V(par)}{N_V(par)}$	
(1) Number, volume, and surface densities may be available anyhow	(1) Must *recognize* all fragments
(2) Disregard orientation for $\overline{v}_N$	(2) Overprojection and other problems
	(3) Rather large CE
Cavalieri, $\overline{v}_N = \dfrac{1}{n}\sum\limits^{n} v_i;\ \ v_i = \bar{t}\cdot a(p)\cdot\sum P_i$	
(1) Observe complete distribution	(1) Exhaustive sectioning; t *known*
(2) Disregard orientation	(2) Over- and/or underprojection
(3) $CV_N(v)$ can be estimated without knowing t	(3) Surface cannot be estimated
Selector. $\overline{v}_N = \dfrac{1}{n}\cdot\dfrac{\pi}{3}\cdot\sum\limits^{n}\overline{l_{0,i}^3}$ and $\overline{v}_V = \dfrac{\pi}{3}\cdot\overline{l_0^3}$	
(1) No need for a nucleus	(1) Exhaustive sectioning
(2) No need to know t	(2) Whenever applicable, nucleator/rotator > selector
(3) $SD_N(v) = \sqrt{\overline{v}_N\left(\overline{v}_V - \overline{v}_N\right)};\ \overline{v}_N$ and $\overline{v}_V$	(3) Isotropic test lines
closely related	
Point-sampled intercept, $\overline{v}_V = \dfrac{\pi}{3}\cdot\overline{l_0^3}$	
(1) Just one section	(1) *Careful*! it is *not* $\overline{v}_N$; but then $\overline{v}_V = \overline{v}_N\left(1 + CV_N^2(v)\right)$
(2) Overprojection small	(2) Must *recognize* all fragments
	(3) Isotropic test lines

* $\overline{v}_N$, Number-weighted mean volume; SD, standard deviation; n, number of observations; v_i volume of individual cells; CV, coefficient of variation; P, no. of test points; $\overline{v}_V$, volume-weighted mean volume; t, section thickness; $a(p)$, area per test point; CE, coefficient of error; l, intercept length.

This estimator requires that isotropic uniform random (IUR) sections are used or the assumption that the tissue deformation in the z-axis is equal to the tissue deformation in the x- and y-axes. Another labor-intensive approach for estimating tissue deformation is to weigh a brain piece, process it, and cut it up exhaustively. The weight of the brain piece before processing is transformed to a volume and compared with the volume of the cut-up piece estimated by the principle of Cavalieri. Considering the theoretical and practical difficulties in estimating tissue deformation, it is important to recognize that deformation of tissue is markedly reduced in plastic embedding materials when compared with paraffin (Hanstede and Gerrits 1983; Miller and Meyer 1990). This probably also holds true for carefully prepared vibratome and frozen sections in the x-y plane (Dorph-Petersen 1999; Andersen and Gundersen 1999).

8 A note on estimation of neuron and glia cell volume

The estimation of neuron or glia cell volume has been improved quite significantly with design-based stereology. It is nowadays not necessary to conduct time-consuming serial reconstruction to estimate neuron or glia cell volume. Five different general methods (Table 8.1) are available. They are based on disector-sampling to provide the ordinary number-weighted mean volume of neurons or glia cells. The only exception is the point-sampled linear intercept method which is performed on a single section and provides the volume-weighted mean volume of neurons or glia cells. The advantages and disadvantages of the various methods are shown in Table 8.1.

References

Andersen, B.B. and **Gundersen, H.J.G.** (1999). Lost caps and anisotropic deformation of thick sections; with a note on disector counting of nucleoli in eucaryotic cells. *J. Microsc.* **196**, 69–73.

Braendgaard, H., Evans, S.M., Howard, C.V., and **Gundersen, H.J.G.** (1990). The total number of neurons in the human neocortex unbiasedly estimated using optical disectors. *J. Microsc.* **157**, 285–304.

Cruz-Orive, L.M. (1986). Particle number can be estimated using a disector of unknown thickness: the selector. *J. Microsc.* **145**, 121–42.

Dorph-Petersen, K-A. (1999). Stereological estimation using vertical sections in a complex tissue. *J. Microsc.* **195**, 79–86.

Gadeberg, P., Gundersen, H.J.G., and **Tagehoj, F.** (1999). How accurate are measurements on MRI? A study on multiple sclerosis using reliable 3D stereological methods. *J. Magn. Reson. Imaging* **10**, 72–9.

Gundersen, H.J.G. (1988). The nucleator. *J. Microsc.* **151**, 3–21.

Gundersen, H.J.G. and **Jensen, E.B.** (1983). Particle sizes and their distributions estimated from line-and point-sampled intercepts. Including graphical unfolding. *J. Microsc.* **131**, 291–310.

Gundersen, H.J.G. and **Jensen, E.B.** (1985). Stereological estimation of the volume-weighted mean volume of arbitrary particles observed on random sections. *J. Microsc.* **138**, 127–42.

Gundersen, H.J.G. and **Jensen, E.B.** (1987). The efficiency of systematic sampling in stereology and its prediction. *J. Microsc.* **147**, 229–63.

Gundersen, H.J.G., Boysen, M., and **Reith, A.** (1981). Comparison of semiautomatic digitizer-tablet and simple point counting performances in morphometry. *Virchows Archiv* **37**, 317–25.

Gundersen, H.J.G., Bagger, P., Bendtsen, T.F., Evans, S.M., Korbo, L., Marcussen, N., Møller, A., Nielsen, K., Nyengaard, J.R., Pakkenberg, B., Sørensen, F.B., Vesterby, A., and **West, M.J.** (1988). The new stereological tools: disector, fractionator, nucleator and point sampled intercepts and their use in pathological research and diagnosis. *Acta Pathol. Microbiol. Immunol. Scand.* **96**, 857–81.

Hanstede, H.J. and **Gerrits, P.O.** (1983). The effects of embedding in water-soluble plastics on the final dimensions of liver sections. *J. Microsc.* **131**, 79–86.

Janson, A.M. and **Møller, A.** (1993). Chronic nicotine treatment counteracts nigral cell loss induced by a partial mesodiencephalic hemitransection: an analysis of the total number and mean volume of neurons and glia in substantia nigra of the male rat. *Neuroscience* **57**, 931–41.

Jensen, E.B.V. and **Gundersen, H.J.G.** (1993). The rotator. *J. Microsc.* **171**, 35–44.

Mathieu, O., Cruz-Orive, L.M., Hoppeler, H., and **Weibel, E.R.** (1981). Measuring error and sampling variation in stereology: comparison of the efficiency of various methods for planar image analysis. *J. Microsc.* **121**, 75–88.

Mayhew, T.M. and **Olsen, D.R.** (1991). Magnetic resonance imaging (MRI) and model-free estimates of brain volume determined using the Cavalieri principle. *J. Anat.* **178**, 133–44.

Miller, P.L. and **Meyer, T.W.** (1990). Effects of tissue preparation on glomerular volume and capillary structure in the rat. *Lab. Invest.* **63**, 862–6.

Pakkenberg, B. and **Gundersen, H.J.G.** (1997). Neocortical neuron number in humans: effect of sex and age. *J. Comp. Neurol.* **384**, 312–20.

Pakkenberg, B., Boesen, J., Albeck, M., and **Gjerris, F.** (1989). Unbiased and efficient estimation of total ventricular volume of the brain obtained from CT-scans by a stereological method. *Neuroradiology* **31**, 413–17.

Regeur, L. (2000). Increasing loss of brain tissue with increasing dementia: a stereo-logical study of post-mortem brains from elderly females. *Eur. J. Neurol.* **7**, 147–54.

Serra, J. (1982). *Image analysis and mathematical morphology.* Academic Press, London.

Sterio, D.C. (1984). The unbiased estimation of number and sizes of arbitrary particles using the disector. *J. Microsc.* **134**, 127–36.

Tandrup, T., Gundersen, H.J.G., and **Jensen, E.B.V.** (1997). The optical rotator. *J. Microsc.* **186**, 108–20.

Weibel, E.R. (1979). *Stereological methods.* Vol. 1. *Practical methods for biological morphometry.* Academic Press, London.

Yong, T., Nyengaard, J.R., Pakkenberg, B., and **Gundersen, H.J.G.** (1997). Age-induced white matter changes in the human brain: a stereological investigation. *Neurobiol. Aging* **18**, 609–15.

CHAPTER 9

THE NUCLEATOR AND THE PLANAR AND OPTICAL ROTATORS APPLIED IN RAT DORSAL ROOT GANGLIA

TRINE TANDRUP

1 Introduction

Obtaining quantitative data from cells in a brain tissue is not simple. The brain has to be cut into sections and magnified before information can be obtained. During this procedure the overview can be lost and cells can be cut in two or damaged at the cutting surface. In a microscope the measurements that can be performed on the profile of a given cell are the easiest to obtain. However, simple two-dimensional area measurements of profiles are influenced not only by the size but also by shape and orientation of the perikaryon. The correct parameter for describing size of a neuron perikaryon is a volume or a surface estimate. To obtain unbiased estimates of these three-dimensional parameters careful considerations about sampling of information in three-dimensions are necessary. Stereological principles can obtain such three-dimensional information from two-dimensional section planes using statistical sampling theories.

Estimation of mean volume of cells using stereological principles involves careful considerations of:

- sampling of a population of cells;
- problems concerning orientation of the cells and sampling of directions;
- choice of a volume estimator.

It is impractical and inefficient to measure all perikarya and a representative sample of the cell population is therefore required. The sampling of perikarya for volume estimates should be performed so that every cell has the same probability of being included in the sample, for example, by the use of the disector principle. This results in a 'number-weighted' mean volume, which is the way we usually think about mean volume. Another possibility is that the cells are sampled proportionally to their volume, so that larger cells have a higher probability of being included in the estimate. This is the 'volume-weighted' mean volume, which is useful in particular situations, for example, in malignancy grading of various cancer types (Sørensen 1992, 1995).

Most of the stereological volume estimators require isotropic test lines in space. Usually it is only in special cases that the population of perikarya can be considered naturally isotropic. Therefore, in many cases, the vertical section technique (Baddeley *et al.* 1986) is convenient to use because it allows the observer to keep track of one of the dimensions. Another possibility is that of isotropic uniform random (IUR) sections where the tissue is randomly rotated in three dimensions.

The nucleator, the planar rotator, and the optical rotator are different volume estimators. They all belong to the family of local estimators but are derived from mathematically different principles. All three methods can be used in sections of clearly delineated objects associated with a point such as neuron perikarya containing a nucleolus. All three methods exist in versions for both vertical section and IUR section designs. The nucleator and planar rotator only use one plane through the object for measurements, whereas the optical rotator uses several planes for the estimate. The optical rotator is therefore, in principle, more precise than the nucleator and planar rotator but also more laborious. The choice of volume estimator most frequently falls on the planar rotator since it provides an acceptable precise and efficient estimation. Generally, due to the simplicity of the method, the nucleator method would be chosen if the measurements needed to be performed by hand. The optical rotator is chosen if very precise estimates on single cells are necessary.

1.1 The nucleator principle

The nucleator (Gundersen 1988) is a local estimator where the volume is estimated in the section plane containing the unique point. In vertical sections the volume,

v, is estimated from cubed intercept lengths between a unique point and the border of the object multiplied by a constant (Fig. 9.1(I)).

$$v := \frac{4\pi}{3} \cdot \overline{\ell_n^3} \tag{1.1}$$

where ℓ_n is the length of the intercept. If no computer is available the nucleator in vertical sections is relatively easy to perform manually (Fig. 9.1 (III)). Using a IUR design of the nucleator the cells are all ready randomly rotated in three dimensions. The volume is estimated from eqn 1.1 (Fig. 9.1 (II)).

1.2 The planar rotator principle

The planar rotator technique (Jensen and Gundersen 1993) is also performed in the plane containing the unique point. In vertical sections, where randomness is ensured in two dimensions by rotating the objects around the vertical axis, the line separating the half-planes going through the unique point has to be parallel with the vertical axis (Fig. 9.1 (IV)). Intercepts from the line separating the half-planes to the border along the test lines are obtained and v estimated as

$$v := \pi \cdot t \cdot \sum \frac{\ell_+^2 + \ell_-^2}{2} \tag{1.2}$$

where t is the distance between test lines and ℓ_+ and ℓ_- (collectively, $\ell_{+/-}$) are the lengths of the intercepts of the positive and negative half-planes, respectively.

In IUR designs intercepts from the line separating the half planes to the border along the line grid can be obtained (Fig. 9.1 (V)) and v estimated as

$$v := 2 \cdot t \cdot \sum \frac{g_{i+} + g_{i-}}{2} \tag{1.3}$$

where

$$g_{i+/-} = \ell_{+/-} \cdot \sqrt{\ell_{+/-}^2 + a_i^2} + a_i^2 \cdot \ln\left(\frac{\ell_{+/-}}{a_i} + \sqrt{\left(\frac{\ell_{+/-}}{a_i}\right) - 1}\right)$$

t is the distance between test lines, ℓ_+ and ℓ_- (collectively, $\ell_{+/-}$) are the lengths of the intercepts on the positive and negative half-plane and a_i is the distance from the unique point to a given line in the line grid.

In cells that are not convex special rules have to be followed to obtain $\ell_{+/-}$ (Fig. 9.1(VI)). The length of a given intercept can be calculated as

$$\ell_{+/-} = \sum_+ \ell_{+/-}(+) - \sum_- \ell_{+/-}(-) \tag{1.4}$$

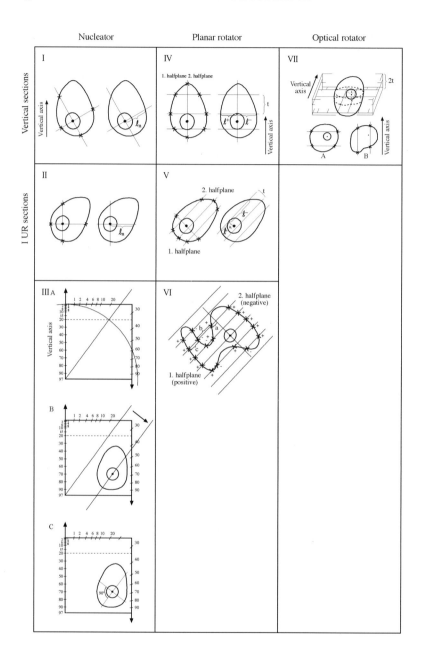

Fig. 9.1 (I) Nucleator measurements on vertical sections. Lines of which the probability of the first line is sine-weighted to the vertical axis. The other line is perpendicular to the first. Intercept lengths are obtained along all systematically random sampled lines. (II) Nucleator measurements on IUR sections. The directions in which to measure can be decided by the observer, as the cells are IUR-rotated themselves. Intercept lengths are obtained along all systematically sampled lines. (III) Nucleator measurements on vertical sections by hand. A counting frame with orientations can be designed. (a) Equidistant numbers (–97) on the y-axis are followed to the point where the circle is hit. This point is extrapolated to the counting frame with a line through 97. (b) When a cell is sampled in the counting frame the next systematic random direction (period 37) is found and the direction is shifted parallel until it is placed through the nucleolus. The second direction is perpendicular to the first. (c) Measurements can be performed along the lines. It can be an advantage to use a non-equidistant ruler with classes.

(IV) Planar rotator on vertical sections. Estimations are obtained in two half-planes separated by a line through the unique point. A line grid is placed perpendicular to the line separating the half-planes. The distance between lines in the line grid is always known but may vary with the size of the cell to increase the efficiency. The distance between the lines in the grid is known (t) and proportionate with the size of the object under study. Intercepts from the line separating the half-planes to the border along the lines in the line grid are measured l_+ and l_-. Equation (1.2) is used for calculation of volume. (V) Planar rotator on cells from IUR sections. The line separating the half-planes going through the unique point can be in any direction but the estimator is most precise if the line is placed perpendicular to the longest axis of the object. (VI) Definition of the sign of the intersection point in rotator measurements. The point on a half-line starting most distant from the line separation of the two half-planes is always +. The line is followed towards the one separating the two half-planes and the intersection point is given alternating signs (– or +). a and c are the line segments of a positive intersection point and b is the line segemnt of a negative intersection point in the positive half-plane. ℓ_+ can be estimated using eqn (1.4), $\ell_+ = a + c - b$.

(VII) Optical rotator on vertical sections. A virtual slice is placed parallel to the histological slice. The virtual slice here contains line grids in two focal planes (A and B). The thickness and the number of systematically randomly placed parallel optical sections within the virtual slice can be decided by the observer. In each of the optical sections a line grid is placed and the distance from the unique point to the intersection between the line grid and the cell membrane is found.

where $\ell_{+/-}(+)$ is the line segment of a positive intersection point and $\ell_{+/-}(-)$ is the line segment of a negative intersection point.

1.3 The optical rotator principle

The optical rotator technique (Tandrup *et al.* 1997) uses information from more than one section through the object and is therefore more precise than both the nucleator and the planar rotator. However, the method is more laborious and the degree of precision on single cells provided with this technique is only necessary in very special cases. Using the optical rotator technique an isotropic virtual slice is placed in the cell parallel to the histological section around the unique point and with reference to the point (Fig. 9.1(VII)).

If the unique point is placed near the center of the cell, which it will most often be, the slice in which measurements are performed will avoid the problematic regions with fussy delineations at the top and the bottom of the cell. The mathematical calculations are complicated and cannot be done by hand—a computer program is indispensable.

2 Worked example: rat dorsal root ganglion

Stereological principles applied to the rat dorsal root ganglion are valuable for the evaluation of structural changes (cell loss, atrophy, etc.) after intervention and treatment in experiments. Diseases or other pathological processes in the peripheral nerves are reflected in the nerve cell bodies lying in the dorsal root ganglion. In conditions leading to degeneration of axons it is important to know whether the whole nerve cell, including the perikarya, is lost or whether the damages are only local with possibility for regeneration. In conditions where the cell body is preserved, changes in volume of the perikarya can provide information about the nature of the damages, about the time course, and the possibility of selective action on a specific neuron subtype.

The dorsal root ganglion is a spindle-shaped enlargement in which the dorsal root becomes the spinal nerve. It is localized just outside the spinal canal and mainly consists of perikarya belonging to the sensory nerves, and their axons, of fibers from the motor root, and of supportive tissue. In normal rat dorsal root ganglion the cell population can be divided into two subtypes at light microscopy (Andres 1961; Lieberman 1976; Duce and Keen 1977; Rambourg et al. 1983). Neurons are characterized as A-cells if the nucleus of the cells is large and light, with one large centrally placed nucleolus only. Cells characterized as B-cells typically have a light nucleus with multiple smaller nucleoli, often located at the periphery of the nucleus. The cytoplasm of the B-cells generally appears darker than the A-cells, and often with an uneven distribution within the perikarya (Fig. 9.2).

In many locations in the central nervous system there can be difficulties distinguishing neurons from glial cells. However, the neurons in the dorsal root ganglion are rather large and, due to their characteristic morphology, readily distinguishable from fibers, satellite cells, connective tissue, and mast cells. The ganglion is also distinct and easy to isolate, which makes it easy to determine which cells belong to the ganglion. This is an advantage since it is one of the requirements in all the quantitative techniques. In nuclei appearing in the central nervous system problems can arise if cells mix with other similar cell populations or fade out in one or more directions without a natural endpoint. The ganglion is connected cranially to

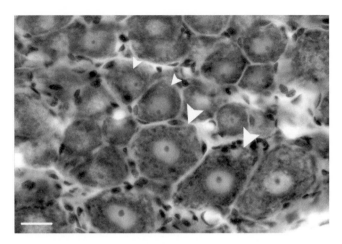

Fig. 9.2 The population of neuron perikarya in the rat dorsal root ganglion consists of A-cells (large arrowheads) and B-cells (small arrowheads). Scale bar, 21 μm.

the roots and caudally to the spinal nerve. However, ganglion cells can be located some distance proximally into the dorsal root. Cutting the ganglion from the root at some distance (about 2–3 mm in rats) from the ganglion, however, solves the problem. The cell population is usually well delineated towards the spinal nerve.

In this section considerations, necessary before estimation of number-weighted mean perikaryal volume in the rat fifth lumber dorsal root ganglion, are described followed by a worked example for the nucleator and the planar rotator.

2.1 Considerations

It is important to know the characteristics of the tissue and on that basis decide on the stereological design before proceeding with the histological process since many of the decisions influence the handling of the tissue. Three major subjects have to be addressed, namely, sampling of perikarya, orientation and anisotropy, and type of volume estimator.

2.1.1 A representative sample of dorsal root ganglion cells

The disector principle is central in many stereological estimates, and it is also extremely useful in the dorsal root ganglion. The disector is a probe that can sample cells with equal probability and therefore be used for estimates of number-weighted mean volume. The principle can be used both for counting cells and for obtaining a representative sample of cells, for example, for estimation of mean perikaryal volume. In the optical disector principle (Gundersen 1986; Braendgaard *et al.* 1990), cells are sampled in a thick section by lowering the thin optical plane down through the tissue and sample cells coming into focus. The optical disector is preferred in the dorsal root ganglion since it is faster and easier to use than the physical disector. Sampling of B-cells in physical disectors followed by an estimation of perikaryal volume can be troublesome due to their many nucleoli. It can be accomplished but several sections may have to be looked through for every sampled cell to find the largest of the nucleoli and with that the section in which the measurements has to be obtained. It is more simple to use the physical disector technique on the A-cell population in which less than 1% has more than one nucleolus. In A-cells the nucleolus could therefore be used as a sampling unit in physical disectors, and the measurements could be obtained on the same pair of sections.

The number of B-cells is approximately twice the number of A-cells in the dorsal root ganglion. If the precision of the estimate of mean perikaryal volume has to be the same for A- and B-cells, different sampling fractions of the cell population have to be obtained for the two cell types, that is, the sampling fraction for the A-cells has to be approximately twice that of the B-cells. This can be accomplished either by having two counting frames of unequal size and the same step length or by using different step lengths and the same counting frame. The solution with two unequal sized counting frames is the most efficient. The sampling scheme in which two independent samples are obtained is the least efficient.

Sampling of perikarya requires a unique counting/sampling unit. In the dorsal root ganglion there are several possibilities. For example, the whole cell can be used as a sampling unit. However, in large cells the nucleus and nucleolus may not be present in the section in which the perikarya is counted. This is rather impractical if the nucleolus is used in a contemporary estimation of volume. Furthermore, the nucleus and the cytoplasm around the nucleus are rather important in the dorsal root ganglion for the histological definition of subpopulations of ganglion cells. In the past studies have used the nucleolus as a counting unit to minimize the problems of splitting cells in two sections (Arvidsson *et al.* 1986; Schmalbruch 1987). However, choosing a smaller counting unit may reduce the bias but does not eliminate it when using previous morphometric methods. The nucleolus can be used as a counting unit in the disector, but some of the cells have more than one nucleolus. One nucleolus could be chosen at random using a random number table; however, it is difficult to separate large chromatin condensations from small nucleoli. Another possibility is to use the largest nucleolus as counting unit, which also requires that the observer is able to visualize all nucleoli in a nucleus. If there are two large nucleoli of the same size in a nucleus, which is rare, one could be chosen at random. The nucleus can also be used as sampling unit in the dorsal root ganglion cells. It is unique meaning that every perikaryon contains exactly one and it is centered providing optimal possibilities for a contemporary morphological description of the cell. However, if, in a study using the nucleus as sampling unit, volume estimations are intended, a unique point inside the cell is required and the largest nucleolus is often chosen for this purpose again requiring that the whole nucleus is available for study.

2.1.2 Orientation

The perikarya are ovoid with their long axis most often orientated in the fiber direction where the main stream is along the long axis of the ganglion. In this case the cells cannot be considered naturally isotropic. In estimates of volume and surface area it is convenient to take advantage of this information about the anisotropy of the cells in the ganglion and design the study so that the variation is minimal. All local estimators require sampling of cells uniformly in isotropic directions in three dimensions. As the cells are not isotropic themselves, IUR sections or vertical section techniques (Baddeley *et al.* 1986) have to be applied. IUR sections in the ganglion are impractical. Either the ganglion has to be cut in smaller pieces, which are rotated randomly creating artificial edges, or the whole ganglion has to be rotated randomly, creating a huge variation in the number of sections cut. In vertical sections the observer can maintain the orientation of the ganglion in a section, have the cells cut at an angle where the anisotropy is maximally presented, and still obtain measurements in isotropic directions. Having decided to cut longitudinal sections the vertical axis can only be chosen along the long axis of the ganglion. In the vertical section technique the ganglion has to be rotated around the vertical axis creating randomness in two dimensions. Randomness in the last dimension is created in the last step of the volume estimator. The rotation of the ganglion around the vertical axis is easily performed around the long axis of the ganglion in agar on a concave surface (Fig. 9.3). The opposite axis would create more sections and not present the anisotropy of the cells as well.

2.1.3 Decision on type of volume estimator

Before using one of the volume estimators, a decision about the type of volume estimator is necessary as they have different requirements for sampling of three-dimensional information. Which type of method it will be convenient to use in a given organ is decided on the basis of the characteristics of the cells. All ganglion cells contain elements that can be used as points and it is therefore possible to use both global and local estimators. Among the advantages of the local estimators in the ganglion are that they are very efficient and in most of them the whole cell need not be present in a section and available for measurements as in global estimators. Some of the A-cells in the rat dorsal root ganglion are so large that very thick sections, where there could be problems with some of the histological stains, are necessary if the section has to contain the entire cell. Furthermore, it may be desireable to avoid the group of global estimators due to the problems with fuzzy delineation of the cell in the top and bottom focal planes (due to the tangentially cut membrane).

It is a requirement for the local estimators that a cell is associated with a point. The element closest to a point is the nucleolus. However, in the dorsal root ganglion cells some cell types have multiple nucleoli and a sampling of the largest

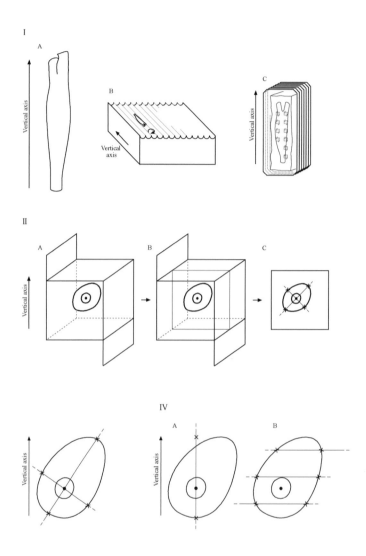

nucleolus is necessary as the point has to be unique for the cell. If two nucleoli are of equal size one of them can be randomly sampled using a random number table. Another problem is that the nucleolus is not a point. However, the error made by placing the point in the middle of the nucleolus is negligible (Moller *et al.* 1990; Tandrup *et al.* 1997).

The choice of estimator among the local estimators depends on the precision needed, the desired workload, and the equipment available. Generally, the local

Fig. 9.3 (I) Vertical sections in the dorsal root ganglion. (a) The long axis of the dorsal root ganglion is defined as the vertical axis. (b) The ganglion is rotated randomly around the vertical axis and embedded in agar. (c) After embedding in glycolmethacrylate serial vertical sections are cut. The vertical axis is along the long axis of the ganglion parallel to the long side of the section. Systematically random sampling of disectors by stepping through a section.

(II) Sampling of dorsal root ganglion cells for estimation of volume. (a) The cells are sampled in optical disectors. (b) The focal plane in which the largest nucleolus of the sampled cells is the sharpest delineated is found. (c) The measurements for the volume estimator are obtained.

(III) Computer-assisted estimation of perikaryal volume on vertical sections using the nucleator. After indication of the center of the largest nucleolus (x) the computer generates sine-weighted lines and the observer marks the intersection between cell boundary and the lines (x).

(IV) Computer-assisted estimation of perikaryal volume on vertical sections using the planar rotator. (a) After indication of the center of the largest nucleolus (x) the computer generates a line through the nucleolus parallel to the vertical axis. The observer marks the extensions of the cell in direction of the vertical axis (x). (b) The computer generates a line grid and the observer marks the intersection between cell boundary and the lines (x).

estimators available for estimation of mean perikarya volume fall into two categories. In one group of estimators only a single focal plane through the reference point is needed, for example, the nucleator and the planar rotator. In the other group of estimators, information from more or varying focal planes in the same cell is used, for example the optical rotator. The precision of the last group is higher than from the group using only a single fixed section through the cell as more three-dimensional information is used (Tandrup *et al.* 1997). However, the workload in methods using more or varying focal planes is also larger and, for most purposes, a very precise estimate of the volumes of single cells is not needed in the ganglion. Among ganglion cells themselves the variation is so large (ranging from 2000 to 200 000 μm^3) that the variation of the volume due to estimation is negligible. For estimation of mean perikaryal volume for the total population and for estimation of volume in the two subpopulations of perikarya, the precision of estimators relaying on only one focal plane through the nucleolus is sufficient—it is also sufficient for determining the distribution of volume. Both local estimators, nucleator and rotator, can be used in the ganglion without problems and with an acceptable efficiency. The times spent on estimation with the two methods are almost identical and the precision is not very different (Jensen and Gundersen 1993). The planar rotator is slightly more precise depending on the cell under study and is therefore the standard estimator of mean perikaryal volume in the dorsal root ganglion in an optical disector design.

2.2 Practical procedures

2.2.1 Histological procedure

The rats were anesthetized with an intraperitoneal injection of a mixture of pento-barbital (50 mg/ml) and diazepam (5 mg/ml). The perfusion was initiated by a pre-rinse with Tyrode's buffer, for 10 seconds, followed by a 10 minute perfusion with 4% glutaraldehyde dissolved in a 0.08M phosphate buffer at a pH of 7.40. After the perfusion the fifth lumbar dorsal root ganglia were removed.

The long axis of the ganglion was defined as the vertical axis of the organ. The ganglia were rotated randomly around the vertical axis and embedded in 7% agar on a tempered aluminum block with a surface of small concaves. The tissues were then dehydrated in graded series of alcohol (2 hours in 70%, 3 hours in 96%, 2 hours in 99%) and embedded in glycolmethacrylate (Technovit, Culser, GmbH). Thick serial vertical sections of 30 μm were cut along the vertical axis on a Supercut 2065, Reichert-Jung, using water on the cut surface and the Ralph glass knife. The curled sections were allowed stretch in water. From the bath the sections were laid on a glass slide and placed in a 60°C oven for a couple of hours. Every third section was sampled for further study using a random starting point. Sections were stained with cresyl violet acetate (Sigma), modified for thick sections (Tandrup 1993). A 0.25 g% solution of the powder in demineralized water was heated to 60°C for 20 minutes. The solution was poured through a coarse filter and, after the solution had attained room temperature, the pH was adjusted to 3.5 with concentrated acetic acid. The sections were stained for 2 hours and 20 minutes in this solution. The sections were differentiated in 1% acetic acid and graded series of alcohol (1 × 70%, 2 × 96%, and 2 × 99%) for approximately 7–10 minutes. Finally, the sections were placed in water briefly to avoid wrinkles. Then the sections were dried in an oven at 60°C. Coverslips were not used.

The best embedding mediums for estimation of volume are materials in which changes in the dimensions of the tissue are as small as possible, for example, glycol-methacrylate. However, for some stains the penetration is only poor in glycol-methacrylate. Giemsa penetrates very well in thick sections of dorsal root ganglion but cresyl violet acetate results in a more detailed staining of the cells. Three-dimensional information obtained in thick sections is very important in the dorsal root ganglion for correct validation of cell types. The central region around the nucleus is especially of value.

2.2.2 Microscope and computer

Sampling of neurons with disectors and measurements of perikarya volume were performed on a microscope/computer set-up (CAST Grid®, Olympus Denmark). A microcator controlled the movements of the focal plane in the vertical direction and a stepping motor moved the sections in the transverse direction for systematic random sampling of counting fields. A 60 × oil immersion lens (numerical aperture

(NA), 1.40; depth of focus, ~0.5 μm) was used, which gave a total magnification of 1700×. With a special stage it was possible to rotate the tissue resulting in the desired orientation of the tissue on the computer screen. The contrast in the microscope can be enhanced with some histological stains using a filter in the light source, for example, a yellow filter with the cresyl violet acetate stain.

2.2.3 Sampling of perikarya in the dorsal root ganglion

Before starting sampling in a section the section must be orientated according to directions in the computer program for orientation of the vertical axis for the volume estimation performed later. A sample of cells in the dorsal root ganglion was obtained using optical disectors (Gundersen 1986; Braendgaard *et al.* 1990) on every third histological section. As the plane of focus was moved 15 μm down through the section, the numbers of neurons appearing within the counting frame (Gundersen 1977) was counted, the counting unit being the nucleus. The disector was started approximately 3 μm below the section surface. Cells present in the initial plan of the optical disector were not counted, whereas all cells in the final optical plan were included. The counting frame for A-cells was 3550 μm² and 1700 μm² for B-cells. Using a step length of 200 μm this sampling scheme resulted in sampling of about 100 of each of the two cell types (Fig. 9.2). A side-effect of this procedure is that nearly all figures necessary for the estimation of total number are obtained as well.

2.2.4 Measurements and estimation

The local volume estimator on the vertical sections was applied on each of the sampled cells during the sampling procedure (Fig. 9.3).

The nucleator

The nucleator technique was applied on a sampled cell during the sampling procedure (Fig. 9.3). Measurements were performed in the focal plane containing the largest nucleolus. The middle of the nucleus was marked as the origin. The vertical axis was visible as the long axis of the ganglion and oriented according to the requirements in the computer program before sampling in the section was initiated. The computer program was set to generate four sine-weighted systematic random directions (two lines) for measurements. The intersection between the lines and the cell border was marked on the screen and the computer automatically calculated the mean volume. The computer estimation could easily be checked using eqn (1.1).

The planar rotator

The planar vertical rotator technique is applied on each of the sampled cells (Fig. 9.3). Measurements with the vertical planar rotator were performed in the focal plane containing the largest nucleolus. The middle of the nucleus was marked as the origin. The vertical axis was visible as the long axis of the ganglion and oriented according to the requirements in the computer program before sampling in the section was initiated. The computer generated a line through the nucleolus parallel to the vertical axis. The extension of the cell in the focal plane along the vertical axis was marked and from these figures the computer generated a line grid suitable for a cell of the approximate size. The number of lines hitting the cell could be adjusted, but usually three lines were sufficient. The computer superimposed the line grid and the intersections between the cell boundary and the line grid were marked. The computer automatically calculated the mean volume. The computer estimation could easily be checked using eqn (1.2).

2.3 Presentation of data

2.3.1 Estimates of the volume of A- and B-cells from one animal

The fifth lumbar dorsal root ganglion in rats contains twice as many B-cells (approximately 11 500) as A-cells. B-cells are generally smaller than A-cells with a considerable overlap between size distributions (Lawson 1979; Tandrup 1993). In the volume distribution about 70% of the B-cell and 50% of the A-cells population are contained in classes where overlapping occur (Fig. 9.4). The size was one of the first ways to describe different types of perikarya in the ganglion (Dogiel 1896). The cells were described as large light or small dark cells. This classification is still used (Lawson 1979). However, due to the overlap of the two size distributions, the use of cell size as a major determinant in classification of subtypes of ganglion cells is causing misclassifications.

2.3.2 Variation in mean volume of A- and B-cells among animals in a group

The relative variation of mean perikaryal volume among animals can be expressed as the observed coefficient of variation, OCV, calculated from the standard deviation between the mean volumes among animals, OCV = SD/mean. This variation is a combination of variation introduced by noise, sampling, counting, and measuring procedures, which can be expressed by the coefficient of error, CE, and the biological variation, CV_{biol}

$$OCV^2 = CV_{biol}^2 + \overline{CE}^2. \tag{2.1}$$

OCV and CE can be estimated from the various measures and CV_{biol} calculated from eqn (2.1). CV_{biol} and CE can be used for evaluation of efficiency. If the CE is larger than the CV_{biol} the efficiency of the stereological procedures is generally sufficient. CE for the perikaryal volume is estimated in every animal and the average is calculated.

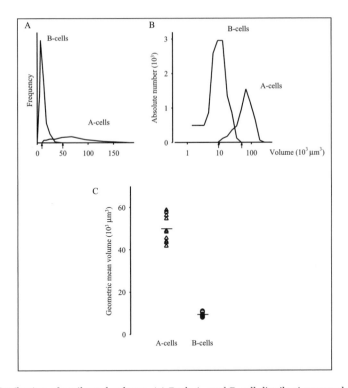

Fig. 9.4 Distribution of perikaryal volume. (a) Both A- and B-cell distributions are skewed to the left. The arrow indicates the mean volume. (b) Due to the skewness of the curves logarithmic transformation and estimation of geometric mean volume (arrow) is of advantage before statistical analysis. (c) Geometric mean volume of A- and B-cell perikarya in a group of normal L5 dorsal root ganglia. Bars indicate mean in the group.

$$CE = \frac{SEM}{mean} \quad where \quad SEM = \frac{SD}{\sqrt{n}} \qquad (2.2)$$

where SD is the standard deviation of the observed volumes within an animal and n the number of observations.

The average CE ($\overline{CE}$) is calculated as

$$\overline{CE} = \sqrt{CE_1^2 + CE_2^2 + \dots CE_i^2}. \qquad (2.3)$$

2.3.3 Change in mean perikaryal volume in an experimental group

Change in mean volume may be caused by many different changes in a neuron population. One underlying cause could be a change of size in all cells. However, it could also be caused by differential changes in the cell population. A subgroup of cells could be unchanged and perhaps even changed in the opposite direction of the mean volume. Changes in mean volume in a population do not even have to be based on changes of volume of the cells but can be caused by a selective cell loss as well. To obtain information on the underlying causes of changes in mean volume it is an advantage to have the possibility of evaluating both number of neurons and the distribution of volume.

One animal model, in which the benefit from information obtained by estimating perikaryal volume has been demonstrated, is chronic acrylamide intoxication. This intoxication has been used as a model of a group of diseases causing peripheral neuropathy (amyotrophic lateral sclerosis, Werdnig–Hoffmann's disease, Frederich's ataxia, thiamine and riboflavin deficiency, alcohol intoxication, tropical neuropathy, and porphyria (Cavanagh 1979)). In rats the acrylamide intoxication is clinically characterized by ataxia and muscle weakness. The intoxication causes degeneration of the axon both in the peripheral and central nervous system. In the dorsal root ganglion Cavanagh (1982) described degeneration with a chromatolytic reaction in the perikarya during acrylamide intoxication. In a stereological study the mean perikarya volume was reduced but no cell loss was detected (Tandrup and Braendgaard 1994; Tandrup and Jakobsen, 2003). The perikaryal shrinkage was furthermore analysed in the two subpopulations of dorsal root ganglion cells. A selective vulnerability of the A-cells with perikaryal shrinkage of 20–28% was demonstrated whereas the volume of B-cells did not change (Fig. 9.5). Later it was shown that atrophy is not restricted to the perikarya alone. The intoxication causes an atrophy of both A-cell perikarya and myelinated axons (Tandrup and Jakobsen 2003).

3 Concluding remarks

Stereology is mostly about sampling and only a little about estimation. The area where it is valuable to use most of the effort in planing a stereological study is in evaluating the possibilities of sampling in all its variations in a tissue—sampling of sections, sampling of perikarya, sampling of directions in which to measure, etc. The last, and the simplest, part is the estimation. Sampling varies from one brain region to another and is highly dependent on characteristics of the brain region. Having chosen an estimator the estimation step is almost the same for all brain regions. However, sampling and choice of estimator is linked, as some estimators require a given kind of sampling. The challenge for the researcher is therefore to choose the combination of sampling and estimator optimal for a given purpose in a given brain region.

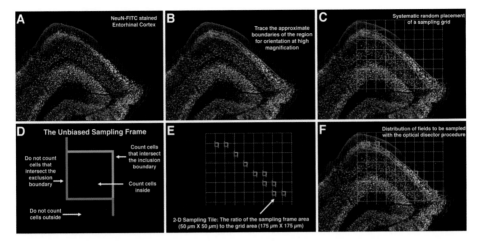

Plate 1 (see also **Fig. 4.2**) Sampling procedures for the optical fractionator using a fluorescent probe does not differ from those described for chromatic stains (West 1993). (a) A section stained with the neuronal marker NeuN detected with the fluorophore FITC (see text for details) is brought into the field of view using a low power objective (4×). (b) The structure to be counted (entorhinal cortex layer II) is outlined with the help of commercial stereology software (StereoInvestigator; MicroBrightField, Inc). The entire region to be sampled does not need to appear in the field of view. If a higher magnification is necessary to define the structure, the software will reposition the stage interactively to define regions outside of the initial, reduced field of view. (c) A regular grid array is placed over the region in a systematic uniform fashion between subsequent samples to ensure that the experimenter does not impose a biased distribution of the sampling. (d) The design of the unbiased sampling frame. Use of the frame ensures that cells visualized in two-dimensions are not over- or undersampled. (e) Grid intersects that overlie the region of interest serve as a positional index for the unbiased sampling frames to be sampled at high magnification (see Fig. 4.3). It is the ratio of these measurements that supplies the value for area sampling fraction (asf) used in the calculation of the optical fractionator result (see Fig. 4.6(b)). (f) All of the sampling elements brought together over the region of interest.

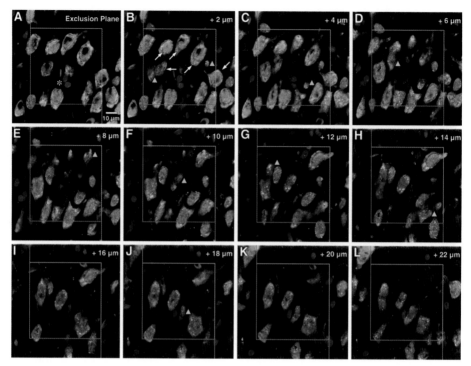

Plate 2 (see also **Fig. 4.3**) The optical disector counting procedure is applied at each sampling site indicated in Fig. 4.2(f). After changing to a suitable high NA immersion objective lens, the software positions the stage at each of the sampling sites in turn. Within each counting frame, sections are sampled with equal probability without bias for cell size, shape, or orientation. All cells were stained with propidium iodide (red) and entorhinal neurons by immunofluorescent labeling against the neuronal marker NeuN (FITC, green) as described in the text. Non-neuronal cells are colored only red (asterisk in panel A) while neurons are labeled with both red and green, which combine to show as yellow to orange depending upon the signal intensity contributed by each primary color. A series of focal planes collected at a sampling site was viewed in sequence and is represented here at discrete 2 μm intervals; in practice, the change of focal planes while counting is a continuous process. The first focal plane (0 μm; panel (a)) imaged after the guard region from the top of the section was used to define an exclusion plane. Any neuron that intersects this plane was excluded from being counted even if it meets the counting criteria in subsequent optical planes (arrows). Subsequent optical planes are viewed and neurons are counted (blue triangles) as they come into view provided they meet the sampling frame requirements (see Fig. 4.2(d)). For example, in (e) a new neuron is first observed and a blue triangle is placed just to the right of it. This neuron is observed in subsequent focal planes (panels (f)–(k)), but it is only the first time it appears that a mark is placed to avoid counting the cell more than once. In this figure, the last focal plane in which a newly visible cell would be counted is at +20 μm (panel (k)). An additional focal plane (l) is shown to illustrate that there should be a guard region at the bottom of the optical disector to evaluate the identity of any new cell that appears in the bottom plane and to verify that it meets the counting rules. Using these counting rules, one counts real cells in a volumetric subsample of the entire tissue. This raw count of cells is used to estimate total neuronal number (see Fig. 4.6).

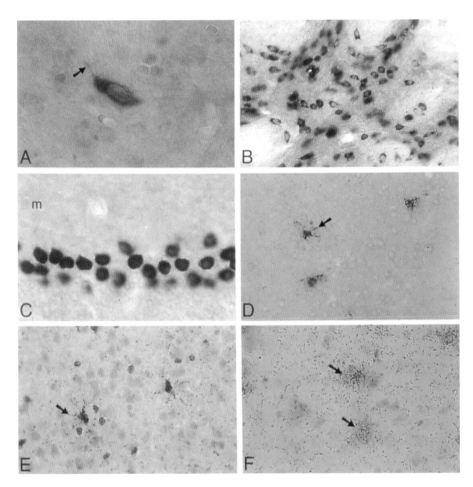

Plate 3 (see also **Fig. 5.2**) High- and medium-power micrographs of ISH reactions performed with (a)–(e) AP-labeled and (f) radiolabeled (^{35}S) probes. (a) Striatal somatostatin (SS) mRNA expressing neuron in adult rat striatum. Counterstained with Neutral red. The arrow marks a SS mRNA containing dendrite. (b) Enkephalin mRNA expressing neurons in adult rat striatum. (c) Calbindin mRNA expressing cerebellar Purkinje cells. (d) Tumor necrosis factor (TNF) mRNA expressing process-bearing (arrow) microglia–macrophages located at the infarct border in mouse subjected to permanent medial cerebral artery occlusion (MCAO). (e) Myelin basic protein (MBP) mRNA expressing oligodendrocytes in the striatum of adult mouse. Note translocation of MBP mRNA out into cellular processes (arrow). (f) Combined ISH for transforming growth factor beta-1 mRNA (black silver grains) combined with immunohistochemistry for microglial-macrophage complement type 3 receptor (red stain). Arrows point to two double-labeled cells located at infarct border in an adult rat subjected to transient MCAO (for details, see Lehrmann *et al.* 1998). The hybridizations shown in (a)–(c) were carried out on 50 μm thick vibratome sections, and those shown in (d)–(f) were carried out on 30 μm (d), (e) or 15 μm (f) thick cryostat sections. m, Molecular layer. Magnification: (a) 1300 ×; (b) 400 ×; (c) 500 ×; (d)–(f) 600 ×.

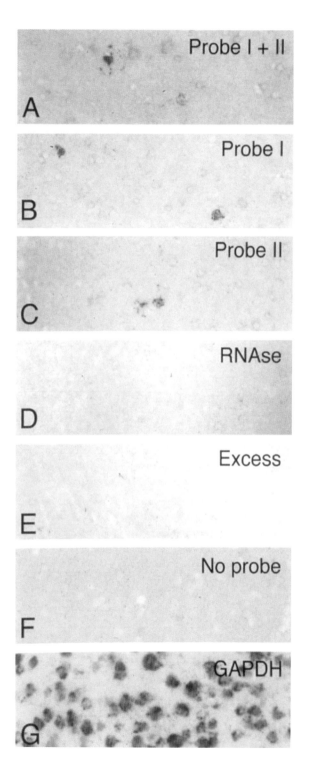

Plate 4 (see also **Fig. 5.3**) Control reactions for specificity of TNF mRNA ISH signal. The controls were performed on parallel sections from a mouse subjected to permanent MCAO. (a) Sections *in situ* hybridized with a probe mixture, consisting of two probes (probe I and II) directed against two different, non-overlapping stretches of TNF mRNA, showed a stronger ISH signal than (b), (c) sections hybridized with each individual probe. (d) Section pretreated with RNAse A prior to ISH shows no signal. Neither do sections hybridized with (e) an excess of unlabeled probe I and II or (f) no probe. (g) The GADPH control demonstrates that mRNA is abundant in the tissue section. Magnification, 400 ×.

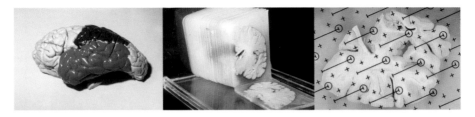

Plate 5 (see also **Fig. 8.1**) The Cavalieri principle is illustrated on the human neocortex in the three parts of this figure. (Left) The four neocortical regions, that is, frontal, temporal, parietal, and occipital, are delineated on the pial surface using different colors (Brændgaard *et al.* 1990; Pakkenberg and Gundersen 1997). The frontal (left) and occipital (right) regions are unstained. The parietal region is dark blue and the temporal region is red. (Middle) One hemisphere is embedded in agar and cut in 7 mm thick slabs. A known fraction (1/4) of the slabs is taken by systematic uniformly random sampling. (Right) On each sampled slab a counting grid with test points is placed at random and test points hitting neocortex are counted. The volume of neocortex equals the multiplication of the distance between the tops of adjacent sampled slabs, the area associated to each test point, and the total number of test points hitting neocortex.

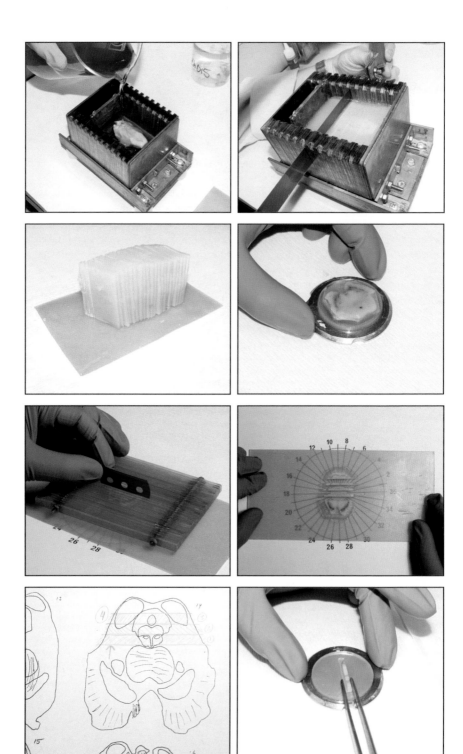

Plate 6 (see also **Fig. 10.3**) (Top four pictures) A brainstem is agar-embedded in a cutting-guide-box and sliced into 2.93-mm-slabs. The resulting set of slabs as well as one slab mounted on the vibratome adaptor is shown. (Bottom four pictures) From the remainder of a slab 2.28 mm wide bars are cut systematic, uniformly random using a razor blade and a randomly rotated cutting guide. Only the bars hitting the dorsal raphe are cut. The exact position of each bar is recorded in maps of the transversal sections. The bars are mounted ' side up' on the vibratome adaptor before the vertical sections are cut.

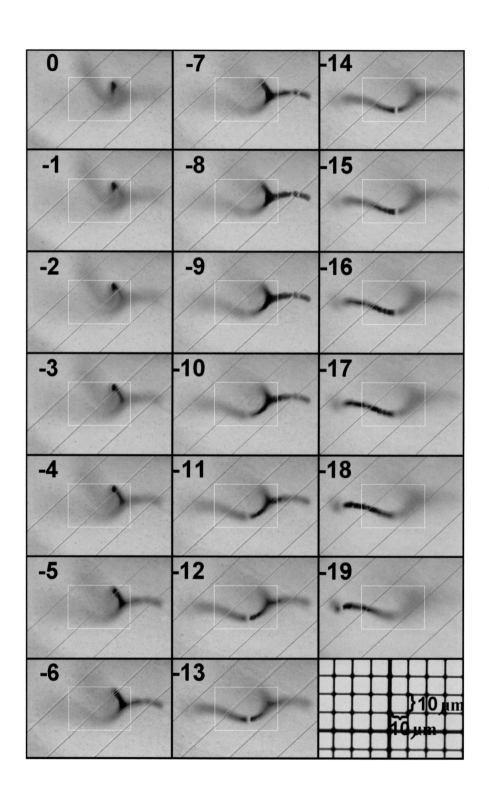

Plate 7 (see also **Fig. 13.7**) A graphical illustration of the counting rule with 'spatial' two-dimensional disectors in the tilted virtual plane exemplified by various events. Three successive focal planes are shown. The upper focal plane contains the look-up exclusion line and a white line showing the projected position of the next disector inclusion line. The middle focal plane contains the inclusion line (green) situated inside the sampling box with the exclusion extension (red, towards the observer) outside the sampling box, and the line of indifference (blue), that is, the two-dimensional representation of the guard area lateral to the two-dimensional-disector. h(dis) must be small enough that all events are recognized, but is made large in the figure for illustrative purposes. The lowermost focal plane contains the line representing the look-down guard area below the sampling box (red). The two-dimensional representation of the structure seen in the focal planes is shown in shadowed red, and the part of the structure between the focal planes is red. In (a) the profile exclusively intersect the green inclusion line and one profile is counted, whereas in (b) the profile touches the red exclusion extension and is thus not counted. In (c) the profile is intersected by the look-up exclusion line and stays in touch with the moving line during focusing which means one coherent 'profile' in the virtual plane and no profiles are counted. In (d) the profile that is intersected by the upper look-up exclusion line leaves the line during focusing, that is, it is not present in the virtual plane, but is intersected again by the inclusion line and thus one profile is counted. In (e) the structure is transected twice by the virtual plane as the profiles are connected outside the moving line, that is, two profiles are counted in (e). When multiple profiles located on the inclusion line are disconnected in the focal plane, information from the guard area in succeeding focal planes is needed to determine whether the profiles represent one coherent 'profile' in the virtual plane. One profile is counted in (f) as the profiles located on the inclusion line stay in touch with and merge on the moving line, that is, represent one coherent U-shaped 'profile' in the virtual plane. In (g) where the two profiles located on the inclusion line merge outside the moving line in succeeding focal planes, the virtual plane contains two profiles. No profiles are counted in (h) where the profiles located on the exclusion extension stay in touch with and merge on the moving line in succeeding focal planes. One profile is counted in (i) where the profiles located on the exclusion extension merge outside the moving line. (Reproduced from Larsen *et al.* (1998) with permission from The Royal Microscopical Society).

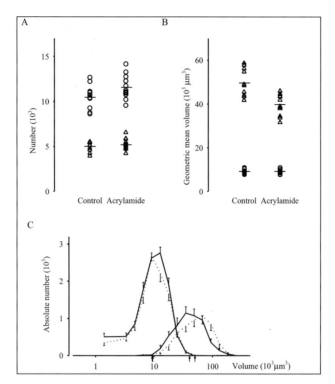

Fig. 9.5 Evaluation of geometric mean volume of L5 dorsal root ganglion cells from rats submitted to chronic acrylamide intoxication (500 mg acrylamide per kg injected intraperitoneally over a period of 7.5 weeks). (a) Number of A- and B-cells. Bars indicate mean in the group. (b) Geometric mean volume of A- and B-cells. Bars indicate mean in the group. (c) Distribution of volume of A- and B-cells. Arrows indicate geometric mean volume. The analysis shows that chronic acrylamide intoxication results in atrophy of A-cell perikarya of all sizes.

References

Andres, K.H. (1961). Untersuchungen über den Feinbau von Spinalganglien. *Z. Zellforsch.* **55**, 1–48.

Arvidsson, J., Ygge, J., and **Grant, G.** (1986). Cell loss in dorsal root ganglia and transganglionic degeneration after sciatic nerve resection in the rat. *Brain Res.* **373**, 15–21.

Baddeley, A.J., Gundersen, H.J.G., and **Cruz Orive, L.M.** (1986). Estimation of surface area from vertical sections. *J. Microsc.* **142**, 259–76.

Braendgaard, H., Evans, S.M., Howard, C.V., and **Gundersen, H.J.** (1990). The total number of neurons in the human neocortex unbiasedly estimated using optical disectors. *J. Microsc.* **157**, 285–304.

Cavanagh, J.B. (1979). The 'dying back' process. A common denominator in many naturally occurring and toxic neuropathies. *Arch. Pathol. Lab. Med.* **103**, 659–64.

Cavanagh, J.B. (1982). The pathokinetics of acrylamide intoxication: a reassessment of the problem. *Neuropathol. Appl. Neurobiol.* **8**, 315–36.

Dogiel, A.S. (1896). Der Bau der Spinalganglien bei den Saugetieren. *Anatomischer Anzeiger* **12**, Centralblatt für die Gesante Wissenschaffliche Anatomie, 140–52.

Duce, I.R. and **Keen, P.** (1977). An ultrastructural classification of the neuronal cell bodies of the rat dorsal root ganglion using zinc iodideosmium impregnation. *Cell Tissue Res.* **185**, 263–77.

Gundersen, H.J. (1977). Notes on the estimation of the numerical density of arbitrary profiles: the edge effect. *J. Microsc.* **111**, 219–23.

Gundersen, H.J. (1986). Stereology of arbitrary particles. A review of unbiased number and size estimators and the presentation of some new ones, in memory of William R. Thompson. *J. Microsc.* **143**, 3–45.

Gundersen, H.J. (1988). The nucleator. *J. Microsc.* **151**, 3–21.

Jensen, E.B. and **Gundersen, H.J.** (1993). The rotator. *J. Microsc.* **170**, 35–44.

Lawson, S.N. (1979). The postnatal development of large light and small dark neurons in mouse dorsal root ganglia. *J. Neurocytol.* **8**, 275–94.

Lieberman, A.R. (1976). Sensory ganglia. In *The peripheral nerve* (ed. D.N. Landon), pp. 188–278. Chapman and Hall, London.

Moller, A., Strange, P., and **Gundersen, H.J.** (1990). Efficient estimation of cell volume and number using the nucleator and the disector. *J. Microsc.* **159**, 61–71.

Rambourg, A.Y., Clermont, T., and **Beaudet, A.** (1983). Ultrastructural features of six types of neurons in rat dorsal root ganglia. *J. Neurocytol.* **12**, 47–66.

Schmalbruch, H. (1987). The number of neurons in dorsal root ganglia L4–L6 of the rat. *Anat. Rec.* **219**, 315–22.

Sørensen, F.B. (1992). Quantitative analysis of nuclear size for objective malignancy grading: a review with emphasis on new, unbiased stereological methods. *Lab. Invest.* **66**, 4–23.

Sørensen, F.B. (1995). Unbiased stereological techniques for practical use in diagnostic histopathology. *Pathologica* **87**, 263–78.

Tandrup, T. (1993). A method for unbiased and efficient estimation of number and mean volume of specified neuron subtypes in rat dorsal root ganglion. *J. Comp. Neurol.* **329**, 269–76.

Tandrup, T. and **Braendgaard, H.** (1994). Number and volume of rat dorsal root ganglion cells in acrylamide intoxication. *J. Neurocytol.* **23**, 242–8.

Tandrup, T. and **Jakobsen, J.** (2003). Long-term acrylamide intoxication induces atrophy of dorsal root ganglion A-cells and of myelinated sensory axons. *J. Neurocytol.* **31**, 79–87.

Tandrup, T., Gundersen, H.J., and **Jensen, E.B.** (1997). The optical rotator. *J. Microsc.* **186**, 108–20.

CHAPTER 10

ESTIMATION OF NUMBER AND VOLUME OF IMMUNOHISTOCHEMICALLY STAINED NEURONS IN COMPLEX BRAIN REGIONS

KARL-ANTON DORPH-PETERSEN,
RABEN ROSENBERG, AND JENS R. NYENGAARD

1 Introduction

One of the most important methods for studying the central nervous system is by microscopy of histological slides. The methods for generating sections are multiple and the staining techniques are becoming more and more sophisticated. Stains based upon immunohistochemical methods have become a standard way to obtain information about subtypes of cells that were indistinguishable in classical non-specific stains. The more advanced histological techniques have put increasing demands upon the tissue processing and, especially in post-mortem studies of the human brain where the investigator has no control of a range of parameters, care has to be taken to make sure that findings are not confounded beyond interpretation (Lewis 2002). However, no matter how sophisticated the histological techniques, the final sections are still nothing but a sample of the original organ of interest. And in order to gain robust information about the brain—and not only about the sections—attention has to be given to how the sections were sampled (see Fig. 10.1). This is the essence of stereology: how to sample the region of interest in such a way that the final sections are directly representative of the original organ.

This chapter explains how to manage a worst-case scenario: the use of immuno-histochemical techniques in thick vibratome sections from post-mortem human brainstems. The sections show anisotropic shrinkage and a significant impact of lost caps. In addition, the relevant stereological techniques demand that the sections be cut with a random position and rotation without losing track of the exact position in the complex brainstem.

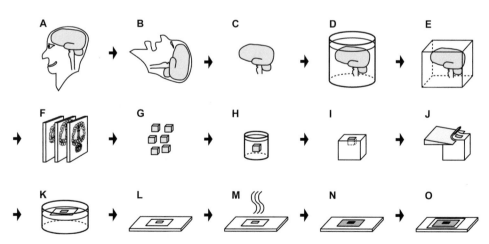

Fig. 10.1 The problem in a nutshell. We are interested in the living brain, but all we have is a slide! However, the final slides do contain direct information about the original organ if every step in the tissue processing is carefully managed. (Illustration from Dorph-Petersen *et al.* (2001*a*) with permission.)

1.1 Background

Since the advent of the method for unbiased number estimation—the disector technique—in the mid-1980s (Sterio 1984), several modifications of the method have been developed: the optical disector as well as the fractionator (Gundersen 1986), the optical fractionator (West *et al.* 1991), and, most recently, the smooth fractionator (Gundersen 2002). However, although there exist plenty of methods for estimating the number of, say, neurons, some specific hurdles need to be addressed in studies based upon immunohistochemical techniques. Antibody sensitivity, specificity, and penetration, as well as antigen stability are of obvious relevance. However, the use of cryostat or vibratome sections adds issues of shrinkage and lost caps. For a review of stereological methods with focus on shrinkage management see Dorph-Petersen *et al.* (2001*a*).

There exist a plethora of stereological methods for cell size estimation. The most used estimators for cell volume are the nucleator (Gundersen 1988) and the rotator (Jensen and Gundersen 1993) methods. Most of these techniques are based upon so-called local stereology (Jensen 1998) and come with an additional requirement (compared to the simple estimation of cell number) of randomness in orientation of the histological sections. However, most of the brain is so complex that sections have to be cut with a particular orientation for consistent and efficient identification of the region of interest, which has limited the (correct) use of these techniques. An approach to the problem has been to assume isotropy of the observed structures and then cut the sections with the preferred orientation. However, in most cases, it is impossible to predict the magnitude of the resulting bias—most neurons are anisotropic (or at least irregular) in shape and oriented in an anisotropic way. An obvious exception is convex, almost spherical particles, such as some cell nuclei, where the bias due to non-rotation of the sections in some cases may be small (Schmitz *et al.* 1999). Nevertheless, there exist deigns where the sections are cut in such a way that both the stereological and the neuroanatomical requirements are met (see Dorph-Petersen 1999).

In this chapter, we estimate the number and size of immunostained, serotonergic neurons in the dorsal raphe (DR) nucleus in a material of human brainstems. The dorsal raphe is one of the major nuclei of the serotonergic system, which is located in the midline of the brainstem. (See the reviews by Törk (1990), Jacobs and Azmitia (1992), and Baumgarten and Grozdanovic (1997) for further details about the serotonergic system in general.) The dorsal raphe nucleus seems to be involved in a range of basic brain functions including the sleep–wake-arousal state of the brain, regulation of cortical blood flow, memory and cognition, motor functions, respiratory and cardiovascular activity, anxiety, and sexual behavior. (For reviews of the extensive literature see e.g. Soubrié (1986) and Jacobs and Fornal (1995).) Because of the complexity and anisotropy of the brainstem, it is an ideal brain region for illustration of the relevant stereological problems and some of their solutions.

2 Material and methods

2.1 Subject sampling and characteristics

Brains from five male subjects were included in the study (see Table 10.1). The subjects were part of the controls analyzed in a depression/suicide study of 20 subjects (Dorph-Petersen *et al.* 2001*b*). The brains were selected from a larger sample collected by the coroner's office in Aarhus (Denmark) and Bergen (Norway) from 1997 to 2000. The subjects were evaluated by an experienced psychiatrist (RR) based upon forensic autopsy reports, police reports (if available), and, if possible, by interviews with the family physician and/or the medical file. In addition the Danish subjects were screened using the central Danish register of psychiatry. The register includes all persons in Denmark whom have been treated at a psychiatric hospital or department. Exclusion criteria were: a history of psychiatric disorders; neurological disorders; significant alcohol or drug abuse; and serious systemic diseases of likely consequences. No absolute restrictions were put on post-mortem interval (PMI), that is, the time from death to fixation of the brain. All subjects were Caucasians of Scandinavian nationality.

2.2 Tissue preparation and sectioning

The brains were immersion fixed in 10% phosphate-buffered formalin (4% formaldehyde in 0.1M phosphate buffer, pH = 7.0) immediately after autopsy. The fixative was changed after approx. one week. The brains were stored in the fixative at room temperature (see storage times in Table 10.1).

Table 10.1 Basic data for the five subjects in the study

Subject number	Age (years)	Height (cm)	Brain weight* (g)	PMI (hours)[†]	Storage time (months)	Cause of death	Pre-death diagnosis[‡]
1	26	167	1550	40	24.1	Car accident	
2	47	174	1650	66	20.5	Cardiovascular	
3	57	189	1425	66	23.1	Cardiovascular	
4	58	179	1325	42	29.5	Cardiovascular	
5	69	180	1450	106	22.3	Cardiovascular	H, NIDD
Mean	51.4	178	1480	64	23.9		
CV[§]	0.32	0.05	0.08	0.42	0.14		

* Brain weight, weight of the fixed brain measured at the time of processing.

[†] PMI, Post-mortem interval (time from death to fixation).

[‡] H, Hypertension; NIDD, non-insulin-dependent diabetes.

[§] CV, Coefficient of variation, i.e. relative standard deviation (CV = SD/mean).

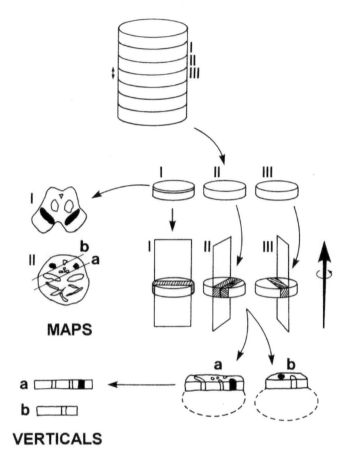

Fig. 10.2 Overview of the 'mapping' design generating both transversal mapping sections for easy identification and counting as well as vertical sections for neuronal somal volume estimation. (Illustration from Dorph-Petersen (1999) with permission.)

Fig. 10.3 (see also **Plate 6**) (Top four pictures) A brainstem is agar-embedded in a cutting-guide-box and sliced into 2.93-mm-slabs. The resulting set of slabs as well as one slab mounted on the vibratome adaptor is shown. (Bottom four pictures) From the remainder of a slab 2.28 mm wide bars are cut systematic, uniformly random using a razor blade and a randomly rotated cutting guide. Only the bars hitting the dorsal raphe are cut. The exact position of each bar is recorded in maps of the transversal sections. The bars are mounted 'side up' on the vibratome adaptor before the vertical sections are cut.

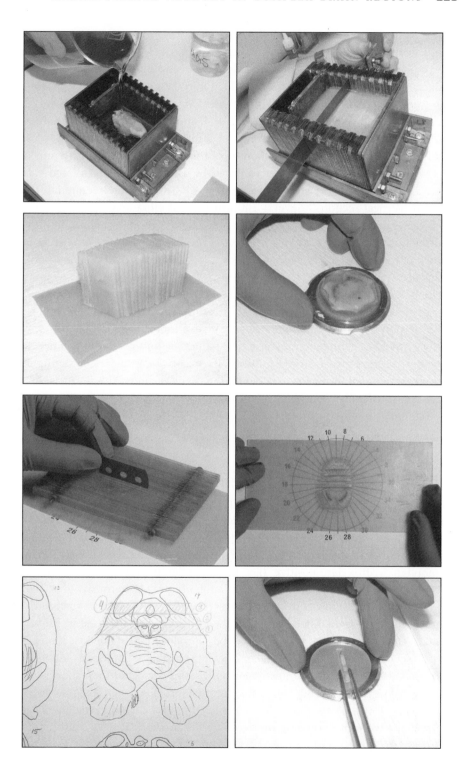

The brains were processed as follows (see Fig. 10.2 for an overview). First the brains were weighed. Then the brainstems were separated from the cerebrum and the cerebellum, and coded to secure blinding of the investigator (KADP). Each brainstem was embedded in 7% agar and cut into slabs perpendicular to its longest axis (see Fig. 10.3). The mean slab thickness was 2.93 mm. Each slab was subsequently re-embedded flatly in 7% agar. Using a calibrated Bio-Rad Polaron H1200 vibratome, several 100 μm transversal mapping sections were cut in parallel to the slab surface. Later, after staining of the mapping sections, 2.28 mm wide bars were cut from the remainder of the slabs using a randomly rotated cutting guide (cf. Fig. 10.3); this step provides the uniform rotation required for subsequent 'vertical sections'. Several 100 μm vertical vibratome sections were cut from each bar parallel to the axis of the brainstem. Although random in position and rotation, the actual positions of these vertical sections in the brainstem were known in relation to the mapping sections. (For a more detailed description of the design see Dorph-Petersen (1999).)

The first complete section (e.g. of full area) from each slab or bar was mounted on a glass slide and air-dried at room temperature. The first complete section is not of full section thickness and therefore not suitable for stereological measurements, but was stored as a guide for aligning the stained sections correctly. The next four vibratome sections were stored individually in small capsules (containing 10% phosphate-buffered formalin) to avoid drying of the sections. This is crucial for the immunohistochemical stain and for reducing the z-axis shrinkage. The capsules were stored at room temperature.

2.3 Staining

From each brainstem two sets of mapping sections and two sets of vertical sections were stained. One set (mapping + vertical) was stained for Nissl substance with toluidine blue. These sections were only used to identify neuroanatomical landmarks determining exact positions in the brainstem. In the second set of sections, tryptophan hydroxylase-positive (i.e. serotonergic) neurons were stained free-floating using PH8 antibody. The PH8 antibody was generously provided by Dr Richard G.H. Cotton, Royal Children's Hospital, Melbourne, Australia. The protocol used was a modified version of the protocol originally used by Törk et al. (1992). Omission of the primary antibody produced unstained sections. The specificity of the antibody has been previously addressed (Haan et al. 1987; Törk et al. 1992). Staining protocols for both stains are listed in Section 6. Extra thin coverslips (thickness 80–100 μm) were used instead of the standard coverslips (thickness ~160 μm) to allow the necessary amount of mounting media over the sections without exceeding the working distance of the 100× oil-immersion objective used in the study.

2.4 Shrinkage

Special attention to tissue shrinkage was necessary in this project. Vibratome sections are known to shrink a lot in thickness (Andersen and Gundersen 1999; Dorph-Petersen 1999), and free-float staining techniques could introduce shrinkage in the xy-plane of the sections.

The special problem of a pronounced and non-uniform shrinkage in the z-axis (the thickness) of the sections was treated in Dorph-Petersen *et al.* (2001*a*), in which shrinkage-robust, stereological methods for unbiased estimation of number and size were described. The current study was based upon these methods. In addition, the staining protocols were optimized in order to minimize the z-axis shrinkage. The final section thickness for the immunostained mapping sections and the vertical sections were 50.6 μm (0.09) and 43.7 μm (0.18), respectively (inter-individual coefficient of variation, CV = SD/mean, in parentheses), that is, the sections collapsed on average 53% along the z-axis. A considerable intraindividual point-to-point variation in section thickness was observed (CV $\approx$ 0.17).

The section areas were monitored for shrinkage by point-counting: a transparency with a fine point grid was superposed over the slabs and bars after sectioning and again over the final sections, and points hitting the tissue counted (see Fig. 10.4). A rather large number of points were counted per set of sections—275 to 400 and 600 to 900 points per set of mapping or vertical sections, respectively. This number of points corresponds to a coefficient of error (CE) of the

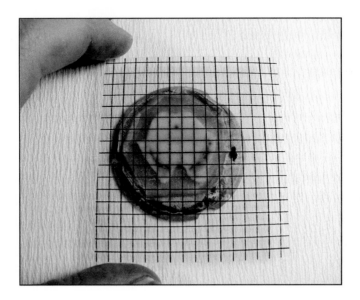

Fig. 10.4 A transparency with a 4 mm × 4 mm square grid is superposed over each tissue slab. A fast, simple, and unbiased estimate of the slab area is made by manually counting the number of crosses hitting the slab.

individual area estimates of about 0.015 (Gundersen and Jensen 1987, fig. 18). The high precision was necessary in order to rule out any significant areal shrinkage. Mean areal shrinkage of 0.9% (2.01) and 3.9% (0.37) was detected for the mapping sections and the vertical sections, respectively (CV in parenthesis). No areal shrinkage correction was used in the study because of the insignificant magnitude of the shrinkage.

A rough estimate of the shrinkage due to fixation was obtained by comparing the fresh brain weights (listed in the forensic autopsy reports) with the fixed brain weights measured at the time of processing. On average, the brain weights increased 3%, suggesting only a minor impact of fixation on global volume.

2.5 Delineation of the dorsal raphe

The dorsal raphe was delineated in the mapping sections as described by Baker *et al.* (1990) and Paxinos and Huang (1995) and facilitated by the antibody stain providing a high contrast to the surrounding, unstained tissue (Fig. 10.5). All positive cells located between or dorsal to the medial longitudinal fascicles and ventral to the cerebral aqueduct (of Sylvius) or fourth ventricle were included. The dorso-

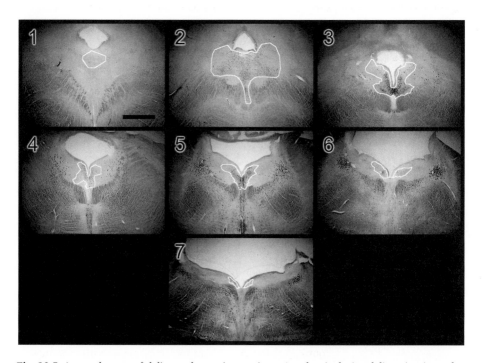

Fig. 10.5 A complete set of delineated mapping sections. A rather inclusive delineation is used including all stained neurons dorsal to the medial longitudinal fascicles. The scale bar is 3 mm, which is also the separation of the sections.

lateral 'wings' of the region, containing few and very scattered positive cells, were fully included guided by the immunostain (but would probably not be detected as a part of the nucleus in a Nissl stain). Rostrally, the first positive neurons were observed at the level of the oculomotor nucleus and caudally defined to end at the level of the caudal pole of the locus coeruleus and the rostral pole of the motor nucleus of the trigeminal nerve. At that level only a few scattered positive neurons, if any, were observed in the sections.

2.6 Microscopy

A modified Olympus BH2 microscope equipped with a MT12 microcator and ND281 read-out (Heidenhain, Germany), a motorized specimen stage (Merzhauser, Germany), and an object rotator (Olympus, Denmark) were used for the microscopy. A CCD camcorder (JAI-2040, Protec, Japan) connected to an IBM PC was mounted on top of the microscope. The computer fitted with a frame grabber (Screen Machine II, Fast Multimedia, Germany) was connected to a 17 inch monitor (IBM, USA) and had the CAST-Grid stereology software (Olympus, Denmark) installed. The initial delineation of the region of interest was done using a 4× objective (Olympus, S-Plan, numerical aperture (NA) 0.13) at a final magnification of 97×. The sections were studied using a 100× oil-immersion objective (Olympus, S-Plan, NA 1.25, working distance 340 μm) at a final magnification of 2420×.

Systematic, uniformly random sampling of fields of view was performed with constant sampling intensity in each set of sections. In each field of view, an unbiased counting frame was superimposed and, by focusing down through the section, optical disector probes were used for sampling of the neurons. The nucleoli of the neurons were used as the sampling unit. No cells with more than one nucleolus were observed. The sampling parameters for the mapping as well as the vertical sections are given in Table 10.2. Because of the pronounced variation in section thickness, a stratified sampling scheme with two different disector heights was used (see details in Section 7). The section thickness was evaluated roughly in all fields of view. The standard disector height (h_1) was used if the local section thick-

Table 10.2 Magnification, X- and Y-step length, area of sampling frame (A(frame)), height of main (h_1) and secondary (h_2) disectors, and finally size of point grid (a(point)) used for shrinkage measurements. The X- and Y-steps were always equal. For the vertical sections the X- and Y-steps were optimized for each subject guided by the corresponding number estimate from the mapping sections

Sections	Magnification	X-step (μm)	Y-step (μm)	A(frame) (μm²)	h_1 (μm)	h_2 (μm)	a(point) (mm²)
Mapping	2420×	260	260	4960	25	10	16.2
Vertical	2420×	180–240	180–240	4960	20	10	0.98

nesses were larger than or equal to 44 μm for the mapping sections and 35 μm for the vertical sections. If the section thickness was smaller than the respective cut-off values, the smaller disector height (h_2) was used. The section thickness was measured precisely in the center of every third 'h_1 frame' and all 'h_2 frames' in the mapping sections and in all the frames in the vertical sections. Using these sampling schemes, an average of 135 (111–164) neurons were sampled in 869 (531–1372) counting frames per set of mapping sections and 189 (179–203) neurons were sampled in 1239 (911–2056) counting frames per set of vertical sections. This implies that only approximately one neuron was sampled for every six counting frames examined. Typically, it is most convenient to sample one to two cells per counting frame. However, in the current project this was not possible due to the very pronounced inhomogeneity in the distribution of the serotonergic neurons in the dorsal raphe.

2.7 Stereological estimation of serotonergic neuron number and volume

The total number N of PH8-positive neurons was estimated for each subject from the mapping sections using the number-weighted mean section thickness based optical fractionator described in Dorph-Petersen *et al.* (2001*a*, equations 7, 9, and 10)

$$N := \frac{1}{\text{ssf}} \cdot \frac{1}{\text{asf}} \cdot \frac{1}{\text{hsf}} \cdot \sum Q^-$$

$$\text{hsf} := \frac{h}{\bar{t}_{Q^-}}$$

$$\bar{t}_{Q^-} := \frac{\sum_i \left(t_i \cdot \bar{q}_i^- \right)}{\sum_i \bar{q}_i^-}$$

where ssf, asf, and hsf are the section sampling fraction, the area sampling fraction, and the height sampling fraction, respectively. $\sum Q^-$ is the total number of the sampled neurons. hsf is estimated from the disector height h and the number-weighted mean section thickness $\bar{t}_{Q^-} \cdot \bar{t}_{Q^-}$ is estimated from the local section thickness t_i and the local neuronal count $\bar{q}_i^-$ in a systematic fraction of the counting frames.

The volume of the containing space, $V(\text{DR})$, was estimated by the Cavalieri method from the area delineated in the mapping sections (Gundersen and Jensen 1987) as

$$V(\text{DR}) := T \cdot \sum A$$

where T is the intersectional distance and $\sum A$ is the sum of the sectional areas.

Volumes of individual PH8-positive neurons were estimated from the vertical sections using the rotator method (Jensen and Gundersen 1993; see also Fig. 10.6).

The geometric mean $\overline{v_N}^g$ and the distribution frequencies were calculated as described in Dorph-Petersen *et al.* (2001*a*, equations 16 and 17):

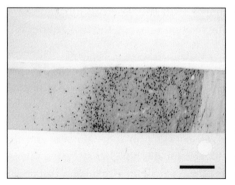
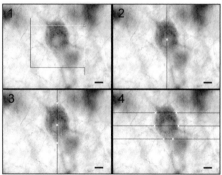

Fig. 10.6 (Left) A vertical section though the dorsal raphe showing numerous PH8-positive neurons. The scale bar is 1 mm. (Right) Volume estimation using the rotator method. (1) A neuronal nucleolus is sampled by an optical disector. (2) The nucleolus is marked. (3) The vertical extent of the neuronal soma (in the focal plane) is marked on the vertical line. (4) The horizontal extent of the neuron is marked. The scale bars are 10 μm.

$$\overline{v_N}^g = \exp\left(\frac{\sum\left(\ln(v_j)\cdot t_j/h_j\right)}{\sum\left(t_j/h_j\right)}\right)$$

$$F_i = N\cdot\frac{\sum\limits_{j=1}^{n_i}\left(t_{ij}/h_{ij}\right)}{\sum\left(t_j/h_j\right)}$$

where v_j is the estimated volume of the jth neuron, and t_j and h_j are the corresponding local section thickness and disector height. The absolute frequency F_i is the estimated number of neurons in the ith size class, from which n_i neurons were sampled. N is the previously estimated total number of neurons. $\sum\limits_{j=1}^{n_i}(t_{ij}/h_{ij})$ is the sum of the thickness/disector height fractions for the neurons belonging to the ith size class, while $\sum(t_j/h_j)$ is the sum of all t_j and h_j fractions (i.e. for all sampled neurons—not only for the neurons belonging to the particular size class). The last equation is a slightly modified version of that cited above (Dorph-Petersen *et al.* 2001*a*, equation 17). The above version estimates the frequencies of the absolute size distribution (in contrast to the relative size distribution).

2.8 Precision of the stereological estimates

The precision of the stereological estimates was evaluated for each type of estimate (total number, total volume, and mean somal volume) by comparing the coefficient of error (CE) of the estimate (due to sampling) with the observed interindividual coefficient of variation, OCV = SD/mean. The relation between the two entities is

$$OCV^2 = CV_{biol}^2 + CE^2.$$

The observed relative variance among subjects equals the sum of the true relative biological variance and the relative variance of the estimate itself. The true relative biological variance, CV_{biol}^2, cannot be observed directly and cannot be changed, as it depends upon the studied population. The CE depends on the estimator and the sample size. To decrease the CE of the individual estimates a higher sampling intensity (i.e. more work) is needed. Depending on this increase in workload a balance of optimal sampling per individual exists. It is evident that the CE should be kept low enough to ensure that a major fraction of the observed relative variance is due to the genuine relative biological variance. As a rule of thumb the CE should at least be kept below 50% of OCV if possible. More work will then only reduce the OCV marginally.

The method used for calculating the CE of the number estimates is described in Section 7. The CE of the reference volume estimates was calculated using the methods described in Gundersen *et al.* (1999). The CV of the somal volume distributions and the CE of the corresponding estimate of geometric mean somal volume was estimated using CV = SD($\ln(v)$) and CE = SEM($\ln(v)$). SD($\ln(v)$) and SEM($\ln(v)$) are the standard deviation and the standard error of the mean, respectively, of the natural logarithms of the individual somal volume estimates.

3 Results

The results of the study are shown in Table 10.3 and Figs 10.7 and 10.8. The distribution of unbiased somal volume estimates was plotted for each subject, but inspection of these (somewhat rough) plots did not reveal any patterns of sig-

Table 10.3 N is the total number of serotonergic neurons, $\overline{v_N}^g$ the geometric mean of the somal volume estimates, and V(DR) the volume of the dorsal raphe. The mean coefficient of error of an estimate is calculated as $\sqrt{\text{mean}(CE^2)}$, and likewise the mean coefficient of variation of the distribution of neuronal volume estimates is calculated as $\sqrt{\text{mean}(CV^2)}$

Subject number	N (CE)	$\overline{v_N}^g$ (μm^3) (CE) [CV]	V(DR) (mm^3) (CE)
1	175 000 (0.09)	14 500 (0.05) [0.72]	174 (0.02)
2	152 000 (0.09)	10 200 (0.07) [1.00]	172 (0.02)
3	112 000 (0.11)	15 900 (0.06) [0.81]	92 (0.05)
4	134 000 (0.09)	12 300 (0.07) [0.97]	130 (0.03)
5	136 000 (0.09)	16 200 (0.05) [0.73]	136 (0.03)
Mean	141 800 (0.09)	13 800 (0.06) [0.86]	141 (0.03)
CV	[0.16]	[0.18]	[0.24]

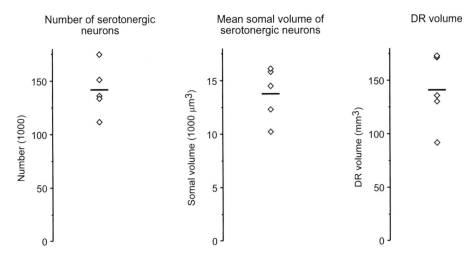

Fig. 10.7 Plots of the results. The horizontal lines indicate the respective means.

nificance. The average absolute distribution of the unbiased somal volume estimates is shown in Fig. 10.8.

Influence of PMI and tissue storage time was evaluated by plotting neuron
number, somal volume, or DR volume as a function of PMI or tissue storage time,
respectively. The plots did not reveal any significant correlation to the two
parameters.

4 Discussion

4.1 Precision of the estimates

Comparing the observed coefficients of variation and the mean coefficients of error
listed in Table 10.3 seems to indicate that the precision of the estimates of mean
somal volume and of DR volume is higher than needed, while the precision of the
individual number estimate could have been higher.

The estimated very low CE (0.03) of the DR volume estimates is an underestimate of the true CE. The very low density of serotonergic neurons in the rostral
cap and dorsolateral wings of DR limits the precision of the exact delineation in the
individual mapping sections. The listed CE does *not* include this extra delineation
variance. The CV of the individual area estimates was evaluated to be in the order
of a few per cent. This suggests a true CE of the volume estimate on the order of
0.05, which is still comfortably low.

At a first glance the absolute size of the CE (0.09) of the number estimate may
seem fine but the low observed CV (0.16) suggests that lowering the CE to ~0.07
would be appropriate. Unfortunately, practical limitations made further reduction

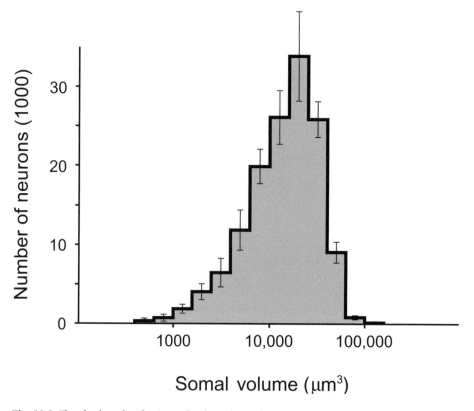

Fig. 10.8 The absolute distributions of unbiased somal volume estimates. Note that 'absolute distribution' indicates that: (1) the area under the graph equals the corresponding unbiased estimate of mean total serotonergic neuron number; (2) the distribution is centered around its (geometric) mean. The graph is based on 943 individual somal volume estimates. The thin bars indicate SEM for each bin.

of the CE difficult. The CE has two main components: the variance due to systematic uniform random sampling of sections, Var(SURS), and the variance due to the counting itself. Eliminating the SURS component would only bring the CE down to ~0.08, and a reduction would require more mapping sections, which is limited by the need for enough material for the vertical sections. Counting more neurons in each subject reduces the CE (with the limit of ~0.055 due to SURS), but it would take a doubling of the count to reduce the CE to ~0.08. This would require on average ~1700 counting frames per set of mapping sections due to the inhomogeneity of DR. It would therefore be rather labor-intensive to increase the number of sampled cells.

In the light of the reasoning above the applied sampling scheme was considered as good as possible given the practical constraints, that is, *de facto* optimal.

4.2 The relevance of post-mortem interval length

The PMI of the subjects included in this study are rather long (mean, 64 hours; CV, 0.42; range, 40–106 hours). This could very well be a severe problem for a study using immunohistochemistry, as the sensitivity as well as the specificity of the stain may be compromised by post-mortem degradation of the antigen of interest. The actual limit depends on the post-mortem stability of the antigen of interest, and may in some cases be very short (Lewis 2002). It is, of course, important that all— or a constant (large) fraction—of the positive neurons are identifiable from the negative background. On the other hand, the absolute intensity and variability of the specific stain are less important for a stereological study, such as the current study, where positive neurons are counted and measured independently of staining intensity.

If the specific stain is compromised due to a too long PMI one or more of the following effects are likely to be observed.

- The estimated total number of positive neurons will be lower than the (unbiased) estimates of total number reported by researchers studying subjects with short(er) PMI.

- The estimated total number of positive neurons will decrease with increased PMI.

- The distribution of estimates of total number of positive neurons among subjects will have an increased variance induced by the variation in PMI itself.

None of these effects were observed in the current study. By contrast, we got larger estimates of total neuron number than described by previous studies with low PMI (Underwood *et al.* 1999) and the variation in our estimates of total neuron number is not remarkable. Based on the observations of the current study we have no reason to believe that the sensitivity of the tryptophan hydroxylase labeling PH8 antibody was compromised by long PMI. In conclusion, the tryptophan hydroxylase appears to have a rather high post-mortem stability.

4.3 Relating results to those of other studies

Table 10.4 lists results of studies of normal subjects using the PH8 antibody technique. The optical disector probe was not used in any of the previous studies, which may therefore be influenced by the effect of lost caps. Cells being lost from the section surface during the sectioning process are the main cause of this effect. At the beginning of the current study a calibration of the problem (in 100 μm vibratome sections of DR) was performed by sampling nucleoli in the full section thickness and plotting the density distribution as a function of vertical position within the section. This calibration indicated that large guard zones above and below the disectors were needed and that underestimates of total number up to

Table 10.4 Results of quantitative studies using the PH8 staining of serotonergic neurons in human DR*

Study	Number of subjects	N(PH8)	V(DR) (mm³)
Baker *et al.* (1991)	7	165 000	
Chan-Palay *et al.* (1992)	3	130 000	
Baker *et al.* (1996)	8	108 000	
Underwood *et al.* (1999)	6	87 000	66
Present study	5	141 800	141

* Some of the data in the study of Baker *et al.* (1996) was previously published by Halliday *et al.* (1993). All the previous studies used 50 μm frozen sections (except for 70 μm sections in the study by Chan-Palay *et al.*).

20% were possible in 50 μm sections (less in thicker sections). This is in agreement with the findings of Andersen and Gundersen (1999). Because the current study used the optical disector with large guard zones, the problem of lost caps was eliminated. The calibration study also confirmed that the antibody fully penetrated our 100 μm sections, as well as linearity of the (pronounced) *z*-axis shrinkage.

The volumes of DR listed by the current study are twice as large as the volumes listed by Underwood *et al.* (1999). This is probably because of a wider delineation of DR in the current study with inclusion of all positive neurons dorsal to the medial longitudinal fasiculi instead of only within the classical boundaries of DR. In the Nissl study reported in Dorph-Petersen (1999), the volumes are of the same magnitude as the volumes reported by Underwood *et al.* (1999).

Only the current study also estimated the somal volume of the neurons in DR. In several of the previous studies different kinds of diameter measures were used to size the neurons. However, diameter measures depend on shape and orientation of the neurons and not only on their size. Therefore diameter measures are not very relevant approximations and are biased by method and—particularly in this anisotropic region of the brain—are less robust than the estimates of the somal volume obtained in the current study, which are based on unbiased principles.

5 Concluding summary

The following take-home-messages are essential in any stereological study based upon immunohistochemistry.

1 *Know your antibody and your tissue*—it is crucial that the antibody sensitivity, specificity, dependence on PMI and storage time, as well as penetration are addressed.

2 *Calibrate for lost caps and antibody penetration*—lost caps and/or lack of antibody penetration may be a significant problem in your material.

3   *Monitor and manage shrinkage*—this includes point counts of sectional areas and multiple section thickness measurements as well as histological techniques optimized for reducing tissue deformation.

4   *Choose the right design*—keep the study as simple as possible but not too simple. A more complex sampling design (e.g. by generating vertical sections) may lead to simpler results and clear-cut conclusions.

6 Appendix 1. Protocol for PH8 staining of 100 μm vibratome sections, free-floating

1   H_2O_2 0.1% (w/w) in tris-buffered saline (TBS; deactivation of endogenous peroxidase) for 1 hour.

2   Triton X-100 0.1% (vol.) in TBS (enhancement of antibody penetration) for 1 hour.

3   Normal horse serum 10% (vol.) in TBS (blocking of non-specific binding sites) for 1 hour.

4   Primary antibody (PH8, raised in mouse) 1:30 000 (vol.) in TBS + 0.1% sodium azide to prevent decomposition due to bacteria and fungi for 45 hours at 4°C.

5   Rinse in TBS, 3×10 min.

6   Biotinylated horse anti-mouse IgG 0.5% (vol.) + 1.5% (vol.) normal blocking serum (horse) in TBS for 5 hours.

7   Rinse in TBS, 3×10 min.

8   ABC (avidin–biotin–peroxidase complex) reagent 4% (vol.) in TBS for 18 hours.

9   Rinse in TBS, 8×5 min.

10 3,3′-diaminobenxidine (DAB) 0.091% + 0.027% H_2O_2 + 0.091% $NiCl_2$ in TBS for a maximum of 3 min.

11 Rinse in TBS, 8×5 min.

12 Mount sections at object glass and air dry at room temperature. Keep the drying time as short as possible (but long enough to leave sections firmly fixed to the slides), that is, 30–45 min.

13 Re-hydration in demineralized water for 15 min.

14 Dehydration:
    ethanol 50% (vol.), 10 min.
    ethanol 70% (vol.), 10 min.
    ethanol 96% (vol.), 2×5 min.
    ethanol 99% (vol.), 2×1 min.

ethanol 99% (vol.), 1 × 2 min.
xylene, 15 min

15   Cover slip with XAM.

In steps 6 and 8 Vectastain *Elite* ABC Kit (peroxidase, mouse-IgG), Vector (PK-6102) is used. In step 10 DAB Substrate Kit, Vector (SK-4100) is used. The standard concentrations were used. XAM is a xylene-based histological mounting medium.

7 Appendix 2. Estimation of the PH8-positive neuron number N and the corresponding CE for subject number 5

The estimate of the total number of PH8-positive neurons in DR is slightly complicated by the fact that a stratified design was used. The highly variable section thickness was measured in every third field of view except when thinner than 44 µm—in that case all section thicknesses were recorded. The section thickness was clearly greater than 44 µm in most fields of view. The estimator of the total neuronal number is actually the sum of the neuronal number estimated in each of the two strata. 'Thick fields of view' correspond to stratum 1 and 'thin fields of view' to stratum 2.

First, the number-weighted mean section thickness is estimated for each stratum using the equation listed in Section 2.7

$$\overline{t_{Q^-_1}} = 51.9 \ \mu m; \quad \overline{t_{Q^-_2}} = 33.0 \ \mu m.$$

The total number of PH8-positive neurons N is estimated from these values and the known constants of the design

$$\text{tsf} := BA/\overline{T} = 100 \ \mu m/2.93 \ mm = 1/29.3$$

$$\text{asf} := A(\text{frame})/(X \text{ step} \times Y \text{ step}) = 4960 \ \mu m^2/(260 \ \mu m \times 260 \ \mu m) = 1/13.63$$

$$\text{hsf}_1 := h_1/\overline{t_{Q^-_1}} = 25 \ \mu m/51.9 \ \mu m = 0.481$$

$$\text{hsf}_2 := h_2/\overline{t_{Q^-_2}} = 10 \ \mu m/33.0 \ \mu m = 0.303$$

$$N := \frac{1}{\text{tsf}} \cdot \frac{1}{\text{asf}} \cdot \left(\frac{1}{\text{hsf}_1} \cdot \sum Q^-_1 + \frac{1}{\text{hsf}_2} \cdot \sum Q^-_2 \right)$$

$$= 29.3 \cdot 13.63 \cdot \left(\frac{1}{0.481} \cdot 131 + \frac{1}{0.303} \cdot 21 \right) = 136314 \approx 136000.$$

tsf is the tissue sampling fraction, asf the area sampling fraction, and hsf_1 and hsf_2 the height sampling fractions. BA is the block advance, that is, the amount of tissue removed from the block for each section by the (calibrated) vibratome. (See Dorph-Petersen *et al.* (2001*a*) for discussion of BA.) $\overline{T}$ is the mean slab thickness.

Table 10.5 The counts for each stratum in each of the eight sections of DR

$t \geq 44$ μm Section number	$Q^-_{i,1}$	$Q^-_{i,1} \cdot Q^-_{i,1}$	$Q^-_{i,1} \cdot Q^-_{i+1,1}$	$Q^-_{i,1} \cdot Q^-_{i+2,1}$
1	0	0	0	0
2	70	4900	3360	280
3	48	2304	192	192
4	4	16	16	12
5	4	16	12	8
6	3	9	6	0
7	2	4	0	
8	0	0		
Sum	131	7249	3586	492
	Noise$_1$	A_1	B_1	C_1

$t < 44$ μm Section number	$Q^-_{i,2}$	$Q^-_{i,2} \cdot Q^-_{i,2}$	$Q^-_{i,2} \cdot Q^-_{i+1,2}$	$Q^-_{i,2} \cdot Q^-_{i+2,2}$
1	0	0	0	0
2	3	9	0	18
3	0	0	0	0
4	6	36	42	6
5	7	49	7	0
6	1	1	0	4
7	0	0	0	
8	4	16		
Sum	21	111	49	28
	Noise$_2$	A_2	B_2	C_2

The CE is calculated based upon the principles explained by Gundersen *et al.* (1999) modified for the stratified version of the fractionator. The numbers calculated in Table 10.5 are the intermediate values used in the CE estimator.

A CE estimate for each stratum is calculated

$$\text{Noise}_1 := \sum Q^-_1 = 131$$
$$\text{Var}(\text{SURS}_1) := \left(3(A_1 - \text{Noise}_1) - 4B_1 + C_1\right)/240 = (21354 - 14344 + 492)/240$$
$$= 31.26$$
$$\text{CE}_1 := \sqrt{\text{Noise}_1 + \text{Var}(\text{SURS}_1)}\Big/\sum Q^-_1 = \sqrt{131 + 31.26}/131 = 0.097$$

$$\text{Noise}_2 := \sum Q^-_2 = 21$$
$$\text{Var}(\text{SURS}_2) := \left(3(A_2 - \text{Noise}_2) - 4B_2 + C_2\right)/240 = (270 - 196 + 28)/240 = 0.425$$
$$\text{CE}_2 := \sqrt{\text{Noise}_2 + \text{Var}(\text{SURS}_2)}\Big/\sum Q^-_2 = \sqrt{21 + 0.425}/21 = 0.220$$

where Var(SURS) is the estimator variance due to systematic, uniformly random sampling and Noise is the variance due to noise of the counting itself.

The two CE estimates are then combined (by weighted addition) into one CE for the estimate of the total N:

$$W_1 := \left(\sum Q_1^-\right)^2 \Big/ \left(\sum Q_1^- + \frac{\text{hsf}_1}{\text{hsf}_2}\cdot\sum Q_2^-\right)^2 = 131^2 \Big/ \left(131 + \frac{0.481}{0.303}\cdot 21\right)^2 = 0.636$$

$$W_2 := \left(\sum Q_2^-\right)^2 \Big/ \left(\frac{\text{hsf}_2}{\text{hsf}_1}\cdot\sum Q_1^- + \sum Q_2^-\right)^2 = 21^2 \Big/ \left(\frac{0.303}{0.481}\cdot 131 + 21\right)^2 = 0.041$$

$$\text{CE} := \sqrt{W_1\cdot\text{CE}_1^2 + W_2\cdot\text{CE}_2^2} = \sqrt{0.636\cdot 0.097^2 + 0.041\cdot 0.220^2} = 0.089 \approx 0.09.$$

This estimate of the CE for the stratified version of the fractionator is conservative, as it is assumed that thin fields of view are distributed randomly. In case particular sections are thinner than average, the result is a bad shape and therefore high CE estimates for each strata. However, the CE of the combined estimator will depend upon the overall shape of the dorsal raphe which is smoother than each of the two strata alone. The estimator therefore in some cases overestimates the CE.

Acknowledgments

Professor Hans Jørgen G. Gundersen is thanked for his invaluable guidance and thoughtful inputs. The skilful technical assistance of Maj-Britt Lundorf is greatly appreciated. The Institute of Forensic Medicine at the University of Aarhus and Professor Inge Morild, Department of Pathology, Gades Institut, Haukeland Sygehus in Bergen are thanked for collecting human brains. Dr Richard G.H. Cotton, Murdoch Institute, Royal Children's Hospital, Melbourne, Australia is thanked for generously donating a batch of PH8 antibody. This study was supported by grants from Århus Universitetshospitals Forskningsinitiativ, P. Carl Petersens Fond, Psykiatrisk Forskningsfond, Fonden til Lægevidenskabens Fremme, Overlæge, dr.med. Einer Geert-Jørgensen og hustru Ellen Geert-Jørgensens Forskningslegat, Pulje til Styrkelse af Psykiatrisk Forskning, Eli Lilly Danmarks Psykiatriske Forskningsfond and Lægeforeningens Forskningsfond.

References

Andersen, B.B. and **Gundersen, H.J.G.** (1999). Pronounced loss of cell nuclei and anisotropic deformation of thick sections. *J. Microsc.* **196 (1)**, 69–73.

Baker, K.G., Halliday, G.M., and **Törk, I.** (1990). Cytoarchitecture of the human dorsal raphe nucleus. *J. Comp. Neurol.* **301**, 147–61.

Baker, K.G., Halliday, G.M., Hornung, J.-P., Geffen, L.B., Cotton, R.G.H., and **Törk, I.** (1991). Distribution, morphology and number of monoamine-synthesizing and substance p-containing neurons in the human dorsal raphe nucleus. *Neuroscience* **42** (3), 757–75.

Baker, K.G., Halliday, G.M., Kril, J.J., and **Harper, C.G.** (1996). Chronic alcoholics without Wernicke–Korsakoff syndrome or cirrhosis do not lose serotonergic neurons in the dorsal raphe nucleus. *Alcohol Clin. Exp. Res.* **20** (1), 61–6.

Baumgarten, H.G. and **Grozdanovic, Z.** (1997). Anatomy of central serotoninergic projection systems. In *Serotoninergic neurons and 5-HT receptors in the CNS* (ed. H.G. Baumgarten and M. Göthert), pp. 41–89. Springer, Berlin.

Chan-Palay, V., Höchli, M., Jentsch, B., Leonard, B., and **Zetzsche, T.** (1992). Raphe serotonin neurons in the human brain stem in normal controls and patients with senile dementia of the Alzheimer type and Parkinson's disease: relationship to monoamine oxidase enzyme localization. *Dementia* **3**, 253–69.

Dorph-Petersen, K.-A. (1999). Stereological estimation using vertical sections in a complex tissue. *J. Microsc.* **195**, 79–86.

Dorph-Petersen, K.-A., Nyengaard, J.R., and **Gundersen, H.J.G.** (2001*a*). Tissue shrinkage and unbiased stereological estimation of particle number and size. *J. Microsc.* **204**, 232–47.

Dorph-Petersen, K.-A., Nyengaard, J.R., Rosenberg, R., and **Gundersen, H.J.G.** (2001*b*). No change in serotonergic neuron number and volume in the human dorsal raphe in depression but observation of a pronounced sexual dimorphism—a stereological study using unbiased principles. *Soc. Neurosci. Abstr.*, Vol. 27, Program no. 111.6.

Gundersen, H.J.G. (1986). Stereology of arbitrary particles. A review of unbiased number and size estimators and the presentation of some new ones, in memory of William R. Thompson. *J. Microsc.* **143** (1), 3–45.

Gundersen, H.J.G. (1988). The nucleator. *J. Microsc.* **151**, 3–21.

Gundersen, H.J.G. (2002). The smooth fractionator. *J. Microsc.* **207** (13), 191–210.

Gundersen, H.J.G. and **Jensen, E.B.** (1987). The efficiency of systematic sampling in stereology and its prediction. *J. Microsc.* **147** (3), 229–63.

Gundersen, H.J.G., Jensen, E.B.V., Kiêu, K., and **Nielsen, J.** (1999). The efficiency of systematic sampling in stereology—reconsidered. *J. Microsc.* **193** (3), 199–211.

Haan, E.A., Jennings, I.G., Cuello, A.C., Nakata, H., Fujisawa, H., Chow, C.W., Kushinsky, R., Brittingham, J., and **Cotton, R.G.H.** (1987). Identification of serotonergic neurons in human brain by a monoclonal antibody binding to all three aromatic amino acid hydroxylases. *Brain Res.* **426**, 19–27.

Halliday, G., Ellis, J., Heard, R., Caine, D., and **Harper, C.** (1993). Brainstem serotonergic neurons in chronic alcoholics with and without the memory impairment of Korsakoff's psychosis. *J. Neuropathol. Exp. Neurol.* **52** (6), 567–79.

Jacobs, B.L. and **Azmitia, E.C.** (1992). Structure and function of the brain serotonin system. *Physiol. Rev.* **72** (1), 165–229.

Jacobs, B.L. and **Fornal, C.A.** (1995). Serotonin and behavior. A general hypothesis. In *Psychopharmacology: the fourth generation of progress* (ed. F.E. Bloom and D.J. Kupfer), pp. 461–9. Raven Press, New York.

Jensen, E.B.V. (1998). *Local stereology.* World Scientific, Singapore.

Jensen, E.B.V. and **Gundersen, H.J.G.** (1993). The rotator. *J. Microsc.* **170**, 35–44.

Lewis, D.A. (2002). The human brain revisited: opportunities and challenges in post-mortem studies of psychiatric disorders. *Neuropsychopharmacology* **26** (2), 143–54.

Paxinos, G. and **Huang, X.-F.** (1995). *Atlas of the human brainstem.* Academic press, San Diego.

Schmitz, C., Schuster, D., Niessen, P., and **Korr, H.** (1999). No difference between estimated mean nuclear volumes of various types of neurons in the mouse brain obtained on either isotropic uniform random sections or conventional frontal or sagittal sections. *J. Neurosci. Methods* **88**, 71–82.

Soubrié, P. (1986). Reconciling the role of central serotonin neurons in human and animal behavior. *Behav. Brain Sci.* **9**, 319–64.

Sterio, D.C. (1984). The unbiased estimation of number and sizes of arbitrary particles using the disector. *J. Microsc.* **134** (2), 127–36.

Törk, I. (1990). Anatomy of the serotonergic system. *Ann. NY Acad. Sci.* **600**, 9–35.

Törk, I., Halliday, G.M., and **Cotton, R.G.H.** (1992). Application of antiphenylala-nine hydroxylase antibodies to the study of the serotonergic system in the human brain. *J. Chem. Neuroanat.* **5**, 311–13.

Underwood, M.D., Khaibulina, A.A., Ellis, S.P., Moran, A., Rice, P.M., Mann, J.J., and **Arango, V.** (1999). Morphometry of the dorsal raphe nucleus serotonergic neurons in suicide victims. *Biol. Psychiatry* **46**, 473–83.

West, M.J., Slomianka, L., and **Gundersen, H.J.G.** (1991). Unbiased stereological estimation of the total number of neurons in the subdivisions of the rat hippocampus using the optical fractionator. *Anat. Rec.* **231**, 482–97.

LENGTH AND SURFACE

CHAPTER 11

SECTION INTRODUCTION

STEPHEN M. EVANS AND JENS R. NYENGAARD

1 Introduction

Estimates of length and surface area provide important information regarding the transfer of material and information. The estimators for surface area and length are closely related. Remember from Chapter 1 that surface area, a two dimensional parameter, is estimated using a line probe i.e. a one dimensional probe and length, a one dimensional parameter, is estimated using an area probe i.e. a two dimensional probe. These relationships and the associated fundamental stereological equations arose from the work of Saltykov (1945 and 1946).

A neurobiologist might wish to know how efficiently a particular area of the brain processes information and how this changes with disease. One of the first parameters to measure would be neuron and synapse number but this might not be enough to demonstrate subtle changes. The next logical step would be to look at other parameters that contribute to the processing of information, for example, the neurons' interconnectivity. Two other important parameters to investigate are the average neuronal surface area (since this would influence the receptive area for synapses) and, of course, the length of dendrites per neuron.

Estimators of surface area are not discussed in detail in this book but a brief outline of the theory and some basic methodology is given below. It is included in this section because of the relationship between surface and length estimators. Surface area is an important biological parameter since it provides information on the exchange of materials across a membrane, for example:

- gaseous exchange over the blood-brain barrier, etc.;
- movement of neurotransmitters into and out of synaptic clefts;
- interaction between neurons and astrocytes as a result of the surface area contact of astrocyte podocytes.

Objects which have a measurable length act as conduits e.g. for information (dendrites and axons), nutrients (blood vessels), force (muscle fibres), and length estimators provide important measures on how these parameters conducted. For example, investigation of the microvasculature by estimation of capillary length density would provide information concerning cerebral blood flow and also forms the basis of the theories underlying modern medical imaging. Consequently, several methods are presented in detail. Other methods not described in detail in this section but mentioned in the subsequent chapters include methods using thick sections and isotropic, virtual planes (Larsen *et al.*, 1998) or virtual spheres—'space balls' (Mouton *et al.*, 2002). The direction of sectioning can now be arbitrary since an isotropic probe is generated. If thick vertical uniformly random sections are available, the length density of the linear structures of interest can be estimated from projection images of the probes through thick sections (Gokhale 1990). In a thick vertical section an isotropic, uniformly random (IUR) surface can be generated by projecting a cycloid through the section if the major axis of the cycloid is parallel to the vertical direction. If the thick vertical section contains the entire linear structure then the total length of the linear structure can be directly estimated without needing to know the section thickness (Gokhale 1992).

Strictly speaking lines/curves in three-dimensional space have a clear mathematical definition with no volume and no surface area. In practice this definition breaks down since most biological structures where length estimates are required are, in fact, tubules. However, when estimating the length of a structure it is assumed that it is 'much longer than it is wide'. Tubules are also subject to over- and underprojection effects, but the closer the tubule approaches a true line/curve, that is, with no surface area or volume, the more these effects diminish.

2 Isotropy revisited

In Chapter 1 isotropy and its requirement for the estimation of length and surface were discussed. However, neuroanatomists have been trained to identify structures on sections when the sectioning has been performed in a specific direction in three-dimensional space, for example, coronally. Consequently, if there is a requirement to generate IUR sections (Mattfeldt *et al.* 1990, Nyengaard and Gundersen 1992), which would make the identification of structures difficult, the experiment may become unfeasible. So before talking about length and surface estimators it would be appropriate to revisit the concept of isotropy. An isotropic direction in space can be defined by two angles, which in the case of Fig. 11.1 are angles v and h.

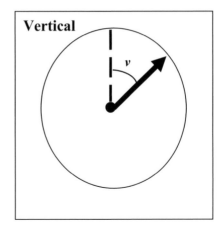

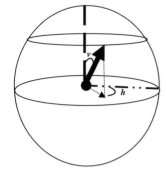

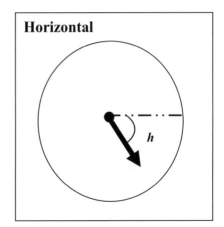

Fig. 11.1 Directions in space are defined by two angles. In the diagram above the direction of the arrowed line is defined by the two angles, *v* and *h*, where *v* represents the angle between a vertical axis and the line in the vertical plane and *h* represents the angle between a horizontal axis and the projection of the line in the horizontal plane. To generate an isotropic direction in space, *v* and *h* would have to be chosen at random and a method to do this is described in Chapter 1. Notice that in this case the length of the line in the vertical plane equals the radius of the sphere but that its length in the horizontal plane varies depending on the angle *v*. Baddeley *et al.* (1986) exploited this in their 'vertical sections' paper to design a method for generating isotropic test lines in three-dimensional space with a preferred direction of sectioning. Only one of the angles, in this case *h*, needs to be random but the length of the line should then be varied randomly on the section. To be more mathematically precise, the length density of the test lines in the vertical sections needs to be sine-weighted. Baddeley *et al.* also pointed out that there is a shape in mathematics that represents this random variation in length, namely, a cycloid, and it is a 'sine-weighted' straight line.

One way to generate isotropic test lines in three-dimensional space is to place test lines of equal length on an IUR section. Another way is to use either cycloids or 'sine-weighted' straight lines on vertical sections. There are a few other ways of generating a random test line in three-dimensional space and that is to do a three-dimensional reconstruction and then place the probe at a random orientation in the three-dimensional reconstruction, but this is rather time-consuming.

3 Surface area

Consider a sphere introduced with a random position into a room full of lines. The probability with which the lines will intersect the sphere's surface will depend on the surface area of the sphere and the total length of line present in the room. Or, put another way, if the number of test lines that intersect the sphere's surface and the total length of the lines are known, then it is possible to estimate the sphere's surface area. Figure 11.2 gives a more intuitive insight as to why lines are used to estimate surface area but the reader is also invited to review Satlykov's work and the method developed by Baddely *et al.* (1986).

The lines are thrown at the sphere by sectioning it first and then placing a grid of test lines on the section. Since the sphere is the **ONLY** simple object to have an isotropic surface this is the sole case where the orientation of the test lines is un-important. For all other objects the assumption of an isotropic surface is not valid and the test lines must have an isotropic orientation in three-dimensional space. For example, instead of using a sphere try using a disk. An estimation of the surface area of an object can therefore be found by using either IUR or vertical sections to cope with this problem of isotropy.

The estimate of surface area is obtained from the relationship

$$S(struct) := 2 \cdot \frac{\sum I(struct)}{l(p) \cdot \sum P(ref)} \cdot V(ref)$$

where ΣI(struct) is the total number of intersections of the grid lines with the structure, ΣP(ref) the total number of points of the integral grid that hit the reference space, and l(p) the length of the test line associated with each point in the grid. Therefore, l(p) $\cdot$ ΣP(ref) will be the total length of the test line used to sample the structure (see Michel and Cruz-Orive (1988) for a clever combination of the Cavalieri principle and vertical sections used to estimate lung volume and pleural surface area).

Other methods for estimating surface area include: the Spatial Grid (Sandau 1987), (which uses a probe consisting of a three dimensional IUR grid of lines to estimate the surface area from the number of intersections between the grid and the surface); and the Surfactor (a method based on the nucleator principle) see Jensen & Gundersen (1987).

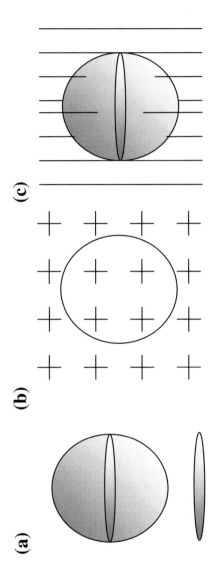

(a)

(b)

(c)

Fig. 11.2 (a) A sphere with radius R is placed on a grid of points and a parallel beam of light is shone from above. A circular shadow (the projected area) with radius R will be produced. The relationship between the projected area of the sphere and the total surface area of the sphere will exist since the projected area of the sphere is πR^2 and the total surface area of the sphere is $4\pi R^2$. (b) The projected area of the sphere can be estimated by counting the number of points that fall inside the projected area and multiplying this by the area associated with the point. If this is multiplied by 4 then this provides an estimate of the surface area of the sphere. In fact, this method would work for any convex object provided the object has a random orientation in three-dimensional space, that is, all possible manifestations of the projected area have equal probability of occurring. The sphere will always give the same projected area no matter the orientation because it is equiaxial in all directions and hence has an inherent isotropy. (c) Another way of estimating the surface area is to imagine that from each point on the grid a test line is projected. Now the surface area of the sphere will be related to the number of intersections with the test line and the length of the test line and this provides the basis for a method of estimating surface area that will work for any object, convex and non-convex, provided the object has an isotropic orientation with respect to the test lines.

4 Length

Consider a twisted curve of finite length in a three-dimensional space. The three-dimensional curve will appear as a profile on any two-dimensional section that passes through the curve. A test frame is the geometric probe that is used to sample the length of an object. The probability that a profile will fall inside the test frame will be proportional to the area of the frame and the length of the curve provided the frame is positioned and orientated randomly in three-dimensional space. The simplest way to implement this procedure is to use an unbiased test frame (Gundersen 1977) placed on an IUR section. The length of the twisted curve of finite length can be estimated from

$$L(struct) := 2 \cdot \frac{\sum Q(struct)}{a(frame) \cdot \sum P(ref)} \cdot V(ref)$$

where $\sum Q(struct)$ is the total number of profiles that fall inside the test frame and do not touch any of the forbidden lines, $\sum P(ref)$ the total number of points associated with the test frame that fall within the reference space, and $a(frame)$ the area of the test frame. Therefore, $a(frame) \cdot \sum P(ref)$ will be the total amount of test area used to sample the structure. For a strongly *ani*sotropic structure, such as neocortical capillaries, the variance can be substantially reduced by taking two extra sections from the tissue block which are orthogonal to the first, that is, the 'Ortrip' (orthogonal triplet probe; Mattfeldt *et al.* 1985).

5 Length estimation using vertical sections

A final method for estimating length that is based on sampling objects on the basis of their surface area is presented. If an IUR line probe is used to sample an object it will do so on the basis of the object's surface area. Objects with a larger surface area will have a higher probability of being sampled. Properties that are measured from this sample will therefore be surface-weighted. In a vertical section design the test lines are sine-weighted and a classical example of such a line is a cycloid.

In this design each time a cycloid hits a tube the diameter is measured, that is, if a cycloid hits a tube twice, the tube diameter is measured twice. This provides a surface-weighted diameter, $\bar{d}_s$, and the ordinary length-weighted diameter, $\bar{d}$, is calculated by computing the harmonic mean of the surface-weighted diameter measurements (Clausen *et al.* 2000)

$$\bar{d} := \overline{d_s^h} = \frac{1}{\frac{1}{m} \cdot \sum \frac{1}{d_i}}$$

where m is the total number of times tubes were hit by cycloids and d_i are the observed diameters. The diameter of a tube in IUR or vertical sections is measured as the longest diameter of the tube profile perpendicular to the longest axis of the tube profile. For tube profiles that are figure-8-shaped, two diameters are measured. Fuzzy tube profiles that represent grazing cuts should be discarded from diameter measurements because they will underestimate the diameter.

When the length-weighted diameter, $\bar{d}$, is calculated the length of the tubes is estimated from

$$L := S / \left(\bar{d} \cdot \pi \right).$$

6 Capillary Length Density and 'diffusion' radius

The cross-sectional area, a, of a 'typical' tissue cylinder around a 'typical' capillary is:

$$a := 1/L_v,$$

where Lv is the length density of capillaries. The estimated, global average cross-sectional area is the correct average of the tissue cross-sectional area, independent of further assumptions. Thus the radius of this typical tissue cyclinder ('diffusion radius' = r) including the radius of the capillary itself is then,

$$r := \frac{1}{\sqrt{\pi \cdot L_v}}.$$

The 'diffusion radius' is an approximation depending on capillary shape and their 3-D distribution.

7 Concluding remarks

A classical stereological concept is when estimating three dimensional parameters the dimensions of the parameter and those of the probe should sum to three. Hence a surface is estimated using a line probe and length is estimated using a (surface) area probe. These are difficult concepts and what is presented here is but a brief and superficial outline of their theory. Alhough estimators of surface area are not discussed in detail in this book, surface area is an important biological parameter. It can be estimated using a variety of stereological methods and the references below provide a good starting point for further reading. The theory, practical implementation, and sources of potential biases of various length estimators are discussed in more detail in Chapters 12 and 13.

References

Baddeley, A.J., Gundersen, H.J.G., and Cruz-Orive, L.M. (1986). Estimation of surface area from vertical sections. *J. Microsc.* **142**, 259–76.

Clausen, H.V., Gundersen, H.J.G., and Larsen, L.G. (2000). Stereological estimation of tubular length from thin vertical sections. *Image Anal. Stereol.* **19**, 205–08.

Gokhale, A.M. (1990). Unbiased estimation of curve length in 3-D using vertical slices. *J. Microsc* **159**, 133–41.

Gokhale, A.M. (1992). Estimation of length density L_V from vertical slices of unknown thickness. *J. Microsc.* **167**, 1–8.

Gundersen, H.J.G. (1977). Notes on the estimation of the numerical density of arbitrary particles: the edge effect. *J. Microsc.* **11**, 219–23.

Jensen, E.B. and Gundersen, H.J.G. (1987). Stereological estimation of surface area of arbitrary particles. Proceedings of the Seventh International Congress for Stereology (ed.: J.L. Chermant). *Acta Stereol.* **6** Suppl. II, 105–22.

Larsen J.O., Gundersen, H.J.G., Nielsen J. (1998). Global spatial sampling with isotropic virtual planes: estimators of length density and total length in thick, arbitrarily orientated sections. *J. Microsc.* **191**, 238–48.

Mattfeldt, T., Mall, G., Gharehbaghi, H., Müller, P. (1990). Estimation of surface area and length with the orientator. *J. Microsc.* **159**, 301–17.

Mattfeldt, T., Möbius, H-J., and Mall, G. (1985). Orthogonal triplet probes: an efficient method for unbiased estimation of length and surface of objects with unknown orientation in space. *J. Microsc.* **139**, 279–89.

Michel, R.P. and Cruz-Orive, L.M. (1988). Application of the Cavalieri principle and vertical sections method to lung: estimation of volume and pleural surface area. *J. Microsc.* **150**, 117–36.

Mouton, P.R., Gokhale, A.M., Ward, N.L, and West, M.J. (2002). Stereological length estimation using spherical probes. *J. Microsc.* **206**, 54–64

Nyengaard, J.R. and Gundersen, H.J.G. (1991). The isector: a simple and direct method for generating isotropic, uniform random sections from small specimens. *J. Microsc.* **165**, 427–31.

Sandau, K. (1987). How to estimate the area of a surface using a spatial grid. *Acta Stereologica* **6/3**, 31–6.

Saltykov, S.A. *Stereometric metallography*, 1st edn. (In Russian). State Publishing House for Metals Sciences, Moscow (1945).

Saltykov, S.A. The method of intersections in metallography (In Russian). Zavodskaja laboratorija, 12, 816–25 (1946).

CHAPTER 12

LENGTH ESTIMATION OF NERVE FIBERS IN HUMAN WHITE MATTER USING ISOTROPIC, UNIFORMLY RANDOM SECTIONS

YONG TANG AND JENS R. NYENGAARD

1 Introduction to length estimation

Unbiased estimation of the length of linear biological structures such as nerve fibers in brain white matter has important implications in neuroscience research. The billions of neurons in the central nervous system (CNS) are connected to one another by means of nerve fibers and synapses to form a complicated network of pathways and neuronal circuits for the transmission of nervous impulses. In order to understand how the brain functions and to explain the causes of various brain diseases, it may be critical to investigate the nerve fibers in a brain. The length of nerve fibers in brain white matter can be estimated using two-dimensional sections through the white matter. When a section plane transects a fiber, the intersections between the section plane and the nerve fiber will be a function of the nerve fiber length. However, it is intuitively clear that the number of intersections is not only directly proportional to the nerve fiber length, but also to the direction of the section plane. Therefore, when estimating the length of linear nerve fibers, there is a strong requirement that the test probe, the section plane, should intersect the nerve fibers with a random direction. There are two ways to ensure that the test probe intersects nerve fibers with a random direction. The first way is to randomize the orientation of nerve fibers in brain white matter, that is, making the biological structures isotropic in three-dimensional space through rotation. This is performed by isotropic, uniformly random (IUR) sections and vertical uniformly random projections. The second way is to randomize the orientation of the test probe. This is done by virtual isotropic planes and virtual isotropic spheres. This chapter will focus on IUR sections.

1.1 Isotropic, uniformly random (IUR) sections

A twisted curve in three-dimensional space will appear as profiles on a two-dimensional section. Therefore, a two-dimensional section is the geometric probe that is used to sample the length of a linear structure. That is, by observing how often a linear structure is hit on a section, that is, counting the number of structural profiles, the length of the linear structure of interest can be estimated. In order to obtain unbiased estimation of the linear structural length by counting the number of the structural profiles on two-dimensional sections, it is required that the profiles of linear structure should be counted on IUR sections unless the linear structures are isotropic themselves. Isotropy implies that the structures are equally represented in all directions of the three-dimensional space. However, biological structures always have a preferred orientation ('anisotropy'). For this reason, we cannot assume for convenience that biological structures are isotropic. The use of IUR sections ensures that every direction of the section plane in three-dimensional space has an equal probability of being sampled and that any bias caused by a possible preferential orientation or distribution of linear structures is avoided. There are two ways of producing IUR sections, the orientator described by Mattfeldt *et al.* (1990) and the isector described by Nyengaard and Gundersen (1992). Another mild requirement for length estimation is that the length of the linear structure studied should be much larger than its diameter (Mattfeldt *et al.* 1985; Osterby and Gundersen 1988). Contrary to general belief, it is not a requirement that the structure under study should be straight or cylindrical.

On IUR sections the length density of the linear structure of interest in the reference space, L_V (struct/ref), is estimated from (Smith and Guttman 1953)

$$L_V(\text{struct/ref}) = 2 \cdot Q_A$$

where Q_A is the number of profiles per unit area of test probe. In order to make this length density estimation formula more intuitively obvious, we can imagine that the linear structures passing perpendicularly through an area, A, are all straight and cut parallel. A box of the tissue of interest with one end area of A and with depth d has a volume $V = A \times d$. The total length of the linear structures, L, in this box would be the number of the structural profiles multiplied by d, and the 'length density', that is, the total length of the linear structures per volume of the tissue, would be

$$\text{Length density} = \frac{L}{V} = \frac{\text{Number of linear structural profiles} \times d}{A \times d}$$
$$= \frac{\text{Number of linear structural profiles}}{A}$$

Since, in most cases, linear biological structures are not parallel this formula underestimates the length density of the linear structures of interest. The constant '2' in the correct length estimation formula is a consequence of the 50% chance of linear structures in three-dimensional space being intersected by a uniform randomly orientated plane.

The total length of the linear structures of interest in the reference, L(struct, ref), is estimated from

$$L(\text{struct, ref}) = L_V(\text{struct/ref}) \cdot V(\text{ref})$$

where V(ref) is the volume of the reference space (usually the volume of the organ or region of interest).

This approach is straightforward to implement. However, the investigator cannot select a convenient section direction (e.g. coronal section, horizontal section), and, therefore, this approach cannot be used for estimating the length density of a linear structure of interest in a region where regional anatomy has to be preserved for the recognition of the region.

1.2 Alternative length estimation methods

There are three other direct stereological methods for length estimation.

1 If thick, vertical, uniformly random sections are available, the length density of the linear structures of interest can be estimated from projection images through thick sections (Gokhale 1990). In a thick vertical section an IUR surface can be generated by projecting a cycloid through the section if the

major axis of the cycloid is parallel to the vertical direction. When the length density of the linear structure is estimated with this approach the section thickness has to be known. If the thick vertical section contains the entire linear structure then the total length of the linear structure can be directly estimated without needing to know the section thickness (Gokhale 1992).

2 Virtual isotropic planes use thick tissue sections in which the parallel virtual planes generated by computer software are randomly rotated and superimposed over magnified linear structures (Larsen et al. 1998). As the image is focused along the Z-axis, the virtual planes move across the image. In order to ensure that the probe, the virtual plane, is isotropic at each sampling plane, the three-dimensional orientations of the virtual planes are randomly selected.

3 Virtual isotropic spheres also use thick tissue sections and virtual spheres to probe linear features within tissue sections. The isotropic surface of a sphere includes all possible orientations, and, therefore, makes a probe for estimating the length of anisotropic structures (Mouton et al. 2002).

1.3 Choosing a method for length estimation of myelinated nerve fibers in the white matter

The two approaches to randomize the orientations of linear structures, namely, IUR sections and vertical, uniformly random projections, require random rotation of the tissue, which will lead to the loss of anatomical landmarks. This is, however, not a problem in white matter. Using relatively thick tissue sections, the test probe, rather than the linear feature in the tissue, can be rendered isotropic using either virtual planes or virtual spheres. Due to the small diameter of the nerve fibers and the fact that the nerve fibers have been fixed for a long time in formaldehyde so that antibodies cannot be used, thick sections are of no use here, leaving only the thin IUR sections.

2 A worked example. Estimating the total length of the myelinated nerve fibers in the white matter of the human brain using IUR sections

Using a stereological method, Pakkenberg and Gundersen (1997) found that a primary structural change that took place in the aging brain was a 28% atrophy of the white matter. Using magnetic resonance imaging (MRI), several in vivo studies also reveal significant age-related loss of white matter volume compared to a much smaller decline in gray matter volume (Miller et al. 1980; Albert 1993; Christiansen et al. 1994; Guttmann et al. 1998). The age-related white matter loss could be due to reduced myelination, loss of myelin sheaths, loss of nerve fibers, or decreased amount of glial cells and intercellular substance, but this cannot be clarified using MRI. In order to detect the changes of the nerve fibers in brain white matter during aging and in various degenerative brain disorders, it is therefore necessary to undertake quantitative research on the nerve fibers in the brain white matter. Since most nerve fibers in the CNS are myelinated (Hildebrand *et al.* 1993) we chose to study the myelinated nerve fibers in brain white matter.

When estimating the total length of the myelinated fibers in brain white matter, there is no problem in defining the region of interest—the entire brain white matter. However, no exact borders for each lobe of brain exist in the white matter. Therefore, the total length of the myelinated fibers in the white matter of each lobe could not be precisely estimated.

3 Experimental design

3.1 Estimation of brain white matter volume

The brains were fixed in 0.1M sodium phosphate buffered formaldehyde for at least 5 months, the meninges removed, and the cerebellum and brainstem detached at mid pons. Right or left hemispheres were chosen at random beforehand. The hemispheres were embedded in 6% agar, sliced coronally at an average of 7 mm wide intervals, starting randomly at the frontal pole. A transparent counting grid with an area per point of 2.25 cm^2 was placed at random over the occipital cut surface of every brain slice. The points hitting the white matter were counted (Fig. 12.1(a)).

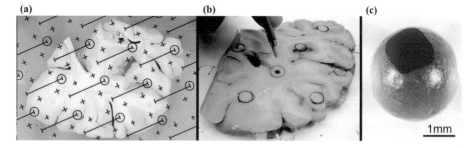

Fig. 12.1 (a) A coronally cut surface of a human brain with a transparent counting grid superimposed at random. When estimating the white matter volume with the Cavalieri principle, all the points (crosses) hitting the white matter were counted. (b) A grid with equidistantly spaced holes was put on top of a slab. Biopsies were sampled where the holes hit white matter. (c) The tissue was embedded according to the isector (Nyengaard and Gundersen 1992), and the 3 mm Epon spheres were rotated randomly before re-embedding. This procedure ensures isotropic, uniformly random sections, so that each tissue block has a uniformly random orientation before being cut.

The volume of the white matter from one hemisphere, V(wm), was estimated using the Cavalieri principle (Gundersen *et al.* 1988)

$$V\left(\text{wm}\right) = \bar{t} \cdot a(\text{p}) \cdot \sum P$$

where V(wm) is the total volume of the brain white matter, $\bar{t}$ the average slice thickness, a(p) the area associated with each point in the grid, and $\sum P$ the total number of points hitting the white matter. The volume of the entire brain white matter was calculated by multiplying V(wm) by 2.

3.2 Uniformly randomly sampling of brain white matter

From the tissue slices cut, on average 25–29 slices per hemisphere, every fourth slice was sampled systematically, the first one being chosen at random among the first four slices. From the slices sampled, tissue blocks were uniformly randomly sampled using a biopsy needle with a diameter of 1.5 mm and a plastic sheet with equidistantly spaced holes. That is, a plastic sheet with equidistantly spaced holes was randomly placed over the frontal face of the sampled slices and the tissue blocks were sampled where the holes of the plastic sheet hit brain white matter using the biopsy needle (Fig. 12.1(b)). From the sampled tissue blocks eight tissue blocks were sampled randomly. The (systematic) uniformly random sampling scheme ensures that all areas of the white matter have the same probability of being sampled and the final sample will therefore represent the entire brain white matter.

The sampled blocks were rinsed for 20 min in maleic acid buffer, postfixed for 1 hour in 1% buffered osmium tetroxide, dehydrated, and stained with uranyl acetate. The tissue was embedded in 3 mm Epon spheres (Fig. 12.1(c), isector; Nyengaard and Gundersen 1992), and the spheres were rotated randomly before being re-embedded. This procedure ensures isotropic, uniformly random sections, so that each tissue block has a uniformly random orientation before being cut. This is essential for avoiding methodological bias of the length measurements due to the anisotropic orientation of the myelinated nerve fibers in brain white matter. One section with a thickness of 0.1 μm was cut from each Epon block using an LKB Historange microtome. The light microscopical sections were stained with 1% toluidine blue.

The estimation of the length density is complicated by a problem—the identification of the profiles of small myelinated fibers in a light microscopical image—since the diameters of the smallest fibers are at the maximum resolution of the light microscope. In this study, we cut very thin sections (0.1 μm) and used a very high magnification (5185×). We identified the myelin profiles by the intense myelin sheath stain and the hole inside the profile. Furthermore, to evaluate the criteria for the identification of small myelin fibers, we compared consecutive

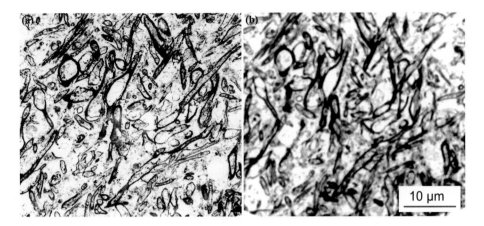

Fig. 12.2 Two consecutive sections were cut from an Epon block for (a) electron microscopy and (b) light microscopy. Coinciding areas on the light microscopic field and electron micrograph were examined. Estimation and comparison of the nerve fiber parameters from light microscopic view fields and electron micrographs were made (modified from Tang and Nyengaard 1997).

sections in light microscopy with electron microscopy with approximately the same magnification and identified the corresponding fibers. Differences of less than 5% were found when comparing estimates of the total length from light and electron microscopy (Fig. 12.2; Tang and Nyengaard 1997).

3.3 Estimation of the length density of the myelinated nerve fibers in brain white matter

The sections were viewed using a modified Olympus BH-2 microscope. An IBM 330–466 DX-2 personal computer and an IBM 14 X monitor were connected to JAI-2040 video camera mounted on top of the microscope. By means of the CAST-Grid software (Olympus, Denmark), the stereological probe, unbiased counting frame, was superimposed upon the video images of tissue sections viewed on the monitor. The area of each counting frame was 165 μm^2 at the tissue level. An oil objective lens (100 ×; numerical aperture (NA) 1.40) and an intermediate lens (5 ×) before the video camera were used when counting. Using a very high light microscopical magnification of 5185×, two fields of view were randomly sampled in each Epon section, that is, the undamaged tissue of the left upper corner and right lower corner were sampled from the IUR sections of the uniformly sampled blocks. The length density of the myelinated nerve fibers in the brain white matter, $L_V(\text{nf/wm})$, was estimated as

$$L_V\left(\text{nf/wm}\right) = 2 \cdot \frac{\sum Q\left(\text{nf}\right)}{\sum A\left(\text{frame}\right)}$$

where 2 is a constant that pertains to IUR section since there is 50% chance for the myelinated nerve fibers in a three-dimensional space to be sectioned when they are cut isotropically. ΣA(frame) is the total area of the unbiased counting frames used when counting the myelinated nerve fiber profiles. ΣQ(nf) denotes the total number of myelinated nerve fiber profiles that were counted in the white matter per hemisphere.

It is obvious that the estimation of the length density of the myelinated nerve fibers in brain white matter reduces to the estimation of the number of the myelinated fiber profiles. The myelinated nerve fiber profiles were counted using the unbiased two-dimensional counting frame (Fig. 12.3; Gundersen 1977). The unbiased counting frame consists of four lines. The upper line and right line are the 'inclusion lines', while the lower line and the left line and their extensions are the 'exclusion lines'. The counting rules for the unbiased frames are: every profile completely or partly inside the counting frame, but not touching the exclusion lines, is counted. In the design of the unbiased counting frame, the higher proba-

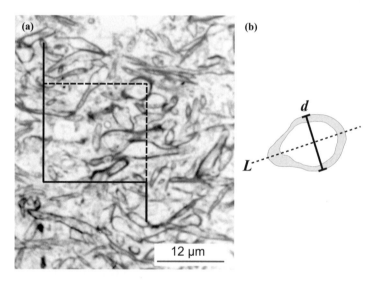

Fig. 12.3 (a) A uniformly randomly sampled field of the white matter from an immersion fixed human autopsy brain is shown to illustrate the estimation of the length density of the myelinated fibers in the brain white matter. The tissue was embedded according to the isector technique to provide an isotropic, uniform random section. The myelinated nerve fiber profiles were sampled by the two-dimensional unbiased counting frame where the area was indicated by the solid lines (exclusion lines) and the dotted lines (inclusion lines). All myelinated nerve fiber profiles completely or partly inside the counting frame, but not touching or intersecting the full-drawn exclusion lines, were counted. (b) A nerve fiber profile is drawn to illustrate the measurement of nerve fiber diameter. Firstly, the longest axis is identified (dotted line and L). Secondly, the nerve fiber profile diameter is measured perpendicular to the longest axis (solid line and d). (Modified from Tang et al. (1997) and Marner et al. (2003).)

bility of larger profiles to appear in a counting frame is taken into account. The exclusion lines are extended so that irregular profiles, for instance, snake-like profiles, will not be overestimated. To fully understand the concept of the unbiased counting frame, one can imagine the entire section filled with counting frames lying side to side. Each profile must only be sampled in one counting frame and must be excluded if appearing in the counting frame below or to the left of the one studied. When using the unbiased counting rules to count the fiber profiles, it is necessary to be able to see the full delineation of the fiber profiles to decide whether parts of the counting frame are being touched. Therefore, an appropriate guard area around the counting frame should be ensured in order to count the myelinated nerve fiber profiles unambiguously in two-dimensions.

3.4 Estimation of the total length of the myelinated nerve fibers in brain white matter

To obtain the total length of the myelinated nerve fibers of the white matter per brain, the length density of the myelinated fibers in the brain white matter was multiplied by the total volume of the white matter per brain.

3.5 Tissue shrinkage

In order to obtain the estimation of the total myelinated nerve fiber length in the brain white matter, the length density of the myelinated nerve fibers in the brain white matter has to be multiplied by the total volume of brain white matter. Therefore, any tissue shrinkage happening between the Cavalieri estimation of the white matter volume and the length density estimation of the myelinated nerve fibers in the white matter should be checked. Two slabs of tissue, 2 mm long and 1.5 mm wide were taken from the white matter of each hemisphere. The dimensions of these tissue slabs were measured carefully before being processed. The tissue cross-sectional area was computed. The tissue was then dehydrated, and embedded together with the rest of the sampled blocks. Sectioning was done carefully to ensure a section perpendicular to the length of the slab. After being stained, the area of the tissue section, A, was measured using point counting

$$A = \Sigma P \times a(\mathrm{p})$$

where ΣP is the number of points hitting tissue, and $a(\mathrm{p})$ is the area associated with each point in the counting grid. The measurements were compared to see if any shrinkage had occurred. The amount of shrinkage was estimated as

Area shrinkage = [(area before) − (area after)]/(area before).

In the current study, the mean area shrinkage induced by the histological processing, was 1.5% in the brains of the subjects, which did not significantly differ from zero.

3.6 Statistics

Although the results with the present stereological method are statistically unbiased, these results represent estimates, and not absolute determinations, of the total length of the myelinated nerve fibers in brain white matter. These estimates have a certain degree of precision (or sampling variance) that is related to the amount of sampling performed. If the amounts of sampling performed were not enough, the estimates would be imprecise, varying widely around the true values. On the other hand, sampling too many regions would be excessively labor-intensive while providing only a minimal increase in precision. Thus, to see if the sampling scheme employed is optimal for the specimens of interest, it is necessary to calculate the sampling variance and analyze the sampling scheme used after a pilot study is done.

Variability within groups is calculated using the dimensionless coefficient of variation (CV = standard deviation/mean). The overall coefficient of variance (CV_{total}^2) of the length estimates of the myelinated nerve fibers among individual brains studied depends on the extent of the estimated stereological sampling variance (CE_{ste}^2) depending on the intensity of the sampling and the inherent biological coefficient of variance (CV_{biol}^2) of the myelinated nerve fiber length in the white matter among brains. They have the relationship

$$CV_{total}^2 = CV_{biol}^2 + CE_{ste}^2.$$

When a pilot study is done and both the total variance (CV_{total}^2) and the computable stereological sampling variance (CE_{ste}^2) are known, the sampling scheme can be optimized. That is, for a quantitative study stereological sampling should be sufficient when the computable stereological sampling variance, CE_{ste}^2, is about half or less than half of the total coefficient of variance, CV_{total}^2.

The CE_{ste} was estimated from the estimated intrabrain coefficient of error (CE) using the relationship

$$CE_{ste} = \sqrt{\text{mean } CE^2}.$$

The total length of the myelinated fibers in brain white matter was calculated from the product of the total volume of the brain white matter and the length density of the myelinated nerve fibers in the brain white matter. Therefore, intrabrain CE has to be calculated from the two-step analyses.

In the current study, the CE of the estimation of the white matter volume, $CE[V(wm)]$, was obtained from another study (Pakkenberg and Gundersen 1997). However, it should be emphasized that the usual formula for estimating CE (standard error of the mean divided by the mean) is only valid for independently derived observations. In the current design, a systematic random sample of slices was used to estimate the white matter volume. Thus, slices were not sampled independently of one another, and therefore estimates were not independently derived observations. Even though the systematic random sampling design is an unbiased sampling design and is more efficient than the independent random sampling design (Gundersen and Jensen 1987), the CE for the dependent observations obtained with the systematic random sampling design cannot be calculated in the ordinary way. The way to estimate the CE for such dependent observations was described by Gundersen and Jensen (1987) and Gundersen *et al.* (1999).

The CE for the estimation of the length density of the myelinated fibers in white matter, CE[L_V(nf/wm)], was calculated according to the formula

$$CE_n\left[\frac{\sum Y}{\sum X}\right] = \sqrt{\frac{n}{n-1}\left[\frac{\sum(X)^2}{\sum X\sum X} + \frac{\sum(Y)^2}{\sum Y\sum Y} - \frac{2\sum(XY)}{\sum X\sum Y}\right]}$$

where $\sum X$ is the number of fields used in each block when estimating the length density of the myelinated nerve fibers in the brain white matter, $\sum Y$ is the number of the myelinated nerve fiber profiles in each tissue block, and n is the number of tissue blocks used when estimating the length density of the nerve fibers in the white matter. The CE of the total length of the myelinated fibers in each brain was calculated from the estimated coefficient of error in the estimation of length density and the estimated coefficient of error in the estimation of white matter volume using the formula

$$CE\left[L(nf, wm)\right] = \sqrt{\left(CE^2\left[L_V(nf/wm)\right] + CE^2\left[V(wm)\right]\right)}.$$

Table 12.1 The stereological estimation of the total length of the myelinated nerve fibers in the white matter of five human brains*

Age (years)	V(wm) (cm³)	CE (%)	ΣQ(nf)	ΣA(frame) (µm²)	L_V(nf/wm) (m/mm³)	CE (%)	L(nf, wm) (10³ km)	CE (%)
18	494	2.4	322	2 640	244	6.3	121	6.7
30	464	2.3	323	2 640	245	4.7	114	5.2
39	488	5.1	328	2 640	249	4.8	122	7.0
45	398	2.6	348	2 640	264	6.7	105	7.2
57	530	2.0	322	2 640	244	5.6	129	5.9
Mean	475	3.1	329	2 640	249	6.7	118	6.4
CV†		10.3				3.4		7.7

* V(wm), the white matter volume per brain; ΣQ(nf), the total number of the myelinated nerve fiber profiles counted per hemisphere; ΣA(frame), the total area of the unbiased counting frames used per hemisphere; L_V(nf/wm), the length density of the myelinated nerve fibers in the white matter; L(nf, wm), the total length of the myelinated nerve fibers in the brain white matter; CE, coefficient of error.
† The observed interindividual coefficient of variation in %.

Table 12.2 Estimating the total length of the myelinated nerve fibers of brain white matter and the coefficient of error of the estimate in a 30-year-old female individual

Section number	Q(nf)
1	46
2	40
3	41
4	37
5	32
6	43
7	48
8	36
$n = 8$	$\Sigma Q(nf) = 323$

Equations used in calculations

(a) $L_V(nf/wm) = 2 \cdot \dfrac{\Sigma Q(nf)}{\Sigma A(frame)} = 2 \cdot \dfrac{323}{2640 \ \mu m^2} = 0.245 \ (\mu m/\mu m^3) = 245 \ (m/mm^3)$

(b) $L(nf, \ wm) = L_V(nf/wm) \cdot V(wm) = 245 \cdot 464 \cdot 10^3 = 114 \cdot 10^6 \ (m) = 114 \cdot 10^3 \ (km)$

(c) $CE\left[L_V(nf/wm)\right] = CE\left[\dfrac{\Sigma Y}{\Sigma X}\right] = \sqrt{\dfrac{n}{n-1}\left[\dfrac{\Sigma(X)^2}{\Sigma X \Sigma X} + \dfrac{\Sigma(Y)^2}{\Sigma Y \Sigma Y} - \dfrac{2\Sigma(XY)}{\Sigma X \Sigma Y}\right]}$

$= \sqrt{\dfrac{8}{8-1}\left[\dfrac{32}{16 \cdot 16} + \dfrac{13239}{323 \cdot 323} - 2 \cdot \dfrac{646}{16 \cdot 323}\right]} = 0.047$

(d) $CE\left[L(nf, \ wm)\right] = \sqrt{\left(CE^2\left[L_V(nf/wm)\right] + CE^2\left[V(wm)\right]\right)} = \sqrt{(0.047)^2 + (0.023)^2}$

$= 0.052$

Comments on equations

Equation (a). Eight isotropic, uniformly random sections from eight uniformly sampled blocks were sampled and two fields of view were randomly sampled in each sampled section. Based on isotropic, uniformly random sections, the length density of the myelinated nerve fibers in brain white matter, $L_V(nf/wm)$, is estimated by multiplying 2 by the ratio between the total number of the myelinated nerve fiber profiles counted per hemisphere and the total area of the counting frames used (see eqn (a)). $\Sigma Q(nf)$ indicates the total number of the myelinated nerve fiber profiles counted per hemisphere and is 323 in eqn (a). $\Sigma A(frame)$ indicates the total area of the unbiased counting frames used per hemisphere, which is calculated by multiplying the area of each counting frame (165 μm2) by the number of the counting frames used. Since eight sections were used per hemisphere and two fields of view were used per section, the total area of the counting frames used was 165 μm² × 16 = 2640 μm² in eqn (a).

Equation (b). The total length of the myelinated nerve fibers in the brain white matter, $L(nf, wm)$, is obtained by multiplying the length density of the myelinated nerve fibers in white matter by the total volume of brain white matter, $V(wm)$ (see eqn (b)). In this individual, $V(wm)$ was 464 cm³, which was obtained from Pakkenberg and Gundersen (1997).

Equation (c). $CE[L_V(nf/wm)]$ indicates the coefficient of error* for the estimation of the myelinated nerve fiber length density in brain white matter. The estimation of the length

Table 12.2 Cont'd

density is a ratio estimator; therefore, the equation in Kroustrup and Gundersen (1983) was used to estimate the CE of this estimate. n is the number of tissue blocks used and is 8. ΣY and ΣX are the numerator and denominator in the length density estimator, that is, the total number of the myelinated fiber profiles counted and the total area of the counting frames used. Since the area of each counting frame is constant ΣX becomes the number of fields of view used in eqn (c).

Equation (d). CE[L(nf,wm)] indicates the coefficient of error* for the estimation of the total myelinated nerve fiber length in brain white matter. CE [V(wm)] is the coefficient of error for the estimate of brain white matter volume. In this individual, CE [V(wm)] was 0.023 (see Pakkenberg and Gundersen 1997). The CE obtained for the volume estimation of brain white matter and that obtained for the length density estimate of the myelinated nerve fibers in the brain white matter were added in the usual way to obtain the total CE (see eqn (d)).

* The coefficient of error of an estimate has no biological significance but is useful when designing and optimizing the sampling scheme.

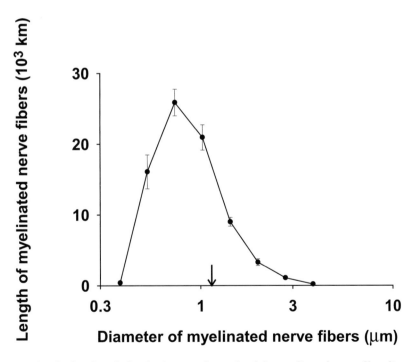

Fig. 12.4 The absolute length distribution on a log-scale of the myelinated nerve fiber diameter in the white matter of five human beings. The class width is 0.2 μm and the error bars are SEM. The arrow indicates the average diameter of the fibers. (Data are from Tang and Nyengaard (1997).)

4 Results

The results from five human brains are presented in Table 12.1 and Fig. 12.4. Using patient number 2 (age 30 years) as an example, Table 12.2 illustrates how the length density and total length of the myelinated nerve fibers in the brain white matter are calculated In addition, the CE for the estimation of the length density of the myelinated fibers in white matter, $CE[L_V(nf/wm)]$, and the CE of the total length of the myelinated fibers in this brain are calculated.

References

Albert, M. (1993). Neuropsychological and neurophysiological changes in healthy adult humans across the age range. *Neurobiol. Aging* **14**, 623–5.

Christiansen, P., Larsson, H.B., Thomsen, C., Wieslander, S.B., and Henriksen, O. (1994). Age dependent white matter lesions and brain volume changes in healthy volunteers. *Acta Radiol.* **35**, 117–22.

Gokhale, A.M. (1990). Unbiased estimation of curve length using vertical slices. *J. Microsc.* **159**, 133–41.

Gokhale, A.M. (1992). Estimation of length density L_V from vertical slices of unknown thickness. *J. Microsc.* **167**, 1–8.

Gundersen, H.J.G. (1977). Notes on the estimation of the numerical density of arbitrary profiles: the edge effect. *J. Microsc.* **111**, 219–23.

Gundersen, H.J.G. and Jensen, E.B. (1987). The efficiency of systematic sampling in stereology and its prediction. *J. Microsc.* **147**, 229–63.

Gundersen, H.J.G., Bendtsen, T.F., Korbo, L., Marcussen, N., Moller, A., Nielsen, K., Nyengaard, J.R., Pakkenberg, B., Sorensen, F.B., Vesterby, A., and West, M.J. (1988). Some new, simple and efficient stereological methods and their use in pathological research and diagnosis. *Acta Pathol. Microbiol. Immunol. Scand.* **96**, 379–94.

Gundersen, H.J.G., Jensen, E.B., Kieu, K., and Nielsen, J. (1999). The efficiency of systematic sampling in stereology—reconsidered. *J. Microsc.* **193**, 199–211.

Guttmann, C.R., Jolesz, F.A., Kikinis, R., Killiany, R.J., Moss, M.B., Sandor, T., and Albert, M.S. (1998). White matter changes with normal aging. *Neurology* **50**, 972–8.

Hildebrand, C., Remahl, S., Persson, H., and Bjartmar, C. (1993). Myelinated nerve fibers in the CNS. *Prog. Neurobiol.* **40**, 319–84.

Kroustrup, J.P. and Gundersen, H.J.G. (1983). Sampling problems in an heterogeneous organ: quantitation of relative and total volume of pancreatic islets by light microscopy. *J. Microsc.* **132**, 43–55.

Larsen, J.O., Gundersen, H.J.G., and Nielsen, J. (1998). Global spatial sampling with isotropic virtual planes: estimators of length density and total length in thick, arbitrarily orientated sections. *J. Microsc.* **191**, 238–48.

Marner, L., Nyengaard, J.R., Tang, Y., and **Pakkenberg, B.** (2003). Marked loss of myelinated nerve fibers in the human brain with age. *J. Comp. Neurol.* **462**, 144–52.

Mattfeldt, T., Mobius, H.-J., and **Mall, G.** (1985). Orthogonal triplet probes: an efficient method for unbiased estimation of length and surface of objects with unknown orientation in space. *J. Microsc.* **139**, 279–89.

Mattfeldt, T., Mall, G., Gharehbaghi, H., and **Moller, P.** (1990). Estimation of surface area and length with the orientator. *J. Microsc.* **159**, 301–17.

Miller, A.K., Alston, R.L., and **Corsellis, J.A.** (1980). Variation with age in the volumes of grey and white matter in the cerebral hemispheres of man: measurements with an image analyzer. *Neuropathol. Appl. Neurobiol.* **6**, 119–32.

Mouton, P.R., Gokhale, A.M., Ward, N., and **West, M.J.** (2002). Stereological length estimation using spherical probes. *J. Microsc.* **206**, 54–64.

Nyengaard, J.R. and **Gundersen, H.J.G.** (1992). The isector: a simple and direct method for generating isotropic, uniform random sections from small specimens. *J. Microsc.* **165**, 427–31.

Osterby, R. and **Gundersen, H.J.G.** (1988). Stereological estimation of capillary length exemplified by changes in renal glomeruli in experimental diabetes. In *Stereology and morphometry in electron microscopy, problems and solutions* (ed. A. Reith and T.M. Mayhew), pp. 113–22. Hemisphere Publishing Corporation, New York.

Pakkenberg, B. and **Gundersen, H.J.G.** (1997). Neocortical neuron number in humans: effect of sex and age. *J. Comp. Neurol.* **384**, 312–20.

Smith, C.S. and **Guttman, L.** (1953). Measurement of internal boundaries in three dimensional structures by random sectioning. *Trans. AIME* **197**, 81–92.

Tang, Y. and **Nyengaard, J.R.** (1997). A stereological method for estimating the total length and size of myelin fibers in human brain white matter. *J. Neurosci. Methods* **73**, 193–200.

Tang, Y., Nyengaard, J.R., Pakkenberg, B., and **Gundersen, H.J.G.** (1997). Age-induced white matter changes in the human brain: a stereological investigation. *Neurobiol. Aging* **18**, 609–15.

CHAPTER 13

VIRTUAL TEST SYSTEMS FOR ESTIMATION OF ORIENTATION-DEPENDENT PARAMETERS IN THICK, ARBITRARILY ORIENTATED SECTIONS EXEMPLIFIED BY LENGTH QUANTIFICATION OF REGENERATING AXONS IN SPINAL CORD LESIONS USING ISOTROPIC, VIRTUAL PLANES

JYTTE OVERGAARD LARSEN, MIA VØNEULER, AND ANNMARIE JANSON

1 Introduction

Global spatial sampling is a *sine qua non* for design-based estimation of surface area and curve length (see e.g. Hennig 1963; Gundersen 1979; Mattfeldt and Mall 1984; Baddeley *et al.* 1986; Gokhale 1990; Cruz-Orive and Howard 1991; Chapter 12, this volume) and local spatial sampling is required for design-based estimation of particle surface and volume (Gundersen 1988; Jensen and Gundersen 1993; Chapter 9, this volume) as well as for estimation of spatial distribution (Chapter 14, this volume). The need for spatial sampling has traditionally been fulfilled by rotating the tissue specimen prior to sectioning to produce either physical isotropic, uniform random (IUR) (Mattfeldt *et al*. 1990, Nyengaard and Gundersen 1992) or vertical, uniform random (VUR) (Baddeley *et al*. 1986) sections. In neurobiology these approaches may conflict with another essential requirement in stereology, namely, the unambiguous identification and delineation of the region of interest. At present two fundamentally different techniques allow for design-based quantification of orientation-dependent parameters in thick arbitrarily orientated sections. One approach is to create an 'image volume' from stacks of stored digitized images and to probe the images with a 'perspective image' of a spatial test system while focusing through the stack of images (Kubínová and Janáček 1998). The other approach, the virtual test system technique, exploits spatial counting rules or measurements applied on to real-time video images of thin focal planes during optically sectioning from top to bottom through a volume in a thick physical section (Larsen *et al*. 1998). The former technique is beyond the scope of this chapter and only the latter will be considered further. Virtual test systems, where the random rotation of the stereological probe is made within volume probes of thick physical sections *after* sectioning by the use of computer-assisted microscopy, ensure correct spatial sampling in a uniform sample of tissue sectioned in the most convenient sectioning direction. Reasons for using arbitrarily orientated physical sections may include interest in a particular sectioning direction in order to ensure proper identification of the reference space or practical constraints associated with the physical cutting procedure. Also, thick sections originally made for other purposes with no requirements for spatial sampling may be used to obtain stereological data. The virtual test systems also offer less important advantages that will be considered later. In this chapter we describe the basic principles governing virtual test systems. The concept is completely general and all kinds of test probes suited for their intended purposes can be implemented. In principle, only computer power and our imagination limit the design of virtual test systems. An in depth description of length estimation using virtual test systems is provided as is a worked example of length quantification of regenerating axons in rat spinal cord lesions using isotropic, virtual planes.

2 Length estimation

The total global length and length density of linear structures extending in a defined volume can, from a practical point of view, be analyzed using different design-based stereological methods (Chapter 12, this volume). When a linear structure is sectioned it will appear as a profile. It is intuitively clear that the number of profiles observed within a test area in the two-dimensional section is proportional to the entire length of the linear structure in the three-dimensional volume (Fig. 13.1). If the structure possesses a preferred orientation the number of profiles observed per unit area will also be dependent on the orientation of the two-dimensional section. There may also be a spatially heterogeneous distribution of the linear structure and thus the number of profiles observed per unit area is also dependent on the position of the two-dimensional-section (Fig. 13.1). Thus all possible orientations in all possible positions in space should have equal probabil-

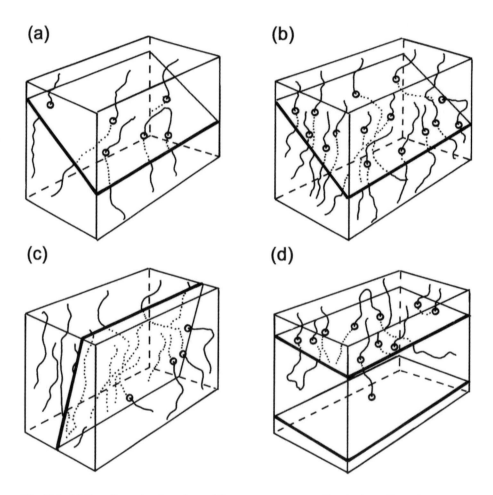

Fig. 13.1 (a) Two-dimensional sections of linear structures extending in three dimensions. Transects between a two-dimensional plane and a linear structure extending in three dimensions will appear as profiles. (b) The number of profiles observed per unit two-dimensional plane is dependent on the length density of the linear structure, but also (c) on the orientation of the plane in cases where the linear structure intrinsically possesses a preferred orientation and (d) on the location of the plane in case of a heterogeneous distribution of the linear structure.

ity of being sampled from the volume in which the estimations are made in order to make reliable quantitative length estimations from two-dimensional-sections. To ensure this the test planes must be isotropic, random in relation to the examined structure (see Fig. 13.2) and systematic, uniform random (see Fig. 13.3).

When all attached requirements are fulfilled the number of profiles observed per unit plane area, Q_A, is related to length density, L_V, by the simple relationship

$$L_V = 2Q_A. \tag{2.1}$$

The total length, L_{tot}, is obtained by multiplying the length density by the volume of the reference space, $V(ref)$,

$$L_{tot} = L_V \cdot V(ref). \tag{2.2}$$

This stereological principle was originally implemented for practical use by analyzing isotropically, rotated thin physical sections with a planar, horizontally orientated test system, the so-called classical length estimator. For the practical generation of these sections, systematic, uniform random isotropic orientations can

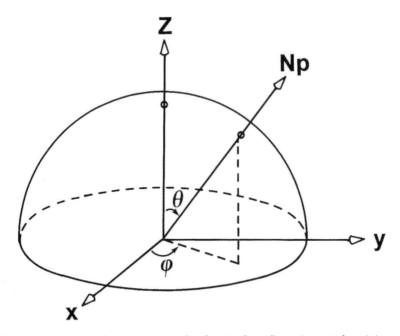

Fig. 13.2 An isotropic random orientation of a plane in three dimensions. A plane is isotropic random when the normal to the plane, N_p, hits any point on the unit hemisphere with equal probability. The orientation of a plane is specified by the spherical polar coordinates of its normal, N_p, and, to achieve an isotropic plane the angle, ϕ, is randomly rotated around the z-axis through the range 0–2π, whereas the angle, θ, is not uniform random, but an arccosine-weighted random angle in the interval 0–π/2. This is because a uniform, random point is less likely to fall near the pole than near the equator.

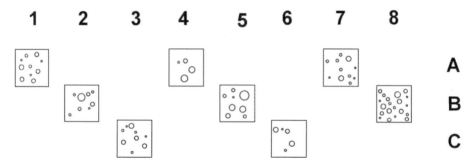

Fig. 13.3 Systematic, uniform randomness. For a given orientation systematic, uniform random sections can be achieved by taking sections systematically equally spaced throughout the organ after taking the first section randomly within the first spacing interval. By choosing a uniform, random start all parts of the object have an equal probability of being sampled. If one systematically samples every third section, that is, one-third of the sections, after randomly selecting the start within the first three sections, one-third of the time one will sample set A (sections 1, 4, and 7), one-third of the time set B (sections 2, 5, and 8), and one-third of the time set C (sections 3 and 6). All sections will thus have an equal probability of being sampled for the analysis.

be achieved by the orientator (Mattfeldt *et al.* 1990) and simple, random isotropic orientations by the isector (Nyengaard and Gundersen 1992). At the resulting sections an estimate of profiles per unit test area can be obtained with any unbiased two-dimensional counting rule, for example, the unbiased counting frame (Gundersen 1977), the associated point counting rule (Miles 1978), or the two-dimensional disector counting rule (Gundersen 1986; Larsen *et al.* 1998).

2.1 Virtual test systems

Imaging techniques that yield three-dimensional volume information from successive two-dimensional focal planes can sustain a test system with full 3D × 4π freedom, i.e. full spatial and rotational. However, a stereological test probe extending in three dimensions is only observed at its intersection with the focal plane, and a line probe is thus seen as a point in the focal plane while a plane probe is seen as a line in the focal plane. The position of the point or, respectively, the line changes during focal plane displacement. For simplicity this description is restricted to computer-assisted microscopic analysis of thick histological sections. Real-time video images of the microscope field are displayed on a screen on to which the stereological test system is superimposed through a computer–video interface, and focal plane displacement microcator recordings can be used as feedback information to control dynamically coordinated visualization of the intersection between the test system and the focal plane. The computer generates a test system extending in three dimensions within the sampled volume probe of the section. As the volume probe is optically sectioned, the intersection between the focal plane and the test system, that is, the 'two-dimensional profile of the test system' is merged with the two-dimensional profile of the structure and sampling/measurements are done using special spatial counting rules or measurements. The two-dimensional profile of the test system moves across the computer screen during focal plane displacement and dynamically maps the test system that exists in three dimensions within the volume probe. Test systems based on isotropic lines in three dimensions are visualized as points that move in successive two-dimensional focal planes, test systems based on isotropic planes in three dimensions are visualized as moving lines, and test systems based on volume probes (e.g. spatial distributions) are visualized as moving areas (see Fig. 13.4).

At present two different test systems have been implemented for length estimation—the original virtual plane technique based on planar areas (Larsen *et al.* 1998) and the 'space ball' technique based on a spherical test system (Calhoun and Mouton 2001; Mouton *et al.* 2002). In the virtual plane technique the computer selects a systematic, uniform random orientation of the virtual plane using an algorithm (Larsen *et al.* 1998). A microcator measuring focal plane displacement along the z-axis, that is, the thickness/depth, is attached to the microscope and continuously provides information about z-axis position as the focal xy-plane is

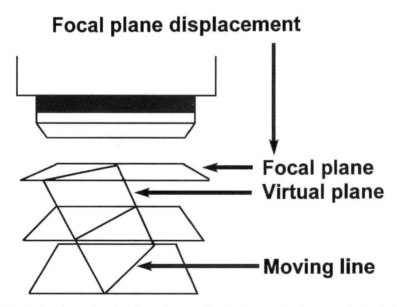

Fig. 13.4 A virtual test plane inside a volume probe. The intersection between the focal plane and an isotropic test plane is a line in the two-dimensional focal plane. The position of the line changes with changing focal planes.

moved through the thick physical section. The computer receives this information and calculates and displays the position of the two-dimensional profile of the test system, that is, a line, during focal plane displacement. The line 'moves' across the computer screen during focal plane displacement and dynamically maps the test system (i.e. the test plane) that exists in three dimensions within the volume probe. The line orientation on the screen depends on the orientation of the virtual plane and the displacement on the screen as a function of z-axis displacement depends on the angle of the virtual plane with respect to the focal plane. This principle is implemented in the CAST-grid software and marketed by The International Stereology Center at Olympus Denmark. The description in the following text refers to this particular implementation.

A set of thick, physical sections that constitute a uniform, random sample of the reference space is analyzed. The section is moved in steps of length d_x and d_y along the x- and y-axes, respectively, in a raster pattern. At each point in the raster pattern the linear structure within a fixed volume, that is, a sampling box is analyzed with isotropic, virtual planes. The volume of the sampling box, v(box), is interactively defined by the area of the sampling box, a(box), which is visualized as a rectangle on the screen, and the depth of the sampling box, h(box), which is the extent of focal plane movement.

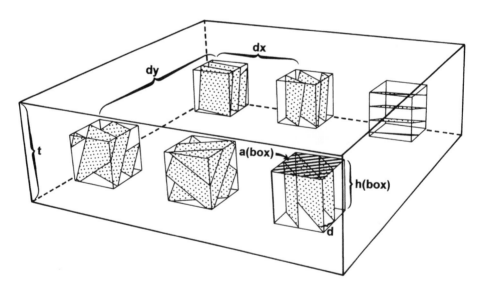

Fig. 13.5 A systematic, random sample of sampling boxes intersected by systematic sets of IUR virtual planes. Within the sampled physical sections of thickness t, each systematically sampled field of vision representing a tissue volume $d_x \times d_y \times t$, is examined with equidistant and parallel IUR, virtual planes in a 'sampling box' of volume a(box) $\times$ h(box). The sampling plane separation is d. (Reproduced from Larsen *et al.* (1998) with permission from The Royal Microscopical Society.)

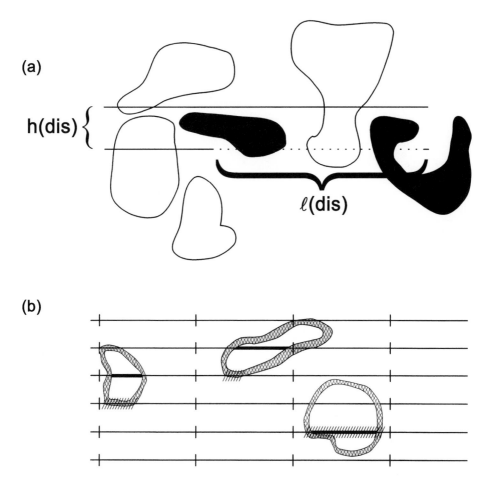

Fig. 13.6 The two-dimensional-disector counting rule. (a) The two-dimensional disector comprises an inclusion line (dashed line) of length, l(dis), with an exclusion extension (lower solid line) separated from an infinite look-up exclusion line (upper solid line) by h(dis). The profiles that exclusively touch the inclusion line are counted (heavily drawn) and profiles that touch the look-up exclusion line or the exclusion extension are excluded from the count. Note that the heavily drawn profile shown to the left extends into the area between the two exclusion lines but without touching the exclusion lines is included in the count. (b) The two-dimensional disector counting rule ensures correct enumeration of profiles that belong to area h(dis) $\times$ l(dis) by imposing an unambiguous two-dimensional lexicographic order of the profiles. That is, each profile of any size, shape, and orientation is counted once and only once, when the dissector line separation is small enough that all profiles are intersected. The inner white profile represents the 'dense core profile' and the one-dimensional intersect of the profile is heavily drawn at the inclusion line in the disector where it is counted. The outer diamond-hatched rim represents 'fuzziness' due to overprojection and the intersect of the profile including its fuzzy rim is hatched at the inclusion line in the disector where it is counted. The symmetry of the two-dimensional disector counting rule provides an unbiased counting rule that is insensitive to overprojection below the degree where two closely apposed structures fail to be recognized as two separate entities.

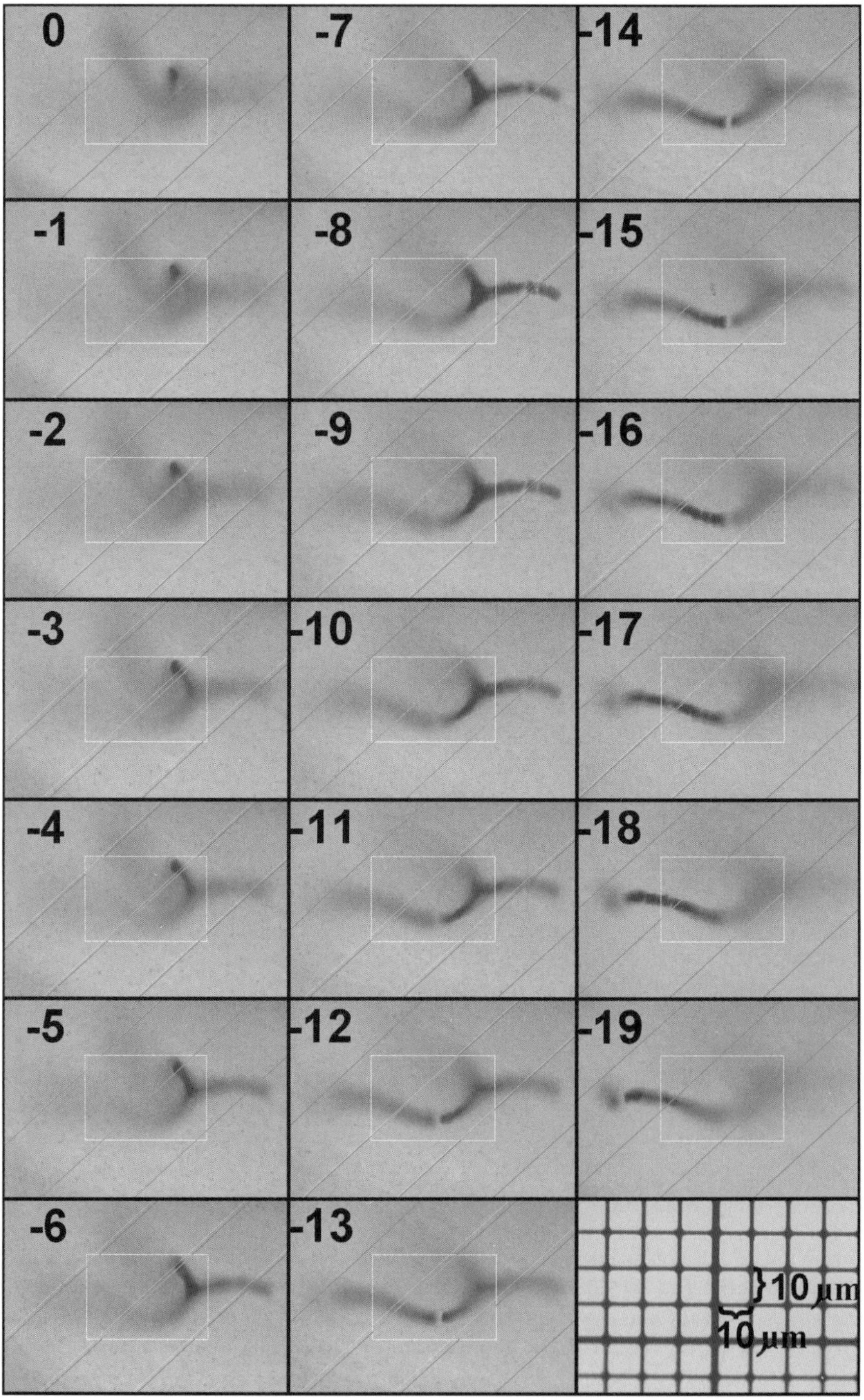
0
-7
-14
-1
-8
-15
-2
-9
-16
-3
-10
-17
-4
-11
-18
-5
-12
-19
-6
-13
10 µm
10 µm

◄ **Fig. 13.7** (see also **Plate 7**) A graphical illustration of the counting rule with 'spatial' two-dimensional disectors in the tilted virtual plane exemplified by various events. Three successive focal planes are shown. The upper focal plane contains the look-up exclusion line and a white line showing the projected position of the next disector inclusion line. The middle focal plane contains the inclusion line (green) situated inside the sampling box with the exclusion extension (red, towards the observer) outside the sampling box, and the line of indifference (blue), that is, the two-dimensional representation of the guard area lateral to the two-dimensional-disector. h(dis) must be small enough that all events are recognized, but is made large in the figure for illustrative purposes. The lowermost focal plane contains the line representing the look-down guard area below the sampling box (red). The two-dimensional representation of the structure seen in the focal planes is shown in shadowed red, and the part of the structure between the focal planes is red. In (a) the profile exclusively intersect the green inclusion line and one profile is counted, whereas in (b) the profile touches the red exclusion extension and is thus not counted. In (c) the profile is intersected by the look-up exclusion line and stays in touch with the moving line during focusing which means one coherent 'profile' in the virtual plane and no profiles are counted. In (d) the profile that is intersected by the upper look-up exclusion line leaves the line during focusing, that is, it is not present in the virtual plane, but is intersected again by the inclusion line and thus one profile is counted. In (e) the structure is transected twice by the virtual plane as the profiles are connected outside the moving line, that is, two profiles are counted in (e). When multiple profiles located on the inclusion line are disconnected in the focal plane, information from the guard area in succeeding focal planes is needed to determine whether the profiles represent one coherent 'profile' in the virtual plane. One profile is counted in (f) as the profiles located on the inclusion line stay in touch with and merge on the moving line, that is, represent one coherent U-shaped 'profile' in the virtual plane. In (g) where the two profiles located on the inclusion line merge outside the moving line in succeeding focal planes, the virtual plane contains two profiles. No profiles are counted in (h) where the profiles located on the exclusion extension stay in touch with and merge on the moving line in succeeding focal planes. One profile is counted in (i) where the profiles located on the exclusion extension merge outside the moving line. (Reproduced from Larsen *et al.* (1998) with permission from The Royal Microscopical Society).

An equal expected sampling area of virtual planes, $E[a(\text{planes})]$, for all orientations, regardless of the x,y,z-dimensions of the sampling box is achieved with Cavalieri's principle by intersecting the sampling box completely with parallel systematic, uniformly random virtual planes with a fixed user-defined plane separation distance, d. The position of the first plane is random within a width equal to the fixed interplane spacing and the succeeding planes are systematically displaced by the fixed plane separation distance with respect to the previous plane (Fig. 13.5). The expected area of the virtual planes, is thus constant

$$E[a(\text{planes})] = v(\text{box})/d. \qquad (2.3)$$

The exact sampling plane area, $a(\text{plane})$, in each systematically sampled box is written automatically on to a log file in the computer. The number of profiles per unit test area of the virtual plane is counted with columns of spatial two-dimensional-disectors (Larsen *et al.* 1998) or by using the associated point-counting rule (Miles 1978); both counting rules require bilateral and lower guard areas. The two-dimensional disector counting rule is explained in Figs 13.6 and 13.7.

An unbiased estimate of the global length density, L_V, is

$$L_V := \frac{2 \cdot \Sigma Q}{\Sigma a(\text{plane})} = \frac{2p(\text{box})}{a(\text{plane})} \cdot \frac{\Sigma Q}{\Sigma p(\text{ref})} \tag{2.4}$$

where ΣQ is the total sum of the number of transects of the linear structure with the virtual plane, and $\Sigma a(\text{plane})$ is the total sum of the sampling plane areas,

$$\frac{1}{p(\text{box})} \Sigma [p(\text{ref}) \cdot a(\text{plane})]$$

where $p(\text{ref})$ is the number of relevant sampling box corners that hit the reference space and $p(\text{box})$ is the number of sampling box corners under consideration, for example, the four topmost corners. For detailed description of counting over natural and artificial edges, see Section 3. Both sums are over all uniformly sampled sections from the reference space.

The total length, L, of the linear structure is estimated as

$$L_1 = L_V \cdot V(\text{ref}) \tag{2.5}$$

where $V(\text{ref})$ is the total volume of the reference space, *or* estimated directly using a fractionator sampling scheme as

$$L_2 = \frac{1}{\text{ssf}} \cdot \frac{1}{\text{asf}} \cdot \frac{1}{\text{hsf}} \cdot \frac{1}{\text{psd}} \cdot 2\Sigma Q \tag{2.6}$$

where ssf is the section sampling fraction, asf the area sampling fraction, hsf the height sampling fraction, and psd the constant probe sampling density,

$$\frac{E[a(\text{plane})]}{v(\text{box})} = \frac{1}{d}$$

$$L_2 := \frac{1}{\text{ssf}} \cdot \frac{d_x \cdot d_y}{a(\text{box})} \cdot \frac{\bar{t}}{h(\text{box})} \cdot d \cdot 2\Sigma Q \tag{2.7}$$

where d_x and d_y are the distances in the *x*- and *y*-directions, respectively, between the systematically sampled fields, and the height sampling fraction, hsf, is the sampling box depth divided by average section thickness, *t*. The section sampling fraction, ssf, is the fraction of uniformly sampled sections from the reference space, for example, 1:10.

Note that L_1 is a ratio estimator with a random variable in the denominator, whereas all denominators in L_2 are design constants, that is, L_2 is a straight-forward unbiased estimator, whereas L_1 is 'only' a consistent or ratio-unbiased estimator.

For the isotropic, virtual plane technique to be applicable, one must be able to make relatively thick, physical sections that constitute a uniform sample of the reference space, and to completely and unambiguously visualize the structure of interest with stains that penetrate the transparent thick, physical sections. Also, the analysis must be performed with a high-power immersion objective and the reference region must be definable at the resulting sections.

By analogy a sphere is dynamically mapped in the space ball technique and all intersections between the structure and the test surface are counted. The advantage of the space ball technique is a simpler counting rule. As shown in Section 5, however, the test surface geometry is biased for three-dimensional linear structures and it offers no counting rule for irregular profiles, that is, it is only applicable to structures that can be assumed to be cylindrical tubes. Furthermore, the symmetry in the counting rule offered by the spatial two-dimensional disector is lost. This symmetry is important for avoiding biases due to overprojection. The space ball technique precludes fractionator sampling and is consequently sensitive to a refractive index mismatch and is only a 'ratio-unbiased estimator'. Of less importance the user may find it difficult to adjust to the fact that the speed of movement as a function of focal plane displacement is not constant within the individual sampling window.

3 A worked example

A liquefied spinal cord lesion is in some respects a suitable region for this type of quantification as the stereological length estimators based on random rotation of the tissue around at least one axis prior to sectioning (Batra *et al.* 1995; Cruz-Orive and Howard 1991; Gokhale 1990; Gundersen 1979) are difficult, if not impossible, to apply. In some types of lesions sectioned in certain orientations there is a risk that the lesion interior is lost in subsequent histological tissue processing and the lesion cavity may collapse. We have tried different section angles and found cross-sections to give the best preservation of the lesion cavity (von Euler *et al.* 1998). Moreover, anatomical orientation is easy and the lesion demarcated and easy to delineate. The trace of the cavity border is clearly visible at all the resulting sections and can thus be delineated. We chose to immunohistochemically label axons containing 200 kDa neurofilaments (NF). The monoclonal antibody had a good penetration and thus we could fulfill the requirement of thick physical sections with well-stained material, though the final section thickness was thinner than desired (see below). We have chosen to fresh freeze the tissue and post-fix the cryosections used for the evaluations as we have observed that the interior of the spinal cord lesion has a tendency to be lost in perfused tissue. Immersion-fixed tissue would probably work equally well.

The spinal cord lesion was caused by photochemically induced thrombosis at thoracic level 8 in rodent spinal cord (see von Euler *et al.* 1997). At time of sacrifice the animal was decapitated, the spinal cord segments containing the entire lesion taken out in one piece, and immediately frozen in a mixture of dry ice and isopentane and stored at −70°C. The spinal cord segments were cut in cross-sections (Microm HM 500 OM, Walldorf, Germany) with a cryostat advance setting of 40 μm and allowed to dry on the glass for at least an hour in room temperature. There is a risk that the section will fall off the glass in subsequent tissue handling but, by allowing the section plenty of time to dry and stick to the glass, this risk was reduced. Before further histological handling the sections were fixed in 4% phosphate buffered paraformaldehyde (pH 7.4). Following removal of endogenous peroxidases with 3% H_2O_2, a systematic, random sample of every sixth section was selected (sections were selected at six-section intervals with a random start in the first six sections) and immunohistochemically stained with monoclonal anti-NF protein 200 antibodies (1:500, IgG concentration of 0.69 μg/ml, Sigma Immuno Chemicals, St. Louis, MO, USA) to label NF-containing axons. For visualization avidin–biotin peroxidase (Vectastain ABC-kit, Vector, Burlingame, CA, USA) followed by vector SG (Vector, Burlingame, CA, USA) was used. The specificity of the NF immunohistochemistry was checked in negative controls omitting the primary antibody or using a negative control serum of the same Ig subclass. The antibody stain penetrated well into the spinal sections and worked excellent in fresh frozen tissue.

The stereological analysis was carried out using an Olympus BH-2 microscope equipped with a motorized specimen stage, controlling movements along the x- and y-axes, and a microcator (VRZ 401, Heidenhain, Traunreut, Germany) monitoring the movements along the z-axis with a resolution of 0.5 μm. The trace of the lesion area was delineated at 144 × using a 4 × objective that was paracentered with the 100 × objective used for the actual counting procedure and the lesion area was recorded. Two-dimensional uniform random sampling in the delineated area was made with precise interactively defined x- and y-steps (here, $d_x = 150$ μm; $d_y = 150$ μm). The two-dimensional representation of the isotropic, virtual planes was software-controlled as previously described (CAST-Grid system, version 1.10, Olympus, Albertslund, Denmark). For each sampled field of vision the NF-positive fibres inside the sampling box were analyzed at a final magnification of 3600 × using a 100 × SPLAN oil immersion objective (numerical aperture (NA), 1.4). The sampling box was defined by the area of the rectangle on the screen, that is, the sampling box area 1120 μm² and the fixed height of focal plane displacement, that is, the sampling box height 10 μm. A set of virtual planes, with virtual plane separation distance 15 μm, was mapped inside the sampling box by random lines that moved across the screen as the focal plane was moved through the section. Intersections between the isotropic, virtual planes and the NF-positive fibres inside the sampling box were counted with spatial two-dimensional disectors. The computer systematically selected a new isotropic, random orientation of the virtual plane for each new field of vision. In each spinal cord lesion 200–500 NF-positive fibre intersections with the isotropic virtual planes were counted. Counting was also performed in the sampling boxes that were intersected by the natural edges. In these sampling boxes, the sampling plane area of the box was included with a weight corresponding to the number of sampling box corners that hit the lesion region. For example, if five sampling box corners hit the reference space, the area of virtual planes in that sampling box is weighted with 5/8. The section thickness, t, was measured with the microcator in all systematically sampled microscope fields. The mean section thickness, $\bar{t}$, for the sections from each spinal cord lesion was used to calculate the total volume ($V_{processed}$) of the lesion after histological processing $V_{processed} = \Sigma A \cdot \bar{T}$, where ΣA is the sum of the lesioned areas in all sampled sections, $\bar{T}$ is the mean distance between sampled areas, that is, $\bar{T} := \bar{t}/\text{ssf}$, where ssf is the section sampling fraction.

There was a high degree of shrinkage along the z-axis in the cryostat sections, most pronounced within the lesion. The reason for this is probably the high water content of the lesion interior. The final histological section had local variations in the section thickness, which required that the data handling be divided into two different procedures: (1) estimation of length density and reference space volume (L_1) was applied for sampling boxes located at places where the section thickness was above the sampling box height (> 10 μm), and (2) a modified (slightly biased) fractionator design (L_2) was applied for the sampling boxes located at places where the section thickness was below the sampling box height (≤ 10 μm).

L_1 was estimated using the equation

$$L_1 := L_v \cdot V(\text{ref}) = 2 \cdot Q_A \cdot V(\text{ref})$$

where $V(\text{ref}) := N(\text{sampling boxes} > 10\ \mu\text{m}) \cdot d_x \cdot d_y \cdot T(L_1)$. N(sampling boxes > 10 μm) is the total number of sampling boxes with a tissue thickness > 10 μm, d_x and d_y are the step lengths between sampled microscope fields, and $T(L_1)$ is the distance between sampled areas, that is, mean section thickness in sampling boxes used for the L_1 estimates multiplied by the inverse section sampling fraction.

L_2 was estimated using the equation

$$L_2 := (2/\text{ssf}) \cdot (d_x \cdot d_y/a(\text{box})) \cdot d \cdot \Sigma Q,$$

where $a(\text{box})$ is the rectangle on the screen representing the sampling box (1120 μm²), d is the virtual plane separation distance (15 μm), and ΣQ is the observed total number of intersections between NF-positive fibres and the isotropic virtual planes inside sampling boxes with section thickness ≤ 10 μm.

The total length estimation (L_{tot}) is the sum of the two above-mentioned estimates

$$L_{tot} := L_1 + L_2.$$

3.1 Calculations from one animal

For each point in the raster pattern a data set comprising data for the number of profiles counted in the sample box, the actual area of virtual test plane in the sample box, and the section thickness at that particular position is obtained. The data are tied for each data set, sorted according to section thickness. Data sets obtained at points in the raster pattern where the section thickness is >10 µm are handled in a density times reference volume setting (Table 13.1) and those where the section thickness is ≤10 µm are handled in a fractionator model (Table 13.2).

In the modified fractionator model used in this example (Table 13.2), counting was performed through the entire section thickness, so the only variable entering the equation is the total number of transects counted. However, a guard area is needed below the sampling box for the counting rule to be unbiased, and so the sampling box height and the mean section thickness also enter the correct formula.

Table 13.1 L_1: length density times reference volume model

Number of profiles counted	Exact sampling plane area (µm²)	Section thickness (µm)
3	949	43.5
1	637	38.5
0	743	37
2	609	32.5
0	912	31.5
5	647	30
1	738	29.5
.	.	.
.	.	.
.	.	.
5	770	11.5
0	598	11
4	833	11
2	769	11
3	631	11
2	724	11
0	657	10.5
0	675	10.5
$\Sigma Q = 384$	$\Sigma a(\text{plane}) = 156\ 531\ \mu m^2$	$t = 17.8\ \mu m$

Calculations based on these data

Constants: $d_x = d_y = 150\ \mu m^2$; section sampling fraction, ssf = 1/12.

$N(\text{sampling boxes} > 10\ \mu m) = 221$

$Q_A := 384/156\ 531\ \mu m^2 = 0.00245$ profiles per μm^2

$L_V := 2 \times 0.00245 = 0.00491\ \mu m^{-2}$

$V(\text{ref}) := 221 \times 150\ \mu m \times 150\ \mu m \times 12 \times 17.8\ \mu m = 939\ 553\ 200\ \mu m^3 = 0.94\ mm^3$

$L_1 := 0.00245\ \mu m^{-2} \times 939\ 553\ 200\ \mu m^3 = 4\ 609\ 802\ \mu m = 4.61\ m$

Table 13.2 L_2: Fractionator model (modified)

Number of profiles counted	Section thickness (μm)
0	10
1	10
0	10
0	10
0	10
4	10
0	10
.	.
.	.
.	.
0	2.5
0	2.5
0	2.5
0	2.5
0	2.5
0	2
0	2
0	2
$\Sigma Q = 24$	

Calculations based on these data

Constants: $d_x = d_y = 150$ μm; $d = 15$ μm; $a(\text{box}) = 1120$ μm²

$L_2 = (2 \times 12 \times 150$ μm $\times 150$ μm $\times 15$ μm$/1120$ μm²$) \times 24 = 173\ 571$ μm $= 0.17$ m

$L_{\text{tot}} = L_1 + L_2 = 4\ 783\ 374$ μm $= 4.78$ m

The massive shrinkage along the z-axis rendered this approach impossible, and the modified fractionator, which is inherently biased, was applied of necessity and should, whenever possible, be avoided. The section thickness should optimally be 40–60 μm for the technique to be efficient. The counting rule is symmetrical so in theory there is no lower limit of the height of the sampling box but, as is the case for the optical disector, fuzziness and overprojection make decisions difficult in a small range (e.g. 0.5 μm) of the total focal plane displacement. The percentage of 'difficult counting' should not be too big. The counting rule requires a guard area in the bottom of the section, even though we can disregard 'lost caps', which must be taken into consideration for number estimation using the optical disector. The length estimator estimates the length of fibres present in the section, and shrinkage of the linear process itself will affect the final result whereas shrinkage and/or deformation of the containing volume is irrelevant. In this example it is not possible to calculate the coefficient of error, as we apply fractionator sampling combined with by the ratio estimator (length density times reference volume).

4 Conclusions

Practical techniques for length estimation of spatial three-dimensional linear structures are now available for isotropically, vertically, and arbitrarily orientated histological sections as well as for a complete non-sectioned specimen in projection. In the quest for more efficient test surface geometries for optical and virtual test systems it is important to avoid a net curvature of the test surface. A net curvature results in bias, the size of which depends on diameter of the test system (and its infinite extension) relative to the diameter of the structure as shown in Section 5. Of equal importance, the test surface geometry must allow an optical or virtual unbiased two-dimensional counting rule to ensure correct enumeration of the profiles observed in a small sample window. A stack of spatial two-dimensional disectors (Larsen *et al.* 1998) provides an unbiased counting rule that can be used for both straight lines and lines with local curvature *when the net curvature of the line and its infinite extensions is zero and the speed of displacement on the screen as a function of z-axis displacement is constant within the observation frame*. The optical counting frame is a special case of that counting rule that can be used for UFAPP (unbiased for all practical purposes) length estimation in vertical sections.

5 Appendix. Test surface geometries valid for length estimation of spatial three-dimensional linear structures

The aforementioned design-based stereological principles for global length estimation are based on stochastic geometry, inferring the length estimate from the *a priori* known probability that one-dimensional lineas features per unit three-dimensional reference space volume are intersected by an isotropic, uniform, random (IUR) two-dimensional test surface. For three-dimensional linear structures of non-zero cross-sectional area (referred to as three-dimensional linear structures in the following text) the measure of length, L, is the axial length. However, a transect between a test surface and a three-dimensional linear structure appears as a two-dimensional profile while the actual zero-dimensional intersect between the one-dimensional axial length and the test surface is imperceptible. Reality, therefore, implies that we are restricted to counting two-dimensional profiles, as opposed to zero-dimensional axial intersects, when estimating length of three-dimensional linear structures. For several, but not all, geometric designs of test surface there is (practically) equality of the probabilities of observing a profile and of an axial intersect (Hennig 1963, Gundersen 1979). The advent of computer-controlled virtual test systems inside thick, physical sections (Larsen *et al.* 1998) justifies considerations about conditions for such an equality. Below we present an overview of basic strategies for length estimation of one-dimensional linear features and the resulting constraints when these principles are applied to length estimation of three-dimensional linear structures.

The probability that a planar IUR test surface T_2 intersects an infinitesimal line element dnL of a stationary one-dimensional lineal feature Y_l in l^3 is directly proportional to the projected length $d_{L'}$ on the normal $\boldsymbol{n}$ to the plane. The isotropy of the test plane ensures that, on average, $d_L = (\pi/4)d_L$ (Saltykov 1946). Gokhale (1990) emphasized that the test surface need not be planar and that a spherical test surface has perfect isotropy and thus, in theory, is applicable for length estimation of one-dimensional linear features. He also realized that, *in projection* through a thick, vertical section (Baddeley *et al.* 1986), intersections between the projected image of the lineal features in the section and cycloid test lines with the minor axis perpendicular to the vertical axis represent (a corrugated) IUR test surface that intersects spatial one-dimensional lineal features. How the number of observable two-dimensional profiles quantitatively relates to the number of axial transects depends on the overall shape and architecture of the linear structure(s) and of the test system (Figs 13.8 and 13.9). For a smooth linear structure probed with an IUR test surface with net-curvature zero, the bias due to the departure from the one-dimensional model reduces to biases related to the end and branching point regions of the linear structure (illustrated in Fig. 13.8). The end-bias has order d/l, where d and l are the diameter and length of the structure(s), respectively

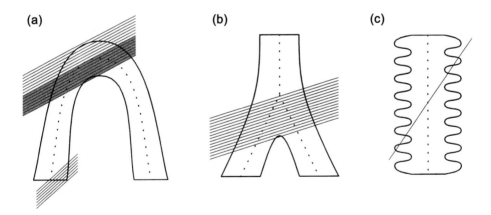

(a)   (b)   (c)

Fig. 13.8 Transects between an IUR test surface with net-curvature zero and a three-dimensional linear structure. (a) It is illustrated in the upper, left corner that the probability of transecting the linear structure (and thereby observing a profile) without intersecting the linear axis is compensated by an identical probability of observing one profile representing two intersects with the lineal axis (gray area), thereby securing equality of the probabilities. At the free ends there is a probability of observing profiles unrelated to the probability of intersecting the lineal axis. (Based on Hennig 1963, fig. 3) (b) In a branching point region the probability of observing one profile representing two intersects with the lineal axis is not offset by any probability of observing one profile representing zero intersects. (c) For a linear structure that is not smooth along the axis there is a probability of observing several profiles that are unrelated to the probability of intersecting the lineal axis.

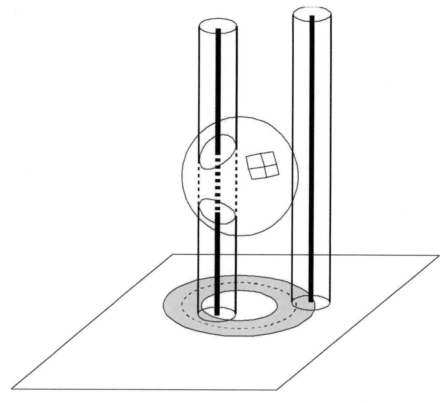

Fig. 13.9 Spherical test surface inserted randomly into a space containing three-dimensional linear structures. The axial length of the three-dimensional linear structure is indicated. It is clear that translation along the z-axis does not change the situation and we can thus reduce our considerations to the situation in the projected two-dimensional image in the xy-plane (perpendicular to the axes of the linear structures). The diamond-hatched area shows the projected two-dimensional image of positions where the intersections between the spherical test surface and the linear structure is observed as one profile. The gray area consists of two annuli (separated by a dashed line) of width $d/2 = r$ on either side of the sphere projection. The area of the zone where one 'profile' is observed is thus $\pi[(R + r)^2 - (R - r)^2]$. In the inner white zone of area, $\pi(R - r)^2$, there is a probability of observing *two* 'profiles', and the aggregate probability of observing a 'profile' is proportional to $\pi[(R + r)^2 - (R - r)^2] + 2\pi(R - r)^2$, whereas the probability of observing *two* axial intersects is proportional to πR^2. 'Profiles' thus outnumber axial intersects by a factor $\frac{[(R + r)^2 - (R - r)^2] + 2(R - r)^2}{2R^2} = \frac{R^2 + r^2}{R^2}$ (based on Gundersen 2001, fig. 4).

(Gundersen 1979). For a continuous three-dimensional linear structure in a network the actual bias is unknown, but it is an underestimate due to branching point regions (Mattfeldt and Mall 1984). These unavoidable biases are generally small and unlikely to differ significantly among the various practical techniques for length estimation. On an IUR test surface with net-curvature zero we may thus *infer* the areal density of axial intersections, I_A, from the areal density of profiles, Q_A, for UFAPP (unbiased for all practical purposes) length estimation, making use of the relationship $L_V := 2Q_A = 2I_A$.

However, for a test surface with positive net curvature the one-to-one relationship between profiles and axial intersects is lost (Gundersen 2001). For simplicity, we consider a spherical test surface to illustrate the problems with positive net curvature because this particular test surface has generic isotropy and we can thus reduce our considerations to translational randomness. As shown in Fig. 13.9, the probability that a test sphere transects a linear structure depends on the diameter D of the sphere relative to the diameter, d, of the linear structure. This results in overcounting and is biased in favor of the part of the linear structure with large diameter. This extra bias due to the net curvature of the test surface is $+r^2/R^2$, where r and R are the radii of the linear structure and the test sphere, respectively (Fig. 13.9 and Gundersen 2001).

References

Baddeley, A.J., Gundersen, H.J.G., and **Cruz-Orive, L.M.** (1986). Estimation of surface area from vertical sections. *J. Microsc.* **142**, 259–76.

Batra, S., König, F., and **Cruz-Orive, L.M.** (1995). Unbiased estimation of capillary length from vertical slices. *J. Microsc.* **178**, 152–9.

Calhoun, M.E. and **Mouton, P.R.** (2001). Length measurements: new develoments in neurostereology and 3D imagery. *J. Chem. Neuroanat.* **21**, 257–65.

Cruz-Orive, L.M. and **Howard, C.V.** (1991). Estimating the length of a bounded curve in three dimensions using total vertical projections. *J. Microsc.* **163**, 101–13.

Gokhale, A.M. (1990). Unbiased estimation of curve length in 3-D using vertical slices. *J. Microsc.* **159**, 133–41.

Gundersen, H.J.G. (1977). Notes on the estimation of the numerical density of arbitrary profiles: the edge effect. *J. Microsc.* **111**, 219–23.

Gundersen, H.J.G. (1979). Estimation of tubule or cylinder Lv, Sv, and Vv on thick sections. *J. Microsc.* **117**, 333–45.

Gundersen, H.J.G. (1986). Stereology of arbitrary particles. A review of unbiased number and size estimators and the presentation of some new ones, in memory of William R. Thomson. *J. Microsc.* **143**, 3–45.

Gundersen, H.J.G. (1988). The nucleator. *J. Microsc.* **151**, 3–21.

Gundersen, H.J.G. (2001). Stereological estimation of tubular length. *J. Microsc.* **207**, 155–60.

Hennig, A. (1963). Length of a three-dimensional linear tract. In *Proceedings of the First International Congress for Stereology*, 44/1–44/4. Bönecke-Druck-Clausthal, Vienna.

Jensen, E.B.V. and **Gundersen, H.J.G.** (1993). The rotator. *J. Microsc.* **170**, 35–44.

Kubínová, L. and **Janácek, J.** (1998). Estimating surface area by the isotropic Fakir method from thick slices cut in an arbitrary direction. *J.Microsc.* **191** (2), 201–11.

Larsen, O.J., Gundersen, H.J.G., and **Nielsen, J.** (1998). Global spatial sampling with isotropic virtual planes: estimators of length density and total length in thick, arbitrarily orientated sections. *J. Microsc.* **191**, 238–48.

Mattfeldt, T. and **Mall, G.** (1984). Estimation of length and surface of anisotropic capillaries. *J. Microsc.* **135**, 181–90.

Mattfeldt, T., Mall, G., and **Gharehbaghi, H.** (1990). Estimation of surface area and length with the orientator. *J. Microsc.* **159**, 301–17.

Miles, R.E. (1978). The sampling, by quadrats, of planar aggregates. *J. Microsc.* **113** (3), 257–67.

Mouton, P.R., Gokhale, A.M., Ward, N.L., and **West, M.J.** (2002). Stereologiscal length estimation using spherical probes. *J. Microsc.* **206**, 54–64.

Nyengaard, J.R. and **Gundersen, H.J.G.** (1992). Short technical note. The isector: a simple and direct method for generating isotropic, uniform random sections from small specimens. *J. Microsc.* **165**, 427–31.

Saltykov, S.A. (1946). *Zavodskaja laboratorija* **12**, 816. (cited in Weibel 1979).

von Euler, M., Sundström, E., Seiger. Å. (1997). Morphological characterization of the evolving rat spinal cord injury after photochemically induced ischemia. *Acta Neuropathol.* **94**, 232–9.

von Euler, M., Larsen, J.O., and **Janson, A.** (1998). Quantitative study of neurofilament-positive fiber length in rat spinal cord lesions using isotropic virtual planes. *J. Comp. Neurol.* **400**, 441–7.

SECOND ORDER STEREOLOGY

CHAPTER 14

SPATIAL DISTRIBUTION

STEPHEN M. EVANS

1 Introduction

Throughout this book a variety of methods have been presented for estimating first-order properties, that is, the means or total quantities, for example, the total number of neocortical neurons or the mean volume of senile plaques in the hippocampus in the brain from a patient with Alzheimer's disease. Stereology also provides tools to look at distribution around the means, second-order properties. For example, neurons come in all shapes and sizes. Using the nucleator/rotator estimators it is possible to estimate the volume of each individual cell sampled. This means that not only can the mean be estimated but also the size distribution around the mean. This chapter presents a method for estimating the variation of number as a function of distance.

The human brain is made up of an ordered three-dimensional array of smaller units, for example, the vertical and horizontal organization of neurons in the cerebral cortex. Knowledge of the structural relationship of these small units should provide more insight into brain function as a whole and how it doesn't function in pathological states. For example, are cerebral neocortex neurons randomly distributed or do they occur in clusters, for example, the clustering of glial cells around neurons, 'satellitosis', in humans during aging? Do transplanted neurons randomly distribute themselves or is there some form of order?

In the estimation of number, the first-order properties describe the mean number of events per unit area as a function of position. Second-order properties describe the variation in the relative frequency of pairs of events as a function of their positions, that is, they quantify spatial patterns of elements of a structure (e.g. clustering, repulsion, etc. between the elements).

The interpretation of a three-dimensional point pattern using only two-dimensional information was previously difficult if not impossible. Braendgaard and Gundersen (1986) give an example using a stack of physical sections where the two-dimensional average nearest-neighbour distance for neurons in rat frontal cortex was estimated to be 50 μm whereas the three-dimensional average nearest-neighbour distance was estimated to be 20 μm, with the largest three-dimensional distance being only 40 μm. Bjaalie and Diggle (1990) give an elegant example of spatial distribution in cat corticopontine neurons in relation to cortical visual field maps and cortical modular organization. However, using physical sections is time-consuming and there is the problem of section alignment. An alternative to physical sections is to use optical sections; see, for example, Baddeley *et al.* (1987) who used a tandem-scanning reflected light microscope to examine the spatial distribution of osteocyte lacunae.

The method described here is unbiased and can directly measure three-dimensional point patterns from measurements done on one section. The method is based on the nucleator principle (Gundersen 1988; Jensen 1998) where the probability that a point in three-dimensional space is hit by the probe is known. One of the unique features of the nucleator is that in one section the three-dimensional distance can be measured from a typical point to everything else that is observable in the section. That is to say, it is possible to estimate the variation of events, in this case the number of points, as a function of distance, that is, the three-dimensional spatial distribution of the points.

2 Theory

A very simple analogy as to how this method works is to imagine a moth flying around a light inside a lampshade. If one were to quickly glance at the lampshade the probability that the moth would be noticed would be related to the size of its shadow on the lampshade, that is, the bigger the shadow the more chance it has of being seen. The shadow that is projected on the lampshade is directly related to the distance of the moth from the light and the size of the moth.

Now, looking at this from a stereological viewpoint, the moth will be frozen in time, that is, it is a stationary process, the light bulb is a point source of light, and the light lampshade is spherical. A section is then taken through the point light source in a random direction in three-dimensional space, that is, an isotropic direction. Again the probability of noticing the moth, that is, sampling it, depends on its size and its distance from the light source. If two sections are used and the disector principle applied, particles can be sampled with equal probability regardless of their size. So now the probability of sampling the moth is directly related to its distance from the light source.

The method can be employed on either isotropic uniform random (IUR) or vertical sections. It provides a map of the variation of the numerical density of events

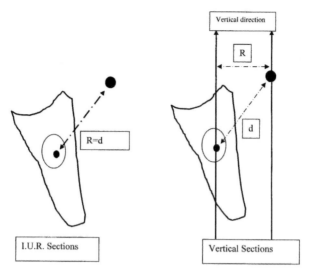

Fig. 14.1 Measurements are made from an arbitrary fixed point, for example, a neuron's nucleolus to an *i*th event, for example, a glial cell, and the numerical density is estimated from the formulae given in the text.

around an arbitrary fixed point at different distances, R, from the fixed point, that is, the spatial distribution of these events around a fixed point. If an event occurs with high probability at a distance R from the fixed point then the numerical density of these events at distance R will also be high. After selecting the arbitrary fixed point, a disector of known height, h_{dis}, is used to select the events that occur around this fixed point. To calculate with what probability the *i*th event occurs at distance R from an arbitrary fixed point it is necessary to measure the distance d_i (see Fig. 14.1). Notice that, for IUR sections, $d = R$.

For vertical sections the formula is

$$\text{est}(N_V, R) = \frac{3 \cdot \pi \cdot \dfrac{\sum d_i}{h_{dis}} + \text{any particles on the vertical axis}}{4 \cdot \pi \cdot R^3}$$

where $\text{est}(N_V)$ is an estimate of the numerical density of events that occur at a distance R from an arbitrary fixed point, d_i is the distance from the *i*th event to the vertical axis that passes through the arbitrary fixed point, that is, an axis that passes through the point and is parallel to the axis of rotation of the vertical section, and h_{dis} is the height of the disector used to sample the events.

For IUR sections the formula simplifies to

$$\text{est}(N_V, R) = \frac{3 \cdot 2 \cdot \dfrac{\sum d_i}{h_{dis}}}{4 \cdot \pi \cdot R^3},$$

but this time $d_i = R$ and is the distance from the ith event to the arbitrary fixed point.

The essential requirements for these formulae to work are the following.

1 The value of R should be large enough so that the numerical density approaches a constant value, i.e. the average numerical density of the events in the structure.

2 The height of the disector must be much less than the distance R.

A frequency distribution map of the variation of the numerical density of events around the fixed point is created by dividing R into a series of classes of equal length. The corresponding measurements d_i are also divided into these classes. The sum of distances, Σd_i, for each class was then averaged for all of the sampled cells so that a three-dimensional map showing the variation of the numerical density of the events as function of distance from the fixed point could be built up. Examples of these frequency distribution maps of the variation of the numerical density of events around the fixed point, the spatial distribution patterns, are shown in Fig. 14.2.

3 A worked example. How to estimate the spatial distribution of glial cells around neurons

The sampling scheme for neocortical tissue was based on the one described by Evans *et al.* (1989) and described in detail by Braendgaard *et al.* (1990). Estimation of the spatial distributions of glial cells around neurons using the nucleator principle is described in Evans and Gundersen (1989). An outline of the method and a worked example is given below.

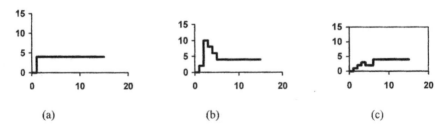

(a) (b) (c)

Fig. 14.2 Some examples of spatial distribution patterns. The numerical density of events is shown on the y-axis and the distance from a fixed arbitrary point is shown on the x-axis. Initially the numerical density is zero, that is, there is volume around the point where the probability of finding an event is zero. (a) The numerical density becomes constant and the spatial distribution could be described as being a Poisson distribution. (b) The numerical density initially increases rapidly in excess of the later constant value, indicating a clustering pattern before settling down to reach a constant-value, Poisson distribution. (c) The numerical density initially increases slowly from zero, then rapidly decreases to below the level of the later constant value. This would indicate a repulsion pattern before settling down to reach a constant-value, Poisson distribution.

Brains with no recorded neurological or psychiatric disorders were removed within 72 hours from time of death and stored in 0.1M sodium phosphate-buffered formaldehyde (pH 7.2, 4% formaldehyde) for at least 6 months, with the fixative being changed every 2 months. The meninges and main arteries were removed and the cerebellum and brainstem detached at a level corresponding to that below the inferior colliculus on the dorsal surface to the rostral border of the pons on the ventral surface. After the cerebrum was divided in half sagittally, one of the hemispheres was weighed and five cortical regions, namely, frontal, temporal, parietal, occipital, and archeo, delineated on the pial surface using Wolbach's Giemsa stain. The hemisphere was then embedded in 6% agar, which provides support during the cutting procedure and preserves the spatial arrangement of the various cortical structures in each coronal slice. The agar-embedded hemisphere was then cut coronally at 7 mm intervals, with a random starting point to fulfill the requirements for the Cavalieri volume estimator, using an instrument illustrated in Fig. 24 in Gundersen and Jensen (1987). This device enabled coronal sections to be cut at a predetermined interval but a variety of devices have been devised to cut parallel sections (see Reed and Howard (1998) for some further examples). Although not strictly necessary for this method, the neocortical regional volume was estimated by point counting and Cavalieri's principle (see Chapter 8). Commencing from the frontal pole, every second coronal slice was sampled for neocortex with the starting slice being chosen at random. Each coronal slice was then placed, with the surface nearest to the occipital pole face up, in the sampling device (shown in Fig. 3 of Braendgaard *et al.* 1990). Briefly, this is a circular plate with a scale around its circumference, from 0 to 60, to which a knife holder could be attached. This allowed perfect straight lines, the thickness of a razor blade, to be cut across the diameter of the plate and perpendicular to its surface. The central points of the plate and slice roughly coincided and the medial side of each coronal slice was aligned parallel to the line, which corresponds to the 0/30 line. The position of the knife holder on the first slice was determined by looking up a number between 1 and 60 in a random number table. The coronal slice was cut only in areas that contained neocortex, moved along 2 positions, that is, 1/30th of a full turn, and another cut made to produce a wedge of neocortex, which was then removed from the slice. The knife holder was then rotated clockwise by 11 positions for sampling in the next coronal slice. Two slabs of 2 mm thickness were cut from the occipital surface of the wedges and these were then cut into 2 mm wide parallel bars, roughly perpendicular to the pial surface. This allows further subsampling while still preserving the wide distribution of the original sample and also it allowed unbiased size estimates to be made using a technique involving 'vertical sections' described by Baddeley *et al.* (1986). The wedges from each region were kept together and their sampling sequence recorded. The same procedure was followed for all the bars and approximately 10 bars per region were systematically sampled. Note that, if any of the bars were damaged during the tissue processing and could not be used, this random destruction would not in any way affect the unbiasedness of the estimate but might slightly increase the coefficient of error due to an increase in the variability of the sample distribution.

As much as possible of the white matter was removed from the subsampled bars, which were then embedded in 4% agar with the aid of an aluminum block with a series of 2 mm wide parallel hemicylindrical indentations in it. The indentations allowed the bars to be given a random rotation around their long axis to fulfill the requirement of vertical sections.

The agar-embedded tissues were dehydrated by immersion in five changes of successively stronger concentrations of ethanol, ($2 \times 75\%$, $2 \times 93\%$, and $1 \times 99\%$, for 1 hour per step), then infiltrated with glycol methacrylate, Technovit©, for 72 hours. The infiltration solution was changed every 24 hours and all solutions were stored in a refrigerator at 4°C before finally being embedded in glycol methacrylate.

The plastic blocks were cut down in a microtome to a common horizontal plane and one 25 μm section taken from each block for use with optical disectors (see Gundersen 1986). An alternative is to take two consecutive 3 μm sections, for use with physical disectors. The sections were stained using a Giemsa stain.

Each bar in each of the sections was sampled in a systematic series of fields of view that were parallel to the long axis of the bar and with the series having a random starting position in the bars. The distance between each field of view was approximately 0.25 mm. Cells were defined as neurons if, primarily, they had a clearly defined cytoplasm. Other factors such as a convex nucleus with clearly defined edges with a diffuse and even chromatin pattern and the presence of a nucleolus were also taken into account. If a cell still had a neuron-like appearance but did not fulfill the above criteria the cell was viewed at a lower magnification so that a comparison could be made with other more easily identifiable neurons in the same area. Neurons were sampled in each field using the disector principle as applied to optical sections.

After the neuron was selected the microscope was focused to the section containing the center of the neuron's nucleolus, from where all measurements were made. If there was no easily recognizable nucleolus, measurements were made from the center of the clearest nuclear profile. Measurements were made from the center of the neuron's nucleolus to the center of the nuclear profile of glial cells in this optical section. Only glial cells whose clearest nuclear profile appeared in this optical section but not in a 'look-up section' 3 μm above the first were used. Two measurements were made because this was a 'vertical sections' sampling design as described in Figs 14.1 and 14.3 (see also Table 14.1).

An estimate of the glial cell numerical density (Table 14.2) when $R = 100$ μm is obtained from

$$\text{est}(N_V, \, R) = \frac{3\pi \cdot \dfrac{\sum d_i}{h_{\text{dis}}} + \text{any particles on the vertical axis}}{4\pi R^3}$$

$$\text{est}(N_V) = \frac{3 \times \pi \times 10\ 707 \div 3}{4 \times \pi \times 100^3} \div 56$$

$$= 4.78 \times 10^{-5} \ \mu m^{-3}.$$

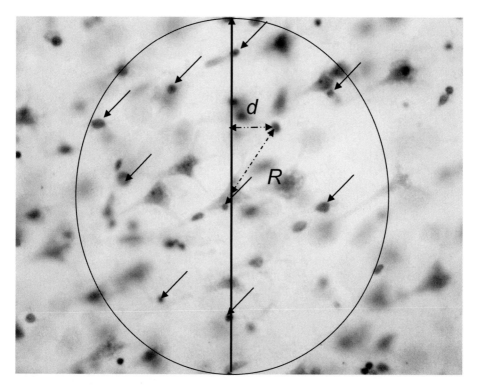

Fig. 14.3 The fixed arbitrary point was a neuron's nucleolus and the measurements were made to selected glial cells (arrowed). The measurements were made to the center of the nuclear profile of glial cells that appeared in this optical section but not in the 'look-up' optical section, 3 μm above this one. Measurements were only made to glial cells that were within a radius of 100 μm (the circle in the figure). This was an arbitrary distance that had been decided upon from previous experiments. An example of the measurements made are given for one glial cell where $d = 25$ μm and $R = 45$ μm and these are entered in Table 14.1.

This is the numerical density of glial cells found around neurons in a spherical volume $4/3\pi R^3$. The numerical density can then be worked out at different distances from the central point. $(R_2)^3 - (R_1)^3$ is the volume of the shell of the sphere at a distance R from the central point, the neuron's nucleolus. It is calculated by subtracting the volume of the larger sphere R_2 from that of the smaller sphere with radius R_1.

An estimate of the numerical density, namely, when $R = 70$ μm, is obtained from

$$\text{est}(N_V) = \frac{3 \times \pi \times 1676 \div 3}{4 \times \pi \times \left(70^3 - 60^3\right)} \div 56$$

$$= 5.89 \times 10^{-5} \ \mu m^{-3}.$$

Table 14.1 All the measurements (in μm) recorded over 56 fields of view are given here. From Fig. 14.3 the measurements for one glial cell where $d = 25$ μm and $R = 45$ μm are shown in bold. The total d for glial cells, in all fields of view, was 10 707 μm

R	d	0
5	+3+2+7+8+5+1+4+5+1+1	0
10	+1+11+4	45
15	+5+20+16+4	16
20	+12+15+1+17+15+12	45
25	+10+14+29+17+28+9+7+13+13+9+8+18+19+28+11+4+17	72
30	+38+37+3+29+2+4+24+12+31+38	254
40	**25**+17+1+7+39+41+37+15+6+40+45+6+13+28+41+30+21+2+5+5	218
50	+35+48+52+34+58+42+32+22+48+36+55+45+33	424
60	+11+64+67+29+46+57+56+59+25+40+64+45+55+33+34+25+44+7+62+59+34+21+24+39+23	540
70	+24+35+17+33+47+12+29+31+56+68+3+27+37+23+12+63+26+27+36+47	1 676
80	+12+24+75+3+68+45+74+23+79+5+4+23+14+30+41+37+9+8+18+19+71+28+11+4+17+69 +12+65	888
90	+11+64+67+29+46+88+35+47+52+89+34+58+42+3+2+7+8+87+8+5+1+4+5+1+13+2+22+48+36 +55+45+33+57+56+59+25+40+38+37+3+29+87+2+4+24+79+12+31+38+64+45+55+33+34+81 +25+44+7+62+59+34+21+24+39+23+24+35+17+33+47+12+29+31+56+68+3+27+37+23+85+12 +63+26+27+36+47+80	3 136
100	+35+47+52+89+34+58+42+3+4+62+3+27+37+23+85+12+63+7+8+87+8+5+91+4+5+1+13+2 +22+48+36+55+45+33+57+56+59+25+40+38+37+3+91+29+87+2+4+24+79+12+51+64+67 +29+46+88+31+38+64+45+55+33+34+81+25+44+7+62+89+34+21+24+39+23+24+35+27+33 +47+12+29+31+56+26+27+36+47+80	3 393

Table 14.2 Estimates of glial cell numerical density at different distances from the central point, $(R_2)^3 - (R_1)^3$

R (μm)	R^3 (μm³)	$(R_2)^3 - (R_1)^3$ (μm³)	$N_V \times 10^5$ (μm⁻³)
5	125	125	0.00
10	1 000	875	22.96
15	3 375	2 375	3.01
20	8 000	4 625	4.34
25	15 625	7 625	4.22
30	27 000	11 375	9.97
40	64 000	37 000	2.63
50	125 000	61 000	3.10
60	216 000	91 000	2.65
70	343 000	127 000	5.89
80	512 000	169 000	2.35
90	729 000	217 000	6.45
100	1 000 000	271 000	5.59

A graph (Fig. 14.4) can then be plotted showing how the numerical density varies with R. The solid line is the variation of glial cell numerical density with distance R and the dotted line is the average numerical density of glial cells.

In making the measurements it can be seen that there are a large number of measurements as R increases. The efficiency of the method could be improved if the sampling of events, glial cells, is reduced as R increases. This can be achieved by making use of the spatial distribution grid whose center is placed on the initial fixed point (Fig. 14.5). The sampling of events, the sampling fraction, is reduced as R is increased. This is done by sampling events that occur close to the center inside a complete sampling circle and then, as R increases, the sampling is done in a semi-circle, then a quarter of a circle, and finally one-eighth of a circle. The measurements of d_i were then multiplied by either 1, 2, 4, or 8 depending on which sampling circle the glial cells were in. After selecting a neuron with the disector, center the grid over the neuron's nucleolus and count all the glial cells that appear in the grid on the section containing the neuron's nucleolus but not in the 'look-up' section. In each field of view the grid should be rotated by one-eighth of a circle to reduce any directional bias.

4 Discussion

Some simple assumptions are made in applying this method. The first is that three-dimensional structures can be reduced to points. In the mathematical derivation of the method it is assumed that d is much greater than the height of the disector. As well as looking at the spatial distribution of points this method can be modified for volume, surface area, and length.

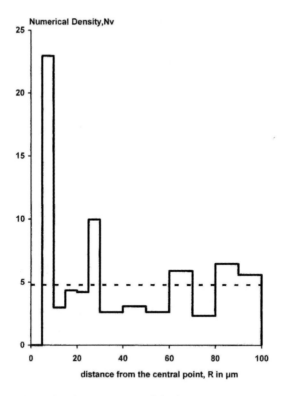

Fig. 14.4 The graph shows that there is an area of dead space at R = 5 µm, namely, inside the neuron where there are no glial cells. This is followed by an area of high density possibly indicating clustering of glia around neurons. Next there is an area of low density possibly indicating an area of repulsion before finally the glial cell numerical density fluctuates around the mean value, 4.78×10^5 µm^{-3}. The density settles to a mean value as more neurons, and their glial cells, are sampled.

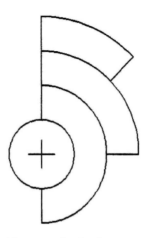

Fig. 14.5 A spatial distribution grid; see text for details.

Although this method was applied in thick sections, using optical sections it can also be applied in thin physical sections. Note how the method does not make use of the reference volume. It produces information about 'local events' and may be very useful in extracting information from samples where the reference volume is difficult, if not impossible, to obtain, for example, brain biopsies.

Second-order properties are at the cutting edge of stereological research. Many of the methods for interpreting them are in their infancy. Having obtained the data, a variety of models can be evoked to study the point pattern with the aim of characterizing the point pattern so that more information can be obtained about the underlying biological process see Diggle (1983) and Stoyan *et al.* (1995). One of the more popular ones is based on the *K* function, which is simply the number of points that lie within a radius of a chosen point divided by the mean number of points per unit volume. However, the *K* function is only a summary statistic and has some shortcomings. For instance Baddeley and Silverman (1984) demonstrated that two quite different point patterns, one a Poisson process and the other containing tight clusters of points, had the same K function.

Although the methods used for directly measuring cell size and the spatial distribution of cells are unbiased they will still be affected by shrinkage which could come from: length of agonia, post-mortem delay, type of fixation, and tissue processing. The spatial distribution method can be used not only to measure the spatial distribution of number but also that of volume, surface area, and length. For instance, it could be used in the study of Alzheimer's disease to show how the senile plaques are distributed about structural elements of the neocortical tissue.

Both these methods produce distributions that are non-standard Gaussian curves. As more data is obtained it will be important to derive some parametric equations for these distributions so that more information about the data can be extracted from the distributions, including methods for estimating error variances.

References

Baddeley, A.J. and **Silverman, B.W.** (1984). A cautionary example of the use of second order methods for analyzing point patterns. *Biometrics* **40**, 1089–93.

Baddeley, A.J., Howard, C.V., Boyde, A., and **Reid, S.** (1987). Three dimensional analysis of the spatial distribution of particles using the tandem scanning reflected light microscope. *Acta Stereol.* **6** (suppl. II), 87–100.

Bjallie, J.G. and **Diggle, P.J.** (1990). Statistical analysis of corticopontine neuron distribution in visual areas 17, 18 and 19 of the cat. *J. Compar. Neurol.* **295**, 15–32.

Braendgaard, H., Evans, S.M., Howard, C.V., and **Gundersen, H.J.** (1990). The total number of neurons in the human neocortex unbiasedly estimated using optical disectors. *J. Microsc* **157**, 285–304.

Braendgaard, H. and **Gundersen, H.J.G.** (1986). The impact of recent stereological advances on quantitative studies of the nervous system. *J. Neurosci. Methods* **18**, 39–78.

Diggle, P.J. (1983). *Statistical analysis of spatial point patterns.* Academic Press, New York.

Evans, S.M. and **Gundersen, H.J.G.** (1989) Estimation of spatial distributions using the nucleator. *Acta Stereol.* **8** (2/2), 395–401.

Evans, S.M., Howard, C.V., and **Gundersen, H.J.G.** (1989). An unbiased estimate of the total number of human neocortical neurons. (Abstract). *J. Anat.* **167**, 249–50.

Gundersen, H.J.G. (1986). Stereology of arbitrary particles. A review of unbiased number and size estimators and the presentation of some new ones, in the memory of William R. Thompson. *J. Microsc.* **143**, 3–45.

Gundersen, H.J.G. (1988). The nucleator. *J. Microsc.* **151**, 3–21.

Gundersen, H.J.G., Bagger, P., Bendtsen, T.F., Evans, S.M., Korbo, L., Marcussen, N., Möller, A., Nielsen, K., Nyengaard, J.R., Pakkenberg, B., Soerensen, F.B., Vesterby, A., and **West, M.J.** (1988). The new stereological tools: disector, fractionator, nucleator and point sampled intercepts and their use in pathological research and diagnosis. *Acta Pathol. Microbiol. Immunol. Scand.* **96**, 857–81.

Jensen, E.B.V. (1998) *Local stereology* ,Advanced Series on Statistical Science and Applied Probability, Vol. 5. See chapter 8.

Stoyan, D., Kendall, W.S., and **Mecke, J.** (1995). *Stochastic geometry and its applications*, 2nd edn. Wiley, Chichester.

CELL CULTURE

UNBIASED MORPHOMETRICAL TECHNIQUES FOR THE QUANTITATIVE ASSESSMENT OF CELLS IN PRIMARY DISSOCIATION CULTURES

H. SCHRÖDER, J. R. NYENGAARD, W. SELBERIS,
E. LAIN, A. KELLER, C. KÖHLER, AND B. WITTER

1 Introduction

The use of *in vitro* systems (cell culture, organotypic culture) is steadily increasing in all areas of modern biomedical research. Cell biology, cancer research, and neuroscience are some of the disciplines in which a growing number of studies depend on the use of cell culture models. Numerous different biochemical techniques are available that provide qualitative and quantitative conclusions from *in vitro* experiments. In addition to the large number of qualitative studies, the quantitative assessment of cell numbers has also become necessary, for example, to quantify the effects of putatively toxic substances on cells or to assess in quantitative terms the impact of experimental conditions on the cell-specific expression of marker proteins.

While the majority of *in vitro* biochemical and molecular biological approaches use state-of-the-art techniques, most *in vitro* studies use crude and in some cases unvalidated quantitative techniques. A frequently applied counting tool is based on a counting grid that has been superimposed on the culture dish. The use of the usual counting grid, in contrast to the two-dimensional unbiased counting frame (Gundersen 1977), does not lead to an unbiased estimate of the number of two-dimensional objects per area.

This poses the following question: How representative may a given grid be for the whole dish? This is by no means a trivial problem given the question as to whether there is a statistical difference in the number of surviving cells following the exposure to a potentially toxic agent as compared to vehicle treatment. A reliable and robust representative result can only be obtained when the sample taken is representative for the whole culture dish. This requirement is only fulfilled when a design-based sampling technique is used.

Based on mathematically very robust theories, stereological techniques were originally developed for the assessment of cells in three-dimensional scenarios. Applicable to *n*-dimensional space, they can easily be adapted to two dimensions. The use of a simple counting grid will not allow for an unbiased assessment of cells and their compartments in cell cultures. Furthermore, stereological techniques offer the advantage of being less time-consuming than the counting grid approach.

To provide an example of an unbiased two-dimensional procedure the present report focuses on a neuroscientific study using a primary neuronal dissociation culture in which it was necessary to obtain reliable data as to whether the induction of potentially toxic conditions would:

(1) affect the total number of neurons, that is, the number of neurons that express the marker protein microtubule-associated protein 2 (MAP2);

(2) affect the expression of certain markers in the cultured neurons, namely, the number of neurons that express $\alpha 4$ nicotinic acetylcholine receptor (nAChR) subunit protein;

(3) affect the co-expression in neurons of different nAChR proteins or phosphorylation markers.

Using immunocytochemical (for (1) and (2)) or immunofluorescent (for (3)) protocols, it was possible to answer these questions on a quantitative level. In this chapter we will provide a description of the use of an unbiased approach for the assessment of the neuron numbers concentrating on points (1) and (2). Finally, we will provide a little note on the estimation of volume/area of cultured neurons.

2 Materials and methods

2.1 Primary hippocampal culture

Hippocampal cell cultures were prepared as described previously (Banker and Cowan 1977; Goslin *et al.* 1998). Embryonic brains were taken from pregnant rats (Wistar) at day E18. They were transferred into a medium (HEPES (20 mM) in Dulbecco's modified Eagle's medium (DMEM) containing 4.5 g/l glucose) and the hippocampi were dissociated enzymatically with 0.05% trypsin/ethylene diaminetetraacetic acid (EDTA) 0.02% for 15 min at 37°C under gentle agitation (60 strokes/min). Subsequently they were homogenized mechanically by pipetting up and down using a Pasteur pipette. After dissociation treatment the tissue was rinsed three times with DMEM to allow for the diffusion of residual trypsin from the tissue. Supernatants of dispersed tissue were recovered by centrifugation at 900 rpm for 5 min. The pellet was filtered and gently resuspended in fresh medium. The amount of viable cells was assessed by trypan blue exclusion in a hemocytometer. The cells were seeded on to 12-mm poly-L-lysine (10 mg/ml) precoated glass coverslips (density about 5.4×10^3 cells/cm^2) and put into 60 mm tissue culture dishes where they were maintained in Neurobasal™ (serum-free medium with B27 supplement, Life Technologies; cf. Brewer *et al.* 1993).

2.2 Aβ_{1-42} preparation and incubation

Aβ_{1-42} was pre-aged in an aliquot of an aqueous 2 μM stock solution for at least 4 days at 37°C, and subsequently added in various concentrations (0.5, 1.0, and 1.5 μM) to cultures that had existed for 7 days and been kept for 3 days under incubation. Controls were performed by omission of Aβ_{1-42}. After incubation, the cells were fixed with 4% paraformaldehyde and 4% sucrose in phosphate-buffered saline (PBS; pH 7.4, 15 min, 37°C). Subsequently, they were rinsed three times with PBS and maintained in a moist chamber at 4°C for further treatment.

Assessment of lactate dehydrogenase (LDH) activity was performed using CytoTox 96 non-radioactive assay (Cytotoxicity Assay kit, Promega®). The viability of the cells after Aβ_{1-42}-treatment was determined by measuring the LDH activity. Upon incubation with 0.5 μM Aβ_{1-42} LDH activity levels increased by approximately 17%, while the application of 1.0 and 1.5 μM Aβ_{1-42} revealed LDH values in the control range. No statistically significant differences in LDH activity levels were observed between the various groups ($p > 0.05$).

2.3 Immunocytochemistry

Immunocytochemistry was performed using the avidin–biotin–peroxidase complex (ABC)-method. Cells were permeabilized using 0.2% Triton X-100 in PBS

(20 min, room temperature (RT)) and subsequently incubated—with washes in PBS (3 × 5 min each) after each step—for 15 min in a 2% hydrogen peroxide solution in PBS and blocked for 1 h in 10% normal rat serum in 1% bovine serum albumin (BSA) and PBS. The primary monoclonal antibodies (mAbs) anti-MAP2 (1:1000, mouse, Sigma) and anti-α4 (mAb 299, rat, at a dilution of 25 µg/ml) were applied overnight at 4°C. The specificity of mAb 299 has been described in detail elsewhere (Peng *et al.* 1994; Whiting and Lindstrom 1988; Schröder *et al.* 2001). After reblocking with 5% goat serum solution and 1% BSA in PBS for 1 h, cultures were incubated in biotinylated secondary antibodies (1:200 in PBS/BSA; goat anti-mouse IgG$_1$ (Amersham) and rabbit anti-rat IgG, 1:400 (Dianova)). Finally, a streptavidin–peroxidase complex (1:200, in 1% BSA in PBS, 40 min (Amersham)) was used. Visualization of the immunoprecipitate was achieved by using diaminobenzidine tetrahydrochloride (DAB) fortified with a glucose oxidase–nickel sulfate peroxidase substrate (5–10 min). Controls performed by omission of the primary antibody during the corresponding incubation revealed negative results. All coverslips were mounted on glass slides using Entellan® (Merck).

2.4 Assessment of neuron numbers

The unbiased stereology-based estimation of total neuron counts was performed using the two-dimensional fractionator (Gundersen 1986) in conjunction with the CAST system (Olympus, Denmark), which allows systematic uniform random sampling using a software-controlable microscope stage. At least eight specimens were evaluated for each of the different incubation conditions (Aβ$_{1-42}$ 0.5, 1.0, and 1.5 µM, vehicle). Means of the individual counting results were used for further statistical evaluation.

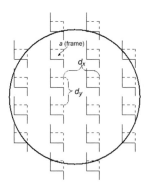

Fig. 15.1 Systematic, uniformly random sampling (SURS) procedure. The cartoon diagrammatically shows the distribution of counting frames across the surface of the culture dish. *a*(frame), Frame area; d_x and d_y, distances between frames in the *x*- and *y*-directions, respectively.

Delineation of the sampling area was performed manually (magnification 1.25 ×) by means of a mouse-driven cursor, the margin of the coverslip (113.3 ± 2.0 mm²) representing the boundary. The distance between the counting frames (frame area: 6976 μm², corrected for magnification) was 600 μm in both the x- and y-directions. The number of counting frames was calculated by a subroutine in the CAST system and automatically superimposed on the video overlay (see Fig. 15.1). The two-dimensional unbiased counting frame was displayed on the computer screen taking up 25% of the screen area. Counting of the neurons (40 × oil immersion objective) was initiated at a random position, continued by moving the computer-controled microscope stage holder stepwise in a meandering pattern over the entire section (see Fig. 15.1). All immunolabeled (MAP2, α4) and morphologically intact appearing neurons sampled by the counting frame were manually marked using a computer mouse. Collected data were saved to a computer file. The total number of neurons sampled was obtained using the formula

$$N = \sum Q_N^- \cdot \frac{d_x \cdot d_y}{a(\text{frame})}$$

$$= \sum Q_N^- \cdot \frac{600 \ \mu m \cdot 600 \ \mu m}{6976 \ \mu m^2}$$

where $\sum Q_N^-$ is the total sum of the neurons counted, d_x and d_y are the sampling steps (μm) in the x- or y-direction, respectively, and $a(\text{frame})$ is the area of the counting frame corrected for magnification (μm²).

2.5 A note on the assessment of neuron volumes

This study did not intend to estimate volumes of neurons but number-weighted mean neuron volume may be estimated in the special case where neurons are cultured and grown on plastic dishes. In the case of a flat support, the number of neurons per area of dish is estimated using the two-dimensional counting frame. The neurons are fixed and embedded in the plastic dish and vertical sections are generated perpendicular to the monolayer. Using light microscopy the mean height of the monolayer, H, is estimated using point counting on the vertical section as

$$H = \frac{\sum P \cdot a(p)}{L}$$

where $a(p)$ is the area associated with each test point in the grid, $\sum P$ is the number of points hitting the neurons, and L is the length of the monolayer base. The mean neuron volume is then the mean height of the monolayer divided by the number of neurons per area of dish (Griffiths *et al.* 1989). The unbiased assessment of neuron volumes is of particular value in those cases where a toxic effect of a given

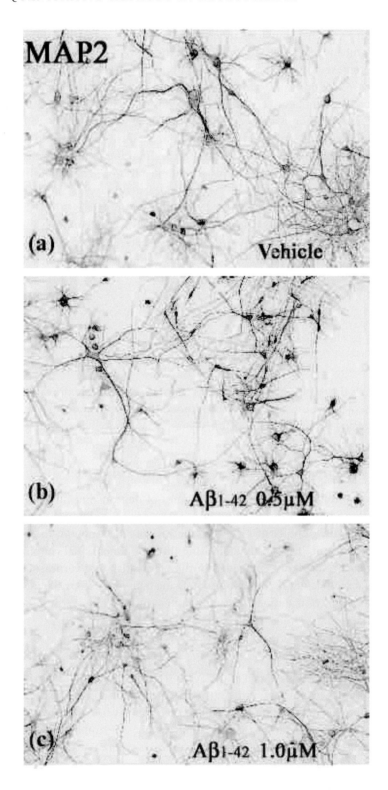

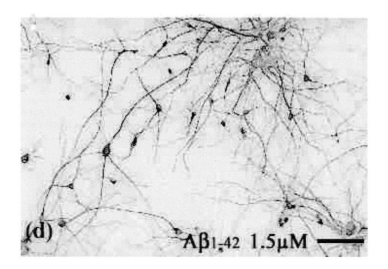

Fig. 15.2 MAP2-immunocytochemistry in hippocampal neuronal dissociation culture following treatment (a) with vehicle and (b)–(d) with $A\beta_{1-42}$ ((b) 0.5 µM; (c) 1.0 µM; (d) 1.5 µM). (a) In the vehicle-treated cultures numerous immunolabeled neurons with their dendritic ramifications are present. (b)–(d) Densities of immunoreactive neurons appear comparable to that shown in (a). Calibration bar, 50 µm.

compound may be assumed which is reflected in volume changes of the exposed neurons (swelling, shrinkage) rather than in changes of their total number. An easier obtainable estimate of 'size' of neurons grown *in vitro* can be generated by determining their average projected horizontal area by means of the two-dimensional nucleator (Gundersen 1988). This can be done in practice using computer-assisted stereology. First, the neuron is sampled by the two-dimensional unbiased counting frame. Second, the approximate center of the neuron is marked. Third, intersections between the boundary of the neuron and a set of radiating test lines are marked and the average projected horizontal area of the neuron is calculated. The average projected horizontal area of the neuron can also be obtained by dividing the areal fraction of neurons per dish, A_A, using point counting, by the number of neurons per area dish, N_A.

2.6 Statistical analysis

Statistical analysis was performed using the SPSS 10.0 software package. Normal distribution of values was assessed by the Kolmogorow–Smirnov test for normality. Statistical comparisons among groups were made by one-way analysis of variance between two means by the grouped Student's *t*-test corrected by the Bonferroni method for multiple comparisons.

Table 15.1   Numbers of immunolabeled neurons upon vehicle and $A\beta_{1-42}$ treatment

| | Number of immunolabeled neurons* upon treatment with | | | |
| | | $A\beta_{1-42}$ | | |
Marker	Vehicle	0.5 µM	1.0 µM	1.5 µM
MAP2	11 911 ± 4 043	11 714 ± 3 969	8 618 ± 3 248	8 530 ± 4 281
α4 nAChR	4 186 ± 1 332	4 038 ± 1 113	3 259 ± 1 329	2 348 ± 698†

* All values given as mean ± standard deviation.

† $p < 0.05$ compared to vehicle treatment; $p < 0.05$ compared to incubation with 0.5 µM $A\beta_{1-42}$.

3   Results

3.1   Total number of neurons

As shown in Fig. 15.2, MAP2-immunoreactive (ir) neurons were present in a relatively high density throughout the culture dishes. Multipolar neurons with a strongly ramified dendritic tree were mostly encountered. As can be seen from Fig. 15.2 (b)–(d), the impact of $A\beta_{1-42}$ on the number of MAP2-ir neurons was only moderate. Reliable data are provided by the unbiased approach showing that the number of MAP2-ir neurones was only slightly decreased in $A\beta_{1-42}$ treated cultures as compared to vehicle-incubated ones (Table 15.1).

3.2   Expression of α4-labeled neurons

In contrast to the effects of $A\beta_{1-42}$ on MAP2-labeled neurons, α4-expressing neurons displayed a clear-cut reduction upon incubation with $A\beta_{1-42}$, reaching statistical significance for those neurons treated with 1.5 µM of the peptide (Table 15.1; Fig. 15.3(a)–(d)).

4   Discussion

In vitro models are important tools for mimicking certain pathophysiological conditions under controled and standardized conditions. The potential of these tools can only be fully reached when quantitative parameters seen *in vitro* are validated with *in vivo* data. Using as an example the potential impact of Aβ on cholinoceptive dysfunction in Alzheimer's disease, mimicked in a primary neuronal dissociation culture, we could show the feasibility of using a modern morphometric technique in order to get reliable quantitative data. This approach has allowed us to obtain clear-cut statistically relevant answers to the questions posed in Section 1, that is, does the treatment with a potentially toxic agent (in this case Aβ) affect:

(1) the number of neurons;

(2) the number of neurons that express $\alpha 4$ nAChR subunit protein?

The total number of neurons—assessed by using MAP2 as an immunocytochemical label—could be shown not to differ significantly between control cultures and those incubated with $A\beta_{1-42}$. With regard to the expression of nAChRs, the quantitative evaluation clearly showed a concentration-dependent significant decrease in number upon treatment for $\alpha 4$ nAChR subunit-expressing neurons.

It would be beyond the scope of this chapter to discuss in detail the biological significance of these findings. In brief, the findings obtained closely resemble those of studies performed on Alzheimer autopsy tissue by means of immuno-histochemistry and Western blotting (e.g. Regeur *et al.* 1994; Schröder *et al.* 1991*a,b*; Wevers *et al.* 1999). The quantitative evaluation of these data on the human brain revealed no difference in the total number of neurons when comparing Alzheimer individuals and age-matched controls. The numbers of $\alpha 4$ nAChR subunit-expressing neurons, however, showed a massive, significant decrease. This finding is corroborated by the results of Western blotting studies (Burghaus *et al.* 2000).

The simple observation of the culture dishes (cf. Figs 15.2 and 15.3) may have raised the suspicion that there is a decrease in the number of $\alpha 4$-expressing neurons while that does not hold true for MAP2-ir neurons. Answers to the question whether the *in vitro* findings can support the above described autopsy findings, however, is only guaranteed by using modern morphometric assessments of neuron numbers. It can now be shown that the *in vitro* behavior of neurons mirrors the *in vivo* situation as shown by Wevers *et al.* (1999).

This feature is an important criterion for the feasibility of using an *in vitro* approach for modeling the *in vivo* events of Alzheimer's disease. This robust methodology gives us a firm base on which to build upon for further studies on the effects of putatively neuroprotective compounds. The impact of these compounds can be assessed using quantitative morphometry in addition to biochemical tools. Currently, we are testing the effects of certain drugs on the phosphorylation status of the tau protein and the associated changes in nAChR expression by means of immunofluorescent double-labeling. By using fluorescence microscopy equipment attached to the computerized microscope, the simultaneous assessment of several cell biological markers becomes possible.

The use of counting grids, which is still widespread in the quantitative evaluation of cell numbers in culture dishes, is hampered by several drawbacks. In the first place, the selection of one or several areas of the culture dishes is not unbiased. It is also highly unlikely that a representative sample of cells can be obtained by this approach. Finally, the simultaneous assessment of different markers, for example, by fluorescent labeling, is almost impossible when using a simple counting grid because it is technically much more difficult and time-consuming. All

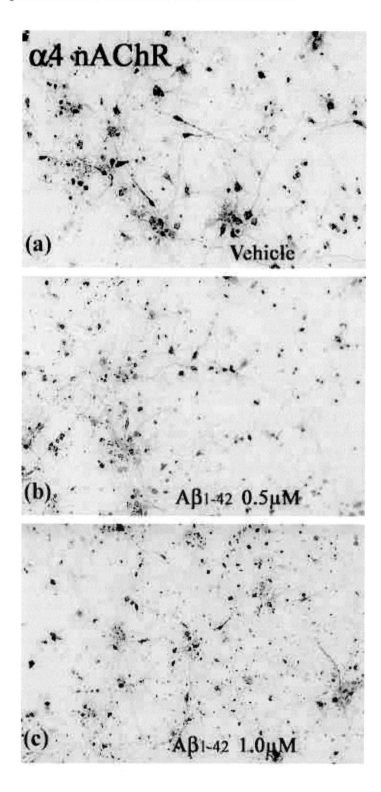

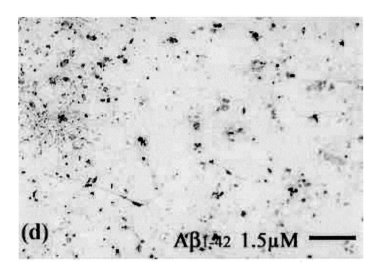

Fig. 15.3 α4 nAChR-immunocytochemistry in hippocampal neuronal dissociation culture following treatment (a) with vehicle and (b)–(d) with $A\beta_{1-42}$ ((b) 0.5 μM; (c) 1.0 μM; (d) 1.5 μM). (a) Numerous strongly immunolabeled neurons are present in untreated cultures. (b)–(d) The density of immunolabeled neurons decreases with increasing concentrations of $A\beta_{1-42}$. Calibration bar, 50 μm.

these problems are circumvented and overcome by use of the technique described in this chapter.

In this study a number of 300 frames per dish was used. In order to restrict the analysis to the minimal necessary time, pilot experiments (determination of the coefficient of variance and coefficient of error) should be performed to try to minimize the number of frames per dish to around 50–75, while still counting about 100–200 cells, by changing the frame area.

In conclusion, while comparisons of autopsy and *in vitro* findings have to be performed with caution, mimicking certain defined aspects of Alzheimer's disease pathology under standardized conditions *in vitro* will lead to a better interpretation of the autopsy results. We believe that the *in vitro* assessment of nAChR expression will be a useful additional piece for a better understanding of the pathophysiological jigsaw of Alzheimer's disease.

Acknowledgements

We are greatly indebted to St. Hösch, S. Hähner and N. Selberis for technical assistance and to I. Koch for the photographic work. We would like to thank Dr W. Bloch for statistical advice and N. Moser for reading the manuscript. The work reported in this chapter was supported by the Deutsche Forschungsgemeinschaft (DFG, Schr 283/18–3, 18–4) and the Fonds der Chemischen

Industrie to H.S., local funding sources to B.W. (Frau Maria Pesch-Siftung, Imhoff-Stiftung).

References

Banker, G.A. and **Cowan, W.M.** (1977). Rat hippocampal neurones in dispersed cell culture. *Brain Res.* **126**, 397–442.

Brewer, G.J., Torricelli, J.R., Evege, E.K., and **Price, P.J.** (1993). Optimized survival of hippocampal neurones in B27-supplemented Neurobasal™, a new serum-free medium combination. *J. Neurosci. Res.* **35**, 567–76.

Burghaus, L., Schütz, U., Krempel, U., de Vos, R.A.I., Jansen Steur, E.N.H., Wevers, A., Lindstrom, J., and **Schröder, H.** (2000). Quantitative assessment of nicotinic acetylcholine receptor proteins in the cerebral cortex of Alzheimer patients. *Brain Res. Mol. Brain Res.* **76**, 385–8.

Goslin, K., Asmussen, H., and **Banker, G.** (1998). Rat hippocampal neurones in low-density culture. In *Culturing nerve cells* (ed. G. Banker and K. Goslin), 2nd edn, pp. 339–70. MIT Press, Cambridge, Massachusetts.

Griffiths, G., Fuller, S.D., Back, R., Hollinshead, M., Pfeiffer, S., and **Simons, K.** (1989). The dynamic nature of the Golgi complex. *J. Cell Biol.* **108**, 277–97.

Gundersen, H.J.G. (1977). Notes on the estimation of the numerical density of arbitrary profiles: the edge effect. *J. Microsc.* **111**, 219–23.

Gundersen, H.J. (1986). Stereology of arbitrary particles. A review of unbiased number and size estimators and the presentation of some new ones, in memory of William R. Thompson. *J. Microsc.* **143**, 3–45.

Gundersen, H.J.G. (1988). The nucleator. *J.Microsc.* **151**, 3–21.

Peng, X., Gerzanich, V., Anand, R., Whiting, P.J., and **Lindstrom, J.** (1994). Nicotine-induced increase in neuronal nicotinic receptors results from a decrease in the rate of receptor turnover. *Mol. Pharmacol.* **46**, 523–30.

Regeur, L., Jensen, G.B., Pakkenberg, H., Evans, S.M., and **Pakkenberg, B.** (1994). No global neocortical nerve cell loss in brains from patients with senile dementia of Alzheimer's type. *Neurobiol. Aging* **15**, 347–52.

Schröder, H., Giacobini, E., Struble, R. G., Zilles, K., and **Maelicke, A.** (1991*a*). Nicotinic cholinoceptive neurones of the frontal cortex are reduced in Alzheimer's disease. *Neurobiol. Aging* **12**, 259–62.

Schröder, H., Giacobini, E., Struble, R.G., Zilles, K., Maelicke, A., Luiten, P.G., and **Strosberg, A.D.** (1991*b*). Cellular distribution and expression of cortical acetylcholine receptors in aging and Alzheimer's disease. *Ann. NY Acad. Sci.* **640**, 189–92.

Schröder, H., Schütz, U., Burghaus, L., Lindstrom, J., Kuryatov, A., Monteggia, L., deVos, R.A., van Noort, G., Wevers, A., Nowacki, S., Happich, E., Moser, N., Arneric, S.P. and **Maelicke, A.** (2001). Expression of the alpha4 isoform of the nicotinic acetylcholine receptor in the fetal human cerebral cortex. *Brain Res. Dev. Brain Res.* **132**, 33–45.

Wevers, A., Monteggia, L., Nowacki, S., Bloch, W., Schütz, U., Lindstrom, J., Pereira, E.F., Eisenberg, H., Giacobini, E., de Vos, R.A.I., Jansen Steur, E.N.H., Maelicke, A., Albuquerque, E.X., and Schröder, H. (1999). Expression of nicotinic acetylcholine receptor subunits in the cerebral cortex in Alzheimer's disease: histotopographical correlation with amyloid plaques and hyperphosphory-lated-tau protein. *Eur. J. Neurosci.* **11**, 2551–65.

Whiting, P.J. and Lindstrom, J.M. (1988). Characterization of bovine and human neuronal nicotinic acetylcholine receptors using monoclonal antibodies. *J. Neurosci.* **8**, 3395–4404.

GLOSSARY

JENS R. NYENGAARD

Bias: a systematic difference between the average estimate and the true value.

Cavalieri principle: an unbiased principle for estimating the volume of any object. With a random start, the object is cut into parallel slices of a known and fixed thickness. The volume is estimated by multiplying the distance between the slices by the total cut area of the same side of all slices.

Coefficient of error (CE): the relative standard error of the mean, CE=SEM/mean, is the dimensionless precision of an estimator, and is important for determining whether the sampling is appropriate. For n independent estimates from the same distribution, SEM:=SD/$\sqrt{n}$. For non-independent estimates, like individual areas of equally spaced sections for e.g. the Cavalieri-estimator of total volume, the situation is much more complex. Even for measurements without measuring noise (independent error) SEM:$\neq$SD/$\sqrt{n}$. For simple sampling schemes, methods based on very recent and advanced sampling theory may provide robust predictions of CE (See 'Systematic, uniformly random sampling'). In practice, almost all measurements and counts have additional, independent noise which also has to be known.

Connectivity: the maximal number of cuts through an object, which does not produce two disconnected objects. These extra or redundant connections are equal to the topologically defined number of units in a network.

Consistent estimator: an estimator where the bias is reduced to an arbitrarily low value by sufficiently increasing the sample size at the level of the independent estimate, i.e. in order to make the bias insignificant the investigator need to sample enough from each study unit (animal/subject). Adding more animals does not reduce the bias. The common example in stereology is estimates of ratios and densities. See 'unbiased estimator'.

Cycloid: a curve describing the trajectory of a point on the periphery of a disc rolling along a straight line. A cycloid arc has a sine-weighted length distribution, i.e. proportional to the sine of the angle to the direction of the minor axis of the cycloid. Cycloids are used for surface estimation on vertical section planes as well as fiber length estimation in thick sections under projection.

Density: many stereological estimators report the ratio of the amount of an object phase to the total reference volume. These ratio estimates are known as densities. Examples are volume density, V_V, surface density, S_V, length density, L_V, and numerical density, N_V. See also 'consistent estimator'.

Disector: a 3-dimensional stereological probe for sampling of objects according to their number. In its simplest form (physical disector), it consists of a pair of (thin) physical sections separated by a known distance and an unbiased 2-dimensional counting frame. The objects are sampled if they are present in the counting frame in one section plane and not in the other plane. In the optical disector (a stack of) thin focal planes within a thick section are used instead of physical sections.

Error variance: the random fluctuation between individual unbiased estimates and the true value. This sampling variance is a random error, with an average of 0, and therefore not a bias.

Euler number: a 0-dimensional measure of structure that can be used to estimate the number of units in a network. A topological definition of a unit, i.e. a loop in the network, is used for the estimation.

Fractionator: a sampling methodology in which a known and usually predetermined fraction of an object is sampled, often in several steps of different sampling fractions. In the final sample, the structural quantity (usually number) is measured. The total structural quantity (e.g. number) is estimated by multiplying the inverse sampling fraction by the structural quantity in the final sample.

Isector: a simple method for generating isotropic section planes by embedding a small object into a spherical mould. Section planes through the spherically embedded and rolled object will be isotropic.

Isotropic design: isotropy means that all directions are equally probable. A design with completely random orientation is therefore an isotropic design, which is necessary if the stereological estimators require isotropic test planes. An isotropic design is usually combined with uniformly random sampling generating isotropic, uniform random (IUR) section planes. See also 'vertical design'.

Lost caps: objects that are physically lost or optically not visible in a thin section or close to the upper and lower surface of a thick section.

Nucleator: a two-step, number-weighted, local size estimator. First, the objects are sampled in proportion to their number using the disector. Secondly, the volume of each object is estimated by measuring the distance along isotropic lines radiating from a unique point, e.g. the nucleolus in a cell, to the boundary. The average estimate is the ordinary, number-weighted, mean volume.

Number-weighted: each object has equal statistical weight equal to 1 regardless of its shape and size.

Noise: the independent variance of an estimator, e.g. point counting and object counting, is called the noise. The sum of the noise and the variance of the systematic sampling procedure is the error variance of the systematic stereological estimators.

Orientator: a two-step procedure for generating isotropic section planes in large objects. First, the object is placed flat on a circle with equidistant divisions along the perimeter. A random number decides the direction of the first cut. Secondly, the isotropic section is obtained using a random number and a circle with cosine-weighted divisions along the perimeter.

Overprojection: an effect of section thickness of transparent slices, which makes all measurements of non-transparent structures larger. The best way to avoid over-projection bias is to make sections as thin as possible.

Point-sampled intercepts: a two-step procedure for estimation of the volume-weighted mean volume. First, the objects are sampled in proportion to their volume using test points. Secondly, the volume of each object is estimated by measuring the length of an isotropic intercept in the object passing through the test point.

Rotator: a number-weighted, local size (volume) estimator performed in two steps like the nucleator except that details of the measurements and computations are different.

Star volume: defined as the average volume of the object that can be seen unobscured from a random point within the object. Star volume is estimated by the use of point-sampled intercepts.

Stereology: the science dealing with sampling of geometric objects. In a more popular form: methods for obtaining 3D information from measures in 2D random sections.

Systematic, uniformly random sampling: sampling performed with a system-atic and a random component. If the goal is to sample approximately 7 slices from a population of 90 numbered slices, the systematic component is the decision of choosing a sampling periodicity of 90/7~12. The random sampling component is to look up a random number between 1 and 12, e.g. 7 for the start of the sample. The following slices are then sampled systematic, uniformly random: 7, 19, 31, 43, 55, 67 and 79.

Unbiased counting frame: a profile is sampled by a counting frame if it is partially or completely inside the frame and does not touch the exclusion line.

Unbiased estimator: an estimator, which on average has the true value. As there is no bias, the precision of the estimate may be increased arbitrarily by increasing the sample size.

Uniformly random sampling: every object in the whole population has an equal probability of being sampled or every position in the object has an equal probability of being hit.

Vertical design: a natural or arbitrary vertical direction is identified or generated in the object, which is rotated at random around this vertical axis. On the properly sampled and uniformly positioned section planes, the vertical direction must be identifiable. All stereological estimators requiring isotropic test lines may then be applied.

Virtual test systems: the use of thick sections and digital light microscopy has made it possible to generate virtual isotropic test planes inside the thick sections for length estimation or virtual isotropic test lines for surface area estimation. The advantage is that the direction of the sections can be arbitrary. The disadvantage is sensitivity to shrinkage in the z-axis (collapse in the thickness of the section).

Volume-weighted: each object has a statistical 'weight' according to its volume. The mean is the volume weighted mean. See also 'point sampled intercepts'.

Ref: The glossary has been modified from Nyengaard, J.R. (1999) Stereologic methods and their application in kidney research. J Am Soc Nephrol. 10(5):1100–1123.

INDEX